P9-DWP-886

Molecular Farming in Plants: Recent Advances and Future Prospects

Aiming Wang • Shengwu Ma
Editors

Molecular Farming in Plants: Recent Advances and Future Prospects

Editors
Aiming Wang
Agriculture and Agri-Food Canada
1391 Sandford Street
London, ON N5V 4T3
Canada
Aiming.wang@agr.gc.ca

Shengwu Ma
Lawson Health Research Institute/Plantigen
Immunology and Transplantation
375 South Street
London, ON N6A 4G5
Canada
sma@uwo.ca

ISBN 978-94-007-2216-3 e-ISBN 978-94-007-2217-0
DOI 10.1007/978-94-007-2217-0
Springer Dordrecht Heidelberg London New York

Library of Congress Control Number: 2011940206

Printed on acid-free paper

Springer is part of Springer Science+Business Media (www.springer.com)

Preface

Molecular farming in plants is an emerging discipline in life sciences. It is also known as molecular pharming or biopharming. Essentially it is using plants including algae to produce pharmaceutical proteins, therapy peptides, vaccine subunits, industrial enzymes and other compounds of interest. The plant system offers practical, biochemical, economic and safety advantages compared with conventional production systems. One successful example is Golden rice that was engineered with the biochemical pathway to biosynthesize beta-carotene, a precursor of pro-vitamin A in the edible parts of rice. This genetically modified rice can be used in areas, particularly in developing countries, where there is a shortage of dietary vitamin A.

This book is aimed at reviewing the principles, current advanced methodology, bottleneck problems and futures of molecular farming in plants in anticipation of providing a textbook for students, teachers, and researchers who are interested in molecular farming in plants and plant biotechnology. This book consists of 12 chapters written by international authorities in the field. Overall current status and future prospective of the discipline is discussed in Chap. 1. In the area of therapeutic strategies using stable transgenic plants, cell suspension cultures (Chap. 3), chloroplasts (Chap. 4), transient expression (Chap. 9), and viral vectors (Chap. 10) are extensively reviewed. In addition to plant leaves that have been commonly used for protein production, promising alternatives including plant seed (Chap. 5) and algae (Chap. 6) are also discussed. From the angle of products, Chaps. 7 and 8 particularly address plant-derived vaccines/therapeutic proteins and industrial proteins, respectively. Downstream processing, as an essential part of molecular farming, is reviewed in Chap. 11. In dealing with public concerns about safety, Chap. 12 provides an updated review in this topic. To make this book concise, basic sciences such as molecular genetics and general plant biotechnology are not included. Interested readers are recommended to read relevant textbooks or reference books for basic knowledge in these areas.

Both the editors wish to express their sincere gratitude to all the authors for accepting their invitation and taking on the challenging task to make valuable contributions to this book. Special thanks go to Jacco Flipsen, Ineke Ravesloot and the other staff at Springer for their strong support and excellent professionalism that make this book possible. Both the editors also thank their families, friends, and colleagues who have provided tremendous support and continuous encouragement from the deliberation of the book to its publication.

Aiming Wang
Shengwu Ma

Contents

Editors

Aiming Wang earned his Ph.D. in plant molecular biology and virology from the University of British Columbia in 1999. He spent his next four years working on plant genomics and biotechnology at the Plant Biotechnology Institute, National Research Council of Canada to work on wheat genomics. In 2003, he assumed a research scientist position at Southern Crop Protection and Food Research Centre, Agriculture and Agi-Food Canada in London, Ontario. In the same year, he was also appointed adjunct professor at Department of Biology, the University of Western Ontario. His research focuses on molecular virus-plant interactions, developing novel antiviral strategies, and potential uses of plant viruses in molecular farming as well as plant biotechnology.

Shengwu Ma completed his Master's degree in Microbiology and Immunology from Nankai University, Tianjin, China, and his Ph.D. in bacterial genetics and plant molecular biology from Carleton University in 1988. He then received further research training in plant-microbial interactions as a post-doctoral fellow at London Centre of Agriculture and Agri-Food Canada. He is now a research scientist at the Lawson Health Research Institute, adjunct professor in Departments of Biology and Medicine, the University of Western Ontario, and co-founder of Plantigen, a "spin off" company focused on the creation of genetically altered plants for use in medical applications. Dr Ma has a strong research interest in molecular farming in plants. He has pioneered research on the use of transgenic plants to express and deliver recombinant autoantigens for the treatment of autoimmune Type 1 diabetes through oral tolerance induction. He has published a number of papers in major scientific journals and is currently on the editorial board of several international scientific journals.

Contributors

Adil Ahmad Southern Crop Protection and Food Research Centre, Agriculture and Agri-Food Canada, 1391 Sandford St, London, ON N5V 4T3, Canada

Didier Breyer Biosafety and Biotechnology Unit, Scientific Institute of Public Health, Rue J. Wytsmanstraat 14, B-1050 Brussels, Belgium

Henry Daniell Department of Molecular Biology and Microbiology, College of Medicine, University of Central Florida, 336 Biomolecular Science Building, Orlando, FL 32816, USA

Adinda De Schrijver Biosafety and Biotechnology Unit, Scientific Institute of Public Health, Rue J. Wytsmanstraat 14, B-1050 Brussels, Belgium

Rainer Fischer Fraunhofer Institute for Molecular Biology and Applied Ecology, Forckenbeckstrasse 6, 52074 Aachen, Germany

RWTH, Worringerweg 1, 52074 Aachen, Germany

Martine Goossens Biosafety and Biotechnology Unit, Scientific Institute of Public Health, Rue J. Wytsmanstraat 14, B-1050 Brussels, Belgium

Christoph Griesbeck Department of Engineering, Environmental and Biotechnology, MCI – Management Center Innsbruck – University of Applied Sciences, Egger-Lienz-Str. 120, 6020 Innsbruck, Austria

Philippe Herman Biosafety and Biotechnology Unit, Scientific Institute of Public Health, Rue J. Wytsmanstraat 14, B-1050 Brussels, Belgium

Elizabeth E. Hood Arkansas Biosciences Institute, Arkansas State University, P.O. Box 639, State University, Jonesboro, AR 72467, USA

Ting-Kuo Huang Department of Chemical Engineering and Materials Science, University of California – Davis, 1031B Kemper Hall, 1 Shields Avenue, Davis, CA 95616, USA

Anthony M. Jevnikar Transplantation Immunology Group, Lawson Health Research Institute, London, ON N6A 4G5, Canada

Jussi Joensuu VTT Technical Research Centre of Finland, Espoo, Finland

Allison R. Kermode Department of Biological Sciences, Simon Fraser University, 8888 University Drive, Burnaby, BC V5A 1S6, Canada

Anna Kirchmayr Department of Engineering, Environmental and Biotechnology, MCI – Management Center Innsbruck – University of Applied Sciences, Egger-Lienz-Str. 120, 6020 Innsbruck, Austria

Shengwu Ma Transplantation Immunology Group, Lawson Health Research Institute, London, ON N6A 4G5, Canada

Department of Biology, University of Western Ontario, London, ON N6A 5B7, Canada

Karen A. McDonald Department of Chemical Engineering and Materials Science, University of California – Davis, 1031B Kemper Hall, 1 Shields Avenue, Davis, CA 95616, USA

Rima Menassa Southern Crop Protection and Food Research Centre, Agriculture and Agri-Food Canada, 1391 Sandford St., London, ON N5V 4T3, Canada

James S. New Department of Molecular Biology and Microbiology, College of Medicine, University of Central Florida, 336 Biomolecular Science Building, Orlando, FL 32816, USA

Zivko L. Nikolov Biological & Agricultural Engineering, Texas A&M University, College Station, TX 77843, USA

Katia Pauwels Biosafety and Biotechnology Unit, Scientific Institute of Public Health, Rue J. Wytsmanstraat 14, B-1050 Brussels, Belgium

Deborah Vicuna Requesens Arkansas Biosciences Institute, Arkansas State University, P.O. Box 639, State University, Jonesboro, AR 72467, USA

Stefan Schillberg Fraunhofer Institute for Molecular Biology and Applied Ecology, Forckenbeckstrasse 6, 52074 Aachen Germany

Richard M. Twyman Department of Biological Sciences, University of Warwick, Coventry CV4 7AL, UK

Aiming Wang Southern Crop Protection and Food Research Centre, Agriculture and Agri-Food Canada, 1391 Sandford St., London, ON N5V 4T3, Canada

Donevan Westerveld Department of Molecular Biology and Microbiology, College of Medicine, University of Central Florida, 336 Biomolecular Science Building, Orlando, FL 32816, USA

Lisa R. Wilken Biological & Agricultural Engineering, Texas A&M University, College Station, TX 77843, USA

Chapter 1
Molecular Farming in Plants: An Overview

Shengwu Ma and Aiming Wang

Abstract Protein-based biopharmaceuticals have become increasingly important due to a combination of their bioreactivity, specificity, safety and overall success rate. *Escherichia coli*, yeast and animal cells have traditionally been used as heterologous expression systems for production of pharmaceutical proteins. However, these conventional expression systems are often limited by high production costs, potential risks of product contamination, and the complexity and difficulty of scale-up to industrial production. Plants have emerged as a promising alternative expression system for production of pharmaceutical proteins because they offer several potential advantages, including low production costs, ease of scale-up to commercial quantities of production and reduced risk of product contamination by mammalian viruses or toxins. Plants are already being used to produce antibodies, vaccines, growth factors and many other proteins of pharmaceutical importance. The use of plants as factories for production of recombinant pharmaceutical proteins, including industrial enzymes, is now more commonly referred to as molecular farming. In this chapter, we discuss the technological basis of molecular farming in plants, with a focus on host systems and approaches/strategies developed to maximize protein yields and to ensure efficient recovery and purification of plant-made recombinant products.

S. Ma (✉)
Transplantation Immunology Group, Lawson Health Research Institute,
London, ON N6A 4G5, Canada

Department of Biology, University of Western Ontario, London, ON N6A 5B7, Canada
e-mail: sma@uwo.ca

A. Wang
Southern Crop Protection and Food Research Centre, Agriculture and Agri-Food Canada,
1391 Sandford St., London, ON N5V 4T3, Canada

A. Wang and S. Ma (eds.), *Molecular Farming in Plants: Recent Advances and Future Prospects*, DOI 10.1007/978-94-007-2217-0_1,

1.1 Introduction

Historically plants have been a primary source for medicinal products and many of the currently available drugs have been directly or indirectly derived from plants. For example, metformin, today's foundation therapy for type 2 diabetes, is derived from galegine, which is naturally found in the herb Goat's rue (G*allega officinalis*) (Witters 2001). In recent decades, advances in molecular biology and genetic engineering have created opportunities to expand the use of plants for a much broader range of products of medical importance. Indeed, molecular farming using genetically engineered plants has emerged as a promising new approach for production of biopharmaceutical products, such as monoclonal antibodies, vaccines, growth factors, cytokines and enzymes. Since the first recombinant protein drug, human insulin, was approved by the U.S. Food and Drug Administration (FDA) 29 years ago, there has been a dramatic increase in the number and frequency of use of therapeutic proteins, largely because protein and peptide drugs, especially antibodies, usually display excellent potency and selectivity for the disease target and have low incidence of toxicity. Presently, a quarter of all FDA approved new drugs are biopharmaceuticals, and by January of 2009, the number of biopharmaceutical drugs licensed by the FDA and the European Medicines Agency (EMA) reached 151 (Ferrer-Miralles et al. 2009). This number is expected to grow considerably over the next few years, because development of new protein and peptide therapeutics is continuing at a rapid pace.

Up to now, most of the clinically available protein and peptide drugs are still derived primarily from mammalian cells, *E. coli*, and, to a lesser extent, yeast and insect cells. While useful, these conventional cell culture-based expression systems are usually limited by high production cost, low efficiency and low yield. Moreover, the use of mammalian cells as a production platform also raises concerns about product safety. The requirement for large quantities of economic and safe therapeutic proteins has fueled a growing interest in the production of recombinant proteins in plant bioreactors.

Plants offer a valuable alternative to traditional fermentation-based expression systems for the commercial production of pharmaceutical proteins, having a number of advantages. First is the ability to perform eukaryotic post-translational modifications such as glycosylation and disulfide bridging that are often essential for biological activity of many mammalian proteins (Ma et al. 2003; Tremblay et al. 2010). Secondly, there is little to no risk of contamination with human pathogens, which is always a concern when using cultured mammalian cells as a bioreactor. Thirdly, plant growth requirements are simple and inexpensive compared to traditional cell culture systems, allowing for inexpensive and nearly unlimited scalability. Plant cell cultures have much simpler growth requirements than mammalian or insect cell cultures, able to utilize light as the main energy source, reducing costs. Last but not least, plant systems are robust and inert, allowing for simplified handling/purification and the ability, in the case of pharmaceutically relevant proteins, to be administered orally with minimal processing. This chapter discusses the technological basis of molecular farming in plants, with a focus on host systems and approaches/strategies developed to maximize protein yields and to ensure efficient recovery and purification of plant-made recombinant products.

1.2 Overview of Plant-Based Expression Systems

There are four major plant-based expression systems available for the production of foreign proteins. Each system has its own characteristics and offers specific benefits. The availability of different plant-based expression systems allows researchers greater convenience and flexibility to choose a more effective system for the production of a particular target protein, given that each protein has its own unique amino acid sequence that will lead to a unique structure, and therefore may require the use of different expression systems in order to achieve maximum protein yield.

1.2.1 Transgenic Plants

Transgenic plants are plants that have been genetically modified by introduction of a foreign gene (transgene) into the nuclear genome of a host plant through the employment of recombinant DNA technology. Thus far, transgenic plants have been the most commonly used expression platform for production of proteins of pharmaceutical importance. One of the main advantages of using nuclear-transformed whole plants as a bioreactor is its flexibility and the efficiency in scaling up the production of recombinant proteins, which can be achieved simply by planting more acres of the biotech crops. Another major advantage is the long-term continuous production of recombinant proteins with little to no external input, because foreign genes are stably integrated into the nuclear genome of the host plant and are inherited in the next generation, and as such, stable and predictable transgene expression can be maintained over many generations (Tremblay et al. 2010). Additionally, the synthesis of foreign proteins in nuclear transgenic plants can be readily targeted to edible plant organs such as leaves, seeds or fruits, allowing them to be delivered orally without the need to perform tedious, expensive downstream processing of expressed products. The disadvantage of transgenic plants as a protein production platform, on the other hand, is that a relatively long period of time (6–9 months) is required to generate plants expressing the target protein at usable levels. There are also some biosafety issues raised with the use of whole plants, as tansgenes could accidently escape from transgenic crops into wild populations through seed mixing or cross-pollination, causing environmental and ecological problems (Pilson and Prendeville 2004).

There are two major types of techniques used for generating transgenic plants: Agrobacterium-medicated transformation and biolistic transformation. The soil bacterium *Agrobacterium tumefaciens,* which causes crown gall disease in many fruit plants, is well known for its ability to infect plants with a tumor-inducing (Ti) plasmid.

A section of the Ti plasmid, called T-DNA, integrates into chromosomes of the plant. Recombinant DNA can be added to the T-DNA and infection by the bacteria containing the recombinant plasmid will provide for transfer of novel genes to plant cells (Gelvin 2003). As not all plants are equally susceptible to *Agrobacterium* transformation, an alternative method to deliver foreign genes into plant cells is through a gene gun, which literally shoots gold or tungsten particles that carry "naked DNA" on their surface into living plant cells,. This method, commonly known as particle bombardment ('biolistics'), has been applied successfully for many cultivated crops, especially monocots like wheat or maize, for which transformation using Agrobacterium is less efficient (Shrawat and Lörz 2006).

1.2.2 Chloroplast Transformed Plants

Chloroplast transformation offers a useful alternative stable expression system to nuclear transformation for the production of therapeutic proteins in plants. A key advantage of expressing foreign genes in plant chloroplasts as compared to their expression in nuclear transformed plants is the high-level accumulation of protein products and transgene containment. The plastid genome is highly polyploid. A typical tobacco leaf cell contains as many as 100 chloroplasts per cell with up to 100 genome copies per chloroplast, and therefore the copy number of any introduced transgene can be amplified by as many as 10,000 per cell, leading to extraordinarily high levels of foreign protein products (Chebolu and Daniell 2009). Indeed, while the level of accumulation appears to vary depending on the expression cassette used and the type of protein being expressed, it is not uncommon to see expression levels in plant chloroplasts ranging from 5% to 20% TSP, an amount that is difficult to achieve using nuclear-transformed plants that give rise to the expression levels within the typical range of 0.001–1% TSP (Chebolu and Daniell 2009; Tremblay et al. 2010). Moreover, because chloroplast genomes are maternally inherited in most plant species, there is minimal chance of the transgene being transferred by cross pollination to wild or related species, thus posing no or lower risk to the environment. Other advantages of chloroplast bioreactors include the precise integration of transgenes by homologous recombination, which is in contrast to nuclear transformation in plants that occurs by the random integration of transgenes into unpredictable locations resulting in varying levels of expression and, in some cases, gene silencing, the ability to express multiple genes in a single transformation event (transgene stacking), as well as the ability to perform the complex post-translational modifications such as disulfide bond formation, protein lipidation, folding and assembly (Verma and Daniell 2007; Chebolu and Daniell 2009; Tremblay et al. 2010). Due to these advantages, transformed chloroplasts have been increasingly used as a bioreactor system in order to achieve high-level expression of pharmaceutical proteins in plants. For example, Ruhlman et al. (2007) reported a 160-fold increase in the accumulation of CTB-insulin fusion protein when expressed in tobacco chloroplasts as compared to its expression in nuclear-transformed tobacco plants. Other examples demonstrating the potential of chloroplast transformation to achieve high yield of foreign proteins in plants include the expression of human immunodeficiency virus (HIV-1) Gag (Pr55gag) polyprotein, a primary HIV vaccine candidate, at expression levels up to 7–8% of TSP in tobacco (Scotti et al. 2009), expression of the p24 and Nef antigens, also HIV vaccine candidates, as a fusion protein in both tobacco and tomato chloroplasts (up to 40% TSP) (Zhou et al. 2008), and the expression of human papillomavirus type 16 (HPV16) L1 structural protein, a vaccine candidate for cervical cancer, in tobacco (up to 24% TSP) (Fernández-San Millán et al. 2008).

A limitation of using plant chloroplasts as an expression platform, however, is that, like bacteria, they are unable to perform glycosylation, a necessity for many pharmaceutical glycoproteins including monoclonal antibodies (Tremblay et al. 2010). Moreover, chloroplast transformation is typically achieved through particle bombardment, and thus far, has been achieved routinely only in a few plant species, although plastid transformation in tobacco was reported over 20 years ago (Svab et al.1990). This may limit the choice of plant species that can be selected as a host system for molecular farming via chloroplast transformation.

1.2.3 Transient Expression

Transient gene expression provides another plant-based platform for production of recombinant pharmaceutical proteins in plants. The system offers a number of advantages over stable gene expression. One of the most important advantages of transient gene expression system is its speed of protein production. Useful amounts of target proteins can be generated within a matter of days or weeks, which is not achievable via stable gene expression. This may be of critical importance in ensuring a rapid and effective response to a sudden outbreak of an infectious disease, such as the 2009 influenza A/H1N1 pandemic that would require a recombinant vaccine product to be made in sufficient quantities within a very short time period in order to meet the needs of a pandemic (Tremblay et al. 2010). The other attractive features of this system include its simplicity and ease of performance, requiring no expensive supplies and equipments. Moreover, transient expression does not depend on chromosomal integration of foreign genes so its expression is not affected by position effects as often observed in nuclear-transformed transgenic plants (Sheludko 2008).

Different approaches have been developed to perform transient gene expression in plants, largely including infiltration of intact plant tissue with recombinant agrobacteria (agroinfiltration) and infection with modified viral vectors. In agroinfiltration, agrobacteria carrying T-DNA harbouring the gene of interest are forced into intact or harvested plant leaf tissues by pressure. This method has now been successfully used for rapid production of clinical grade bio-pharmaceuticals (Pogue et al. 2010). The viral infection method, on the other hand, involves using plant viruses as vectors to deliver foreign genes into plants. The major advantage of virus-mediated transient expression is the rapid and high levels of production of recombinant proteins, as plant viral vectors are able to systemically infect all cells of a plant after inoculation, generating many transcripts of the transgene (Fischer and Emans 2000). Tobacco mosaic virus (TMV) and potato virus X (PVX) are the most commonly used viral vector systems for transient production of pharmaceutically relevant proteins. Using these plant virus-based transient expression systems, a number of pharmaceutical proteins have been produced, including full size antibodies and antibody fragments (Canizares et al. 2005; Lico et al. 2008; Regnard et al. 2010). However, a concern about the use of viral vectors for the expression of proteins is that they may spread into the environment. Furthermore, there are strict limitations on the size of the insert that can be introduced into the virus genomes and successfully assembled into virions (Yusibov et al. 1997). To overcome these disadvantages, Icon Genetics (Halle, Germany) has developed a TMV-based "deconstructed" viral vector system, known as magnICON, for the over-expression of foreign genes. The MagnICON system combines the advantages of *Agrobacterium*-mediated delivery with the speed and expression level/yield of a virus (Gleba et al. 2005). In this system, the TMV genome has been optimized by the removal of cryptic sequences that, if recognized by the plant machineries, could affect viral replication/spreading, and splitted into 5′ and 3′ modules. The 5′ module includes the viral polymerase and the movement protein genes while the 3′ module carries the gene of interest in place of the viral coat protein gene. The two modules, once inserted into binary vectors and delivered to the plant cell nucleus by Agrobacterium mediated infiltration, are then assembled together in vivo by a site specific recombinase delivered by a third *Agrobacterium* cell line (Marillonnet et al. 2004, 2005). This "deconstructed" virus system that lacks the coat protein gene is unable to spread throughout the plant and into the environment, and is not influenced by the dimension of the inserted gene due to the absence of the packaging process, while leads the plant cell machinery to the production of the heterologous protein. This system proved to be very efficient for expression of proteins of interest, with expression levels increased by up to 100-fold. For example, Giritch et al. (2006) reported that by using the MagnICON system, they were able to generate an assembled full-size monoclonal antibody at levels as high as 0.5 g of mAb per kg of fresh leaf biomass.

1.2.4 Plant Cell Suspension Cultures

Plant cell suspension culture provides another attractive plant-based production platform for production of pharmaceutical proteins. The system integrates many of the merits of the whole-plant system with those of microbial and animal cell cultures (Xu et al. 2011). As a production system, plant cell suspension cultures maintain the merits of whole-plant systems, i.e., product safety, easy scale-up, post-translational modifications and ability to synthesize correctly folded and assembled multimeric proteins. On the other hand, plant cells, like bacteria, have relatively rapid doubling times and can be grown in simple synthetic media using conventional bioreactors.

Moreover, plant cells have the potential to secrete the expressed proteins into the culture medium, making product recovery and purification substantially simpler and cheaper than that from plant biomass (Xu et al. 2011). Recombinant glycoproteins derived from cultured plant cells also often have reduced N-glycosylation heterogeneity, owing to the uniformity of cultured plant cell populations (De Muynck et al. 2009). Tobacco suspension cells, particularly those from the closely related cartivars BY-2 and NT-1, are frequently chosen as host cell lines because they exhibit high growth rate and high homogeneity, and are easy to transform by using either particle bombardment or co-cultivation with *Agrobacterium tumefaciens* (Hellwig et al. 2004). Examples of tobacco cell suspension culture-derived pharmaceutical proteins include antibodies (Fischer et al.1999), human interleukin (IL)-2 and IL-4 (Magnuson et al. 1998), and human granulocyte colony-stimulating factor (hG-CSF) (Hong et al.2006). Tobacco cell suspension culture-derived Newcastle disease virus (NDV) vaccine, which was produced by Dow AgroSciences (Indianapolis, IN, USA), has obtained regulatory approval from the US Department of Agriculture (USDA), representing the world's first plant-derived vaccine approved for veterinary application. Cell suspension cultures derived from rice, soybean and tomato have also been used for the production of recombinant proteins (Hellwig et al. 2004). Protalix Biotherapeutics (Carmiel, Israel) has developed the use of suspension cultures of carrot cells to produce recombinant human glucocerebrosidase for the treatment of Gaucher's disease (Shaaltiel et al. 2007), and preclinical and phase I human trials showed that the plant cell-derived protein is safe to use (Aviezer et al. 2009). However, the relatively low protein yields from plant cell suspension cultures remain as a major challenge to be overcome (Hellwig et al. 2004).

1.3 Approaches for Increasing Heterologous Protein Accumulation in Plants

There is now little doubt that the plant production systems offer many benefits over conventional production systems. However, there are still a number of challenges that remain to be overcome before plants can gain widespread acceptance as a mainstream production platform. The relatively low yield poses one of the main challenges in promoting plants as a major protein production system. In the past several years, various molecular approaches have been developed to address the challenge of low product yields in plant-based production systems. Many of these approaches have produced promising results.

1.3.1 Boosting Transcription

Expression of a foreign gene in plants is regulated at different levels. At transcriptional levels, promoter elements can have a dramatic effect on the level of messenger RNA of a transgene, which in turn affects accumulation levels of the protein product. Therefore, increasing the rate of transcription initiation via the use of a strong, constitutive promoter has the potential to enhance recombinant protein yields.

In this respect, the 35S promoter of cauliflower mosaic virus (CaMV) has often been used to drive transgene expression in transgenic plants as well as in transient gene expression systems (other than viral vector-based expression) because it is a strong promoter that is active in all or most cells of many dicot plant species (Twyman et al. 2003; Tremblay et al. 2010). Using the CaMV 35S promoter, a number of pharmaceutically relevant proteins have been produced at relatively high levels in plants, including antibodies, vaccine antigens, autoantigens and growth factors (Tremblay et al. 2010). The CaMV 35S promoter, however, has a much weaker activity in monocots, so alternative promoters are used to drive transgene expression when cereal crops serve as the expression hosts. The promoters of the constitutively expressed rice (*Oryza sativa*) actin 1 gene and corn (*Zea mays*) ubiquitin-1 gene are often the preferred choice (Twyman et al. 2003; Streatfield 2007). In the case that the expressed foreign protein is toxic to the host cells, organ-specific or inducible expression would be necessary. Several regulated promoters, such as organ/tissue- or developmental stage-specific promoters and physically/chemically-inducible promoters, have been developed to meet this need (Twyman et al. 2003; Streatfield 2007). Indeed, tissue- or organ-specific expression of several pharmaceutical proteins, including vaccine antigen HBsAg M, murine single chain variable fragment (scFv) G4 and human interferon-α, has been achieved by using a regulated promoter system (He et al. 2008; De Jaeger et al. 2002).

Several strategies have been followed to further boost transcription over that achieved with plant or plant viral promoters. One strategy is to further optimize promoter activity. For example, incorporation of an additional duplicated enhancer element to CaMV 35S promoter resulted in tenfold higher promoter activity than the natural CaMV35S promoter (Kay et al. 1987). Another method to increase promoter strength is to design novel synthetic promoters combining the most active sequences of multiple well-characterized natural promoters. Indeed, synthetic hybrid promoters consisting of a combination of elements from both the CaMV 35S promoter and the Agrobacterium Ti plasmid mannopine synthase promoter showed activity several-fold higher than either of the two parental promoters (Comai et al. 1990). As polyadenylation influences mRNA stability and translational efficiency, the choice of a poly(A) signal is also important for optimal transgene expression in plants. The poly(A) signal from nopaline synthase (nos) gene of the Ti plasmid of *A. tumefaciens* proved to be highly efficient for increasing mRNA levels of both transiently and stably expressed transgenes, and is often chosen to add to the expression construct (Streatfield 2007). Recently, Nagaya et al. (2010) demonstrated that the heat shock protein 18.2 (HSP 18.2) poly(A) signal of *Arabidopsis thaliana* is more effective than the nos poly (A) signal in enhancing the expression of a transgene in transfected Arabidopsis protoplasts, with more than twofold increase in the transgene mRNA levels. Furthermore, the combination of HSP poly (A) signal and a translational enhancer, the 5′-UTR of the tobacco alcohol dehydrogenase gene (NtADH), resulted in a 60- to 100-fold increase in the amount of the transgene protein compared to the combination of the nos poly (A) signal and NtADH 5′-UTR in transgenic tobacco plants. In plants, many introns enhance gene expression through an increase in the steady-state level of mRNA, referred to as intron-mediated enhancement of gene expression (Rose 2002). Inclusion of one or more introns in a gene construct could therefore be used as a strategy to increase the expression in plants of a transgene that otherwise lacks introns, such as a cDNA or bacterial gene (Koziel et al. 1996). Indeed, Bartlett et al. (2009) demonstrated that the inclusion of an intron at a specific position within the coding sequence of the transgene luciferase significantly increased luciferase activity in transgenic barlay plants. Moreover, the enhanced luciferase activity was stably maintained in the T1 and T2 progeny transgenic lines. Other strategies proven to increase transgene expression include the elimination of cryptic splicing sites and mRNA destabilizing elements to increase RNA stability, and the use of viral suppressors of silencing to prevent or reverse post-transcriptional gene silencing (Desai et al. 2010).

1.3.2 Boosting Translation

In addition to increasing transcription, it is important to maximize the amount of protein translated per unit of mRNA, as mRNA abundance cannot necessarily give rise to high protein levels. By estimation, only 20–40% of protein abundance is determined by the concentration of its corresponding mRNA (Tian et al. 2004; Nie et al. 2006). Protein concentrations depend not only on the mRNA level, but also on the translation rate and the degradation rate. In plants, the efficiency of mRNA translation is the most important determinant of protein abundance (Kuroda and Maliga 2001). Different strategies have been employed to further boost the translational efficiency of transgene mRNAs. One is to add a plant viral 5′ non-translated sequence to the transgene construct to increase the rate of translation initiation. For example, insertion of the 5′-untranslated leader sequence from alfalfa mosaic virus (AMV) RNA 4 between the CaMV 35S promoter and the human respiratory syncytial virus (RSV) fusion protein (F) gene increased viral protein expression by over five fold as compared to its expression from the same construct without this AMV RNA 4 leader sequence in apple leaf protoplasts (Sandhu et al. 1999). The 5′-untranslated leader sequence of tobacco mosaic virus RNA, potato virus X (PVX) RNA or rice seed storage-protein genes has also proven to be effective in boosting transgenic protein expression (Liu et al. 2010). Optimization of code usage of transgenes to match the host's preferred codon usage is another strategy for enhancing foreign protein expression in plants. Transgenes from heterologous species often have a different codon bias to the host plant, which might result in pausing at disfavoured codons and truncation, misincorporation or frameshifting. Such effects can be avoided by introducing silent mutations into the coding region of the transgene by site-directed mutagenesis, which brings transgene codon usage in line with that of the host (Ma et al. 2003). Indeed, Kang et al. (2004) showed that expression of a synthetic CTB gene with codon usage optimized to tobacco plants resulted in approximately 15-fold increase in the protein level compared to its expression using the unmodified native gene. Other strategies for increasing recombinant protein accumulation include optimization of the sequence context of the AUG translation initiation codon, flanking transgenes with nuclear matrix attachment regions, and co-expression with protease inhibitors (Desai et al. 2010).

1.3.3 Subcellular Targeting

Targeting of heterologous proteins to the appropriate subcellular compartment can be critical for obtaining high levels of accumulation, since the structure and stability of the recombinant protein is affected by its route and final destination in the cell (Streatfield 2007). The endoplasmic reticulum (ER) is the port of entry of the protein secretory pathway. Proteins destined for the cell wall, the vacuole or the other compartments of the endomembrane system are first inserted into the ER and then transported to the Golgi complex *en route* to their final destinations. The ER is the compartment where newly-synthesized polypeptides fold, where many multimeric proteins assemble and where glycoproteins acquire their asparagine-linked glycans. The ER also provides a protein quality control function; proteins are usually retained in this compartment until they have acquired their correct conformation. Additionally, plants naturally use the ER to accumulate vast amounts of protein in specific tissues, in the form of large aggregates or oligomers, as part of the developmental process of seed maturation (Herman and Larkins 1999; Vitale and Ceriotti 2004). All these features make the ER a potentially excellent target compartment for overexpression of foreign proteins. To retain a secretory heterologus protein in the ER, a tetrapeptide ER retention signal, KDEL or HDEL, is required at the C-terminus of the target protein. This strategy has been shown to significantly enhance the production yield of antibodies, vaccines, cytokines and many immunoregulatory proteins in transgenic plants (Vaquero et al. 2002; Ko et al. 2003; Ma et al. 2005).

Targeting to other cellular compartments also provides a useful strategy for boosting accumulation of foreign proteins in plants. Wirth et al. (2004) reported that while the expression level of human epidermal growth factor (hEGF) did not exceed 0.001% of TSP when expressed in the cytoplasm of transgenic tobacco plants, the expression level reached about 0.11% of TSP when targeted to the apoplast compartment, resulting in about 100-fold yield increase. Apoplast targeting also facilitates recombinant protein purification. Cheung et al. (2009) showed that the expression of human insulin-like growth factor binding protein-3 (IGFBP-3) can be significantly enhanced by targeting it to protein storage vacuoles in tobacco seeds, reaching 800 μg IGFBP-3/g dry weight of seeds. Targeting to plant chloroplasts has also been shown to increase foreign protein accumulation (Hyunjong et al. 2006; Van Molle et al. 2007). However, it may be necessary, at least initially, to target the protein of interest to different cellular compartments in order to evaluate the best localization for a particular protein.

1.3.4 Fusion Protein-Based Approaches

Fusion protein-based approaches provide another effective means of boosting the production of foreign proteins in plants. Fusion partners have been observed to improve protein yield, to prevent proteolysis, increase solubility and stability, and facilitate isolation and purification of target proteins (Hondred et al. 1999).

Several fusion partners/tags have been identified and used to increase the yield of foreign proteins in plants. A fusion partner that has received considerable attention is Zera, a proline-rich N-terminal domain derived from the maize storage protein γ zein with both self-assembling and protein body formation properties. Moreover, the ability of Zera to form protein bodies is not limited to native maize seeds, but can be extended to nonseed tissues of plants as well as non-plant eukaryotic hosts including cultured fungal, mammalian and insect cells (Torrent et al. 2009). Fusion of Zera to proteins of interest has proven to induce the formation of PBs that facilitate the stable accumulation of recombinant proteins at very high levels. A good example is the expression of F1-V hybrid vaccine antigens with a fused Zera in *N. benthamiana*, *Medicago sativa* (alfalfa) as well as *N. tabacum* NT1 cells, resulting in at least 3 times more recombinant protein than the control F1-V hybrid protein without a fused Zera in all plant hosts tested (Alvarez et al. 2010). F1 and V are the two most immunogenic antigens of *Yersinia pestis*, the bacterium that causes pneumonic and bubonic plague. Another example is the expression of human growth hormone in *N. benthamiana,* with significantly higher levels of the recombinant protein obtained when fused to Zera (Llompart et al. 2010).

The use of elastin-like polypeptides (ELPs) as a fusion partner to increase recombinant protein accumulation in plants has also attracted more attention. ELPs are artificial biopolymers containing repeats of the pentapeptide sequence Val-Pro-Gly-Xaa-Gly (VPGXG), where X can be any naturally occurring amino acid except Pro (Urry 1992). ELP fusions have been shown to significantly enhance the accumulation of a range of different pharmaceutical proteins in plants, including human IL-10 and murine interleukin-4 (Patel et al. 2007), the full-size anti-human immunodeficiency virus type 1 antibody 2F5 (Floss et al. 2008) and anti-foot and mouth disease virus single variable antibody fragment (Joensuu et al. 2009). In general, ELP fusion can increase the expression of recombinant proteins 2- to 100-fold, depending on the individual proteins. The ability of the ELP tag to increase recombinant protein accumulation may be associated with its natural resistance against hydrolysis by proteolytic enzymes, thus reducing the level of protein degradation, and its abilty to increase the solubility of target proteins by protecting against irreversible aggregation and denaturation at high protein concentrations (Conley et al. 2011). Moreover, it was shown that ELP fusion proteins accumulate in PB forms in the leaves and seeds of tobacco, thus insulating them from normal cellular degradation processes and increasing the overall protein yield (Conley et al. 2009). Recently, the potential of hydrophobins as fusion tags to enhance recombinant protein expression in plants has also been investigated. Hydrophobins are small surface-active fungal proteins that have a characteristic pattern of eight conserved Cys residues, which form four intramolecular disulfide bridges and are responsible for stabilizing the protein's structure (Hakanpaa et al. 2004). Joensuu et al. (2010) showed that hydrophobin fusion significantly increased the expression of GFP in transiently transformed *N. benthamiana,* representing 51% of TSP (equivalent to 5.0 mg of the fusion protein per gram of leaf fresh weight). Other fusion partners with demonstrated beneficial effects on the accumulation of target proteins in plants include ubiquitin (Hondred et al. 1999; Mishra et al. 2006), b-glucuronidase (Gil et al. 2001), cholera toxin B subunit (Kim and Langridge 2003; Tremblay et al. 2008), viral coat proteins (Canizares et al. 2005) and human immunoglobulin A (IgA) (Obregon et al. 2006).

1.4 Technologies and Strategies for Recovery of Recombinant Proteins from Plants

Protein recovery and purification are regarded as another critical limiting factor in the successful commercialisation of plant-made recombinant proteins (therapeutic proteins and industrial enzymes), as processing steps represent a significant production cost for recombinant proteins. For example, cost analysis of producing the enzyme β-glucuronidase from transgenic corn revealed that milling, protein extraction and purification operations accounted for ~94% of the production cost (Evangelista et al. 1998), suggesting the importance of downstream processing in determining the economic feasibility of plant-based protein production. Compared to bacterial or yeast systems, protein isolation and purification from leafy plants is more difficult and costly, owing to the complexity of the plant system, and has been considered as a major challenge in plant molecular farming. Therefore, development of simple, low-cost, and reliable methods for purification of target proteins from plant materials is critically important. Toward this end, several methods and strategies have been developed and used for recovery and purification of plant-made recombinant proteins.

1.4.1 Affinity Fusion-Based Protein Purification

The addition of an affinity tag sequence to the genes of interest is the most commonly used method to assist in the purification of a protein. This gene fusion results in the expression of an affinity-tagged fusion protein that can be easily purified via an affinity matrix column. An ideal affinity tag should be small and allow for the rapid and flexible purification from a complex mixture, achieving high yield and purity (Terpe 2003). The poly-histidine tag, usually comprising 6 consecutive histidines (6xHis), is the most widely used affinity tag for purification of recombinant proteins from various expression systems, because it can be performed inexpensively with large volumes and can provide good yields of tagged proteins. The purification of His-tagged proteins is based on the use of a chelated metal ion as an affinity ligand. A commonly used ion is the immobilized nickel-nitrilotriacetic acid chelate [Ni–NTA], which is bound by the imidazole side chain of histidine. While His-tag purification can achieve high levels of purity of tagged proteins in bacterial systems, it only achieves low purity in plants (Sharma and Sharma 2009). StrepII, another frequently used affinity purification tag, consists of eight amino acids (WSHPQFEK), and binds to a streptavidin derivative termed Strep Tactin. StrepII structurally mimics biotin and can be eluted from the StrepTactin column by washing with biotin or desthiobiotin containing buffers. Witte et al. (2004) reported the use of StrepII as an affinity tag to purify recombinant membrane-anchored protein kinase, NtCDPK2, from leaf extracts of transiently transformed *N. benthamiana,* and demonstrated that StrepII-tagged NtCDPK2 could be purified to almost complete homogeneity in one-step using a Strep-Tactin column. Other affinity tags useful for recovering recombinant proteins include FLAG, c-myc, glutathione S-transferase, calmodulin-binding peptide, maltose-binding protein and the cellulose-binding domain (Terpe 2003). For some applications, small tags may not need to be removed following protein purification, as they often do not interfere with protein performance. While affinity fusion-based chromatography methods allow for fast, simple, one-step purification of recombinant products and remain the most efficient protein purification approach, they are generally used only for relatively small-scale purification. For large-scale protein purification, these methods may be too costly. Also, due to their small sizes, most affinity tags do not generally increase the expression of the fusion proteins or enhance their solubility.

1.4.2 Non-affinity-Based Protein Purification

Several non-affinity-based purification methods have also been developed for isolation of plant-made proteins. While elastin-like polypeptide (ELP) provides an efficient fusion protein tag to enhance the production of plant-made recombinant proteins, it also supports the development of a non-affinity-based method for protein purification. ELP consists of repeats of the pentapeptide sequence and undergoes a reversible inverse temperature transition. When the temperature is raised above the inverse transition temperature (T_t), the normally soluble ELP fused polypeptides form insoluble aggregates which can be separated by simple centrifugation. Subsequently, the aggregate can be resolubilized easily by a temperature shift below T_t. This non-chromatographic purification method was termed inverse transition cycling (ITC) (Urry 1992). ITC has been used to purify cytokines (Lin et al. 2006), antibodies (Joensuu et al. 2009), and vaccinal antigens (Floss et al. 2010) from transgenic plant tissues, with typical recovery rates of 30–60%. These rates of protein recovery are, however, lower compared to the recovery rate of 75–95% achieved in bacterial expression systems. The reduced recovery rates of plant-made recombinant proteins by using the same ITC is attributable partly to their much lower expression levels in stable transgenic plants (i.e. <1% TSP) relative to bacteria (i.e. 0.1–1.6 g/L) (Conley et al. 2011).

Like ELP, hydrophobins used as fusion protein tags can not only enhance foreign protein production in plants, but can also aid in their purification by using non-affinity methods. Hydrophobins have been described as the most powerful surface active proteins known and are capable of altering the hydrophobicity of their respective fusion partners, thus enabling efficient purification using a surfactant-based aqueous two-phase system (ATPS) (Linder et al. 2004). Joensuu et al. (2010) reported that by using the ATPS method, they were able to recover up to 91% of the GFP-hydrophobin fusion protein from leaf extracts of transiently transformed *N. benthamiana.* Oleosin fusion provides another method for recovering plant-made proteins via non-affinity methods. Oleosins are hydrophobic proteins localized abundantly in the oil bodies of plant seeds. Oleosin fusion products are capable of integrating into natural oil bodies in transgenic plant seeds, where oil bodies can be easily separated from the rest of the cellular extract by flotation centrifugation (van Rooijen and Moloney 1995). Using this system, hirudin, a pharmaceutical protein commonly used as anticoagulants to prevent thrombosis, has been successfully expressed and purified from seeds of *Brassica napus* (Parmenter et al. 1995).

1.4.3 Soybean Agglutinin Fusion as a Novel System for High-Level Expression and Purification of Plant-Made Proteins

Soybean agglutinin (SBA) is a tetrameric lectin glycoprotein found in soybean seeds. SBA binds to *N*-acetyl-D-galactosamine and is able to induce the agglutination of cells with this glycan on their surface. SBA accumulates to nearly 2% of the soluble protein in soybean seeds and can be isolated from soybean flour with more than 90% yield following one-step affinity purification on beads bearing *N*-acetyl-D-galactosamine (Percin et al. 2009). Therefore, SBA has great potential for exploration as a novel fusion tag to improve the expression and purification of plant-made proteins. To provide a "proof of concept" for the utility of SBA as a new fusion tag for expression and purification of plant-made proteins, we initially expressed SBA as a non-fusion protein in transiently transformed *N. benthamiana,* with expression levels higher than 4% of TSP. As expected, plant-derived rSBA forms stable tetramers, binds specifically to *N*-acetyl-D-galactosamine, and retains the ability to induce red blood cell agglutination. Moreover, plant-made rSBA can be purified from total plant extracts in high purity through a simple one-step N-acetyl-D-galactosamine agarose column (Tremblay et al. 2011a). Following this initial success, we subsequently expressed SBA as a fusion protein with the reporter protein GFP in transiently transformed *N. benthamiana* and achieved expression levels higher than 2% of TSP. Plant-derived SBA-GFP fusion protein was found to assemble into correct tetramers essential for its stability, retain the capacity of SBA to induce red blood cell agglutination and the ability of GFP to fluoresce. More importantly, the fusion protein can be quickly purified to a high degree of purity via the simple one-step N-acetyl-D-galactosamine agarose column (see Figs. 1.1 and 1.2) (Tremblay et al. 2011b). Taken together, these results confirm the potential utility of SBA as a new tool to express and recover recombinant proteins in plants.

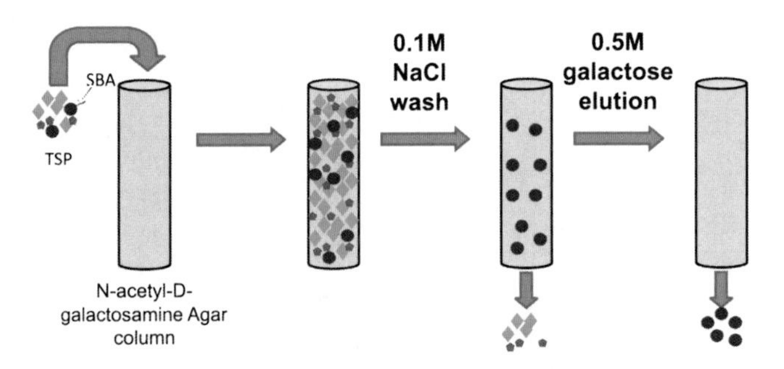

Fig. 1.1 Schematic diagram of one-step purification of plant-made recombinant proteins using an N-acetyl-D-galactosamine-linked agarose column. The total plant protein extract containing a SBA-taged recombinant protein is loaded onto the sugar column. The column is then washed with 0.1 M NaCl solution to remove any non-specific proteins bound to the column. SBA-tagged fusion protein is then eluted from the column by adding 0.5 M galactose

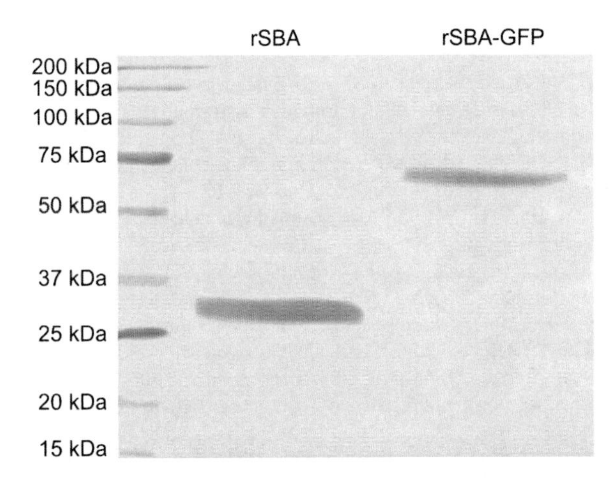

Fig. 1.2 Coomassie Blue stained 12.5% (w/v) SDS-PAGE gel demonstrating purity of rSBA-GFP following purification through an N-acetyl-D-galactosamine-linked agarose column. Purified rSBA from *N. benthamiana* was used as a control. The single band in each lane corresponds to the expected size for rSBA and rSBA-GFP, respectively. Approximinately 12 and 9 μg of purified rSBA and rSBA-GFP were loaded, respectively (Adapted with the permission from Tremblay et al. 2011b)

1.5 Conclusions

Plant-based molecular farming has the potential to become a major new method for low-cost, mass and safe production of biopharmaceuticals. There are several plant-based expression systems that are currently being explored to serve as production platforms, each offering specific benefits. Moreover, a number of technologies and strategies have been developed to further enhance the yield of foreign proteins in plant systems. Several protein purification methods were also developed to isolate recombinant proteins from plant materials. These technological advancements in plant molecular farming have translated into rapid growth in the number of plant-made biopharmaceuticals. It is now possible to use plants to produce a wide range of therapeutic products, including complex molecules such as monoclonal antibodies. Importantly, many plant-made pharmaceuticals have already achieved preclinical validation in a range of disease models following either injections or oral delivery, with some plant-made vaccines in Phase II and Phase III clinical trials (Daniell et al. 2009). While there are still some issues that need to be addressed, such as the possibility of plant-specific glycans inducing allergic responses in humans and regulatory barriers to commercialization, the potential benefit of plant-made pharmaceuticals to human health should not be underestimated.

References

Alvarez ML, Topal E, Martin F, Cardineau GA (2010) Higher accumulation of F1-V fusion recombinant protein in plants after induction of protein body formation. Plant Mol Biol 72:75–89

Aviezer D, Brill-Almon E, Shaaltiel Y, Hashmueli S, Bartfeld D, Mizrachi S, Liberman Y, Freeman A, Zimran A, Galun E (2009) A plant-derived recombinant human glucocerebrosidase enzyme–a preclinical and phase I investigation. PLoS One 4:e4792

Bartlett JG, Snape JW, Harwood WA (2009) Intron-mediated enhancement as a method for increasing transgene expression levels in barley. Plant Biotechnol J 7:856–866

Canizares MC, Nicholson L, Lomonossoff GP (2005) Use of viral vectors for vaccine production in plants. Immunol Cell Biol 83:263–270

Chebolu S, Daniell H (2009) Chloroplast-derived vaccine antigens and biopharmaceuticals: expression, folding, assembly and functionality. Curr Top Microbiol Immunol 332:33–54

Cheung SC, Sun SS, Chan JC, Tong PC (2009) Expression and subcellular targeting of human insulin-like growth factor binding protein-3 in transgenic tobacco plants. Transgenic Res 18:943–951

Comai L, Moran P, Maslyar D (1990) Novel and useful properties of a chimeric plant promoter combining CaMV 35S and MAS elements. Plant Mol Biol 15:373–381

Conley AJ, Joensuu JJ, Menassa R, Brandle JE (2009) Induction of protein body formation in plant leaves by elastin-like polypeptide fusions. BMC Biol 7:4

Conley AJ, Joensuu JJ, Richman A, Menassa R (2011) Protein body-inducing fusions for high-level production and purification of recombinant proteins in plants. Plant Biotechnol J 9:419–433

Daniell H, Singh ND, Mason H, Streatfield SJ (2009) Plant-made vaccine antigens and biopharmaceuticals. Trends Plant Sci 14:669–679

De Jaeger G, Scheffer S, Jacobs A, Zambre M, Zobell O, Goossens A, Depicker A, Angenon G (2002) Boosting heterologous protein production in transgenic dicotyledonous seeds using Phaseolus vulgaris regulatory sequences. Nat Biotechnol 20:1265–1268

De Muynck B, Navarre C, Nizet Y, Stadlmann J, Boutry M (2009) Different subcellular localization and glycosylation for a functional antibody expressed in Nicotiana tabacum plants and suspension cells. Transgenic Res 18:467–482

Desai PN, Shrivastava N, Padh H (2010) Production of heterologous proteins in plants: strategies for optimal expression. Biotechnol Adv 28:427–435

Evangelista RL, Kusnadi AR, Howard JA, Nikolov ZL (1998) Process and economic evaluation of the extraction and purification of recombinant β-glucuronidase from transgenic corn. Biotechnol Prog 14:607–614

Fernández-San Millán A, Ortigosa SM, Hervás-Stubbs S, Corral-Martínez P, Seguí-Simarro JM, Gaétan J, Coursaget P, Veramendi J (2008) Human papillomavirus L1 protein expressed in tobacco chloroplasts self-assembles into virus-like particles that are highly immunogenic. Plant Biotechnol J 6:427–441

Ferrer-Miralles N, Domingo-Espín J, Corchero JL, Vázquez E, Villaverde A (2009) Microbial factories for recombinant pharmaceuticals. Microb Cell Fact 8:17

Fischer R, Emans N (2000) Molecular farming of pharmaceutical proteins. Transgenic Res 9: 279–299

Fischer R, Liao YC, Drossard J (1999) Affinity-purification of a TMVspecific recombinant full-size antibody from a transgenic tobacco suspension culture. J Immunol Methods 226:1–10

Floss DM, Sack M, Stadlmann J, Rademacher T, Scheller J, Stöger E, Fischer R, Conrad U (2008) Biochemical and functional characterization of anti-HIV antibody-ELP fusion proteins from transgenic plants. Plant Biotechnol J 6:379–391

Floss DM, Mockey M, Zanello G, Brosson D, Diogon M, Frutos R, Bruel T, Rodrigues V, Garzon E, Chevaleyre C, Berri M, Salmon H, Conrad U, Dedieu L (2010) Expression and immunogenicity of the mycobacterial Ag85B/ESAT-6 antigens produced in transgenic plants by elastin-like peptide fusion strategy. J Biomed Biotechnol. doi:10.1155/2010/274346

Gelvin BS (2003) Agrobacterium-mediated plant transformation: the biology behind the "gene-jockeying" tool. Microbiol Mol Biol Rev 67:16–37

Gil F, Brun A, Wigdorovitz A, Catalá R, Martínez-Torrecuadrada JL, Casal I, Salinas J, Borca MV, Escribano JM (2001) High-yield expression of a viral peptide vaccine in transgenic plants. FEBS Lett 488:13–17

Giritch A, Marillonnet S, Engler C, van Eldik G, Botterman J, Klimyuk V, Gleba Y (2006) Rapid high-yield expression of full-size IgG antibodies in plants coinfected with noncompeting viral vectors. Proc Natl Acad Sci USA 103:14701–14706

Gleba Y, Klimyuk V, Marillonnet S (2005) Magnifection–a new platform for expressing recombinant vaccines in plants. Vaccine 23:2042–2048

Hakanpää J, Paananen A, Askolin S, Nakari-Setälä T, Parkkinen T, Penttilä M, Linder MB, Rouvinen J (2004) Atomic resolution structure of the HFBII hydrophobin, a self-assembling amphiphile. J Biol Chem 279:534–539

He ZM, Jiang XL, Qi Y, Luo DQ (2008) Assessment of the utility of the tomato fruit-specific E8 promoter for driving vaccine antigen expression. Genetica 133:207–214

Hellwig S, Drossard J, Twyman RM, Fischer R (2004) Plant cell cultures for production of recombinant proteins. Nat Biotechnol 22:1415–1421

Herman EM, Larkins BA (1999) Protein storage bodies and vacuoles. Plant Cell 11:601–614

Hondred D, Walker JM, Mathews DE, Vierstra RD (1999) Use of ubiquitin fusions to augment protein expression in transgenic plants. Plant Physiol 119:713–724

Hong SY, Kwon TH, Jang YS, Kim SH, Yang MS (2006) Production of bioactive human granulocyte-colony stimulating factor in transgenic rice cell suspension cultures. Protein Expr Purif 47:68–73

Hyunjong B, Lee DS, Hwang I (2006) Dual targeting of xylanase to chloroplasts and peroxisomes as a means to increase protein accumulation in plant cells. J Exp Bot 57:161–169

Joensuu JJ, Brown KD, Conley AJ, Clavijo A, Menassa R, Brandle JE (2009) Expression and purification of an anti-Foot-and-mouth disease virus single chain variable antibody fragment in tobacco plants. Transgenic Res 18:685–696

Joensuu JJ, Conley AJ, Lienemann M, Brandle JE, Linder MB, Menassa R (2010) Hydrophobin fusions for high-level transient protein expression and purification in Nicotiana benthamiana. Plant Physiol 152:622–633

Kang TJ, Loc NH, Jang MO, Yang MS (2004) Modification of the cholera toxin B subunit coding sequence to enhance expression in plants. Mol Breed 13:143–153

Kay R, Chan A, Daly M, McPherson J (1987) Duplication of CaMV 35S promoter sequences creates a strong enhancer for plant genes. Science 236:1299–1302

Kim TG, Langridge WH (2003) Assembly of cholera toxin B subunit full-length rotavirus NSP4 fusion protein oligomers in transgenic potato. Plant Cell Rep 21:884–890

Ko K, Tekoah Y, Rudd PM, Harvey DJ, Dwek RA, Spitsin S, Hanlon CA, Rupprecht C, Dietzschold B, Golovkin M, Koprowski H (2003) Function and glycosylation of plant-derived antiviral monoclonal antibody. Proc Natl Acad Sci USA 100:8013–8018

Koziel MG, Carozzi NB, Desai N (1996) Optimizing expression of transgenes with an emphasis on post-transcriptional events. Plant Mol Biol 32:393–405

Kuroda H, Maliga P (2001) Sequences downstream of the translation initiation codon are important determinants of translation efficiency in chloroplasts. Plant Physiol 125:430–436

Lico C, Chen Q, Santi L (2008) Viral vectors for production of recombinant proteins in plants. J Cell Physiol 216:366–377

Lin M, Rose-John S, Grötzinger J, Conrad U, Scheller J (2006) Functional expression of a biologically active fragment of soluble gp130 as an ELP-fusion protein in transgenic plants: purification via inverse transition cycling. Biochem J 398:577–583

Linder MB, Qiao MQ, Laumen F, Selber K, Hyytia T, Nakari-Setala T, Penttila ME (2004) Efficient purification of recombinant proteins using hydrophobins as tags in surfactant-based two-phase systems. Biochemistry 43:11873–11882

Liu WX, Liu HL, Chai ZJ, Xu XP, Song YR, le Qu Q (2010) Evaluation of seed storage-protein gene 5′untranslated regions in enhancing gene expression in transgenic rice seed. Theor Appl Genet 121:1267–1274

Llompart B, Llop-Tous I, Marzabal P, Torrent M, Pallissé R, Bastida M, Ludevid MD, Walas F (2010) Protein production from recombinant protein bodies. Process Biochem 45:1816–1820

Ma JK, Drake PMW, Christou P (2003) The production of recombinant pharmaceutical proteins in plants. Nat Rev Genet 4:794–805

Ma S, Huang Y, Davis A, Yin Z, Mi Q, Menassa R, Brandle JE, Jevnikar AM (2005) Production of biologically active human interleukin-4 in transgenic tobacco and potato. Plant Biotechnol J 3:309–318

Magnuson NS, Linzmaier PM, Reeves R, An G, HayGlass K, Lee JM (1998) Secretion of biologically active human interleukin-2 and interleukin-4 from genetically modified tobacco cells in suspension culture. Protein Expr Purif 13:45–52

Marillonnet S, Giritch A, Gils M, Kandzia R, Klimyuk V, Gleba Y (2004) In planta engineering of viral RNA replicons: efficient assembly by recombination of DNA modules delivered by Agrobacterium. Proc Natl Acad Sci USA 101:6852–6857

Marillonnet S, Thoeringer C, Kandzia R, Klimyuk V, Gleba Y (2005) Systemic Agrobacterium tumefaciens-mediated transfection of viral replicons for efficient transient expression in plants. Nat Biotechnol 23:718–723

Mishra S, Yadav DK, Tuli R (2006) Ubiquitin fusion enhances cholera toxin B subunit expression in transgenic plants and the plant-expressed protein binds GM1 receptors more efficiently. J Biotechnol 127:95–108

Nagaya S, Kawamura K, Shinmyo A, Kato K (2010) The HSP terminator of Arabidopsis thaliana increases gene expression in plant cells. Plant Cell Physiol 51:328–3251

Nie L, Wu G, Zhang W (2006) Correlation between mRNA and protein abundance in Desulfovibrio vulgaris: a multiple regression to identify sources of variations. Biochem Biophys Res Commun 339:603–610

Obregon P, Chargelegue D, Drake PM, Prada A, Nuttall J, Frigerio L, Ma JK (2006) HIV-1 p24-immunoglobulin fusion molecule: a new strategy for plant-based protein production. Plant Biotechnol J 4:195–207

Parmenter DL, Boothe JG, Rooijen GJH, Yeung EC, Moloney MM (1995) Production of biologically active hirudin in plant seeds using oleosin partitioning. Plant Mol Biol 29:1167–1180

Patel J, Zhu H, Menassa R, Gyenis L, Richman A, Brandle J (2007) Elastin-like polypeptide fusions enhance the accumulation of recombinant proteins in tobacco leaves. Transgenic Res 16:239–249

Percin I, Yavuz H, Aksoz E, Denizli A (2009) N-acetyl-dgalactosamine-specific lectin isolation from soyflour with poly(HPMA-GMA) Beads. J Appl Polym Sci 111:148–154

Pilson D, Prendeville HR (2004) Ecological effects of transgenic crops and the escape of transgenes into wild populations. Annu Rev Ecol Evol Syst 35:149–174

Pogue GP, Vojdani F, Palmer KE, Hiatt E, Hume S, Phelps J, Long L, Bohorova N, Kim D, Pauly M, Velasco J, Whaley K, Zeitlin L, Garger SJ, White E, Bai Y, Haydon H, Bratcher B (2010) Production of pharmaceutical-grade recombinant aprotinin and a monoclonal antibody product using plant-based transient expression systems. Plant Biotechnol J 8:638–654

Regnard GL, Halley-Stott RP, Tanzer FL, Hitzeroth II, Rybicki EP (2010) High level protein expression in plants through the use of a novel autonomously replicating geminivirus shuttle vector. Plant Biotechnol J 8:38–46

Rose AB (2002) Requirements for intron-mediated enhancement of gene expression in Arabidopsis. RNA 8:1444–1453

Ruhlman T, Ahangari R, Devine A, Samsam M, Daniell H (2007) Expression of cholera toxin B-proinsulin fusion protein in lettuce and tobacco chloroplasts–oral administration protects against development of insulitis in non-obese diabetic mice. Plant Biotechnol J 5:495–510

Sandhu JS, Krasnyanski SF, Osadian MD, Domier LL, Korban SS, Buetow DE (1999) Enhanced expression of the human respiratory syncytial virus-F gene in apple leaf protoplasts. Plant Cell Rep 18:394–397

Scotti N, Alagna F, Ferraiolo E, Formisano G, Sannino L, Buonaguro L, De Stradis A, Vitale A, Monti L, Grillo S, Buonaguro FM, Cardi T (2009) High-level expression of the HIV-1 Pr55gag polyprotein in transgenic tobacco chloroplasts. Planta 229:1109–1122

Shaaltiel Y, Bartfeld D, Hashmueli S, Baum G, Brill-Almon E, Galili G, Dym O, Boldin-Adamsky SA, Silman I, Sussman JL, Futerman AH, Aviezer D (2007) Production of glucocerebrosidase with terminal mannose glycans for enzyme replacement therapy of Gaucher's disease using a plant cell system. Plant Biotechnol J 5:579–590

Sharma AK, Sharma MK (2009) Plants as bioreactors: recent developments and emerging opportunities. Biotechnol Adv 27:811–832

Sheludko YV (2008) Agrobacterium-mediated transient expression as an approach to production of recombinant proteins in plants. Recent Patents Biotechnol 2:198–208

Shrawat AK, Lörz H (2006) Agrobacterium-mediated transformation of cereals: a promising approach crossing barriers. Plant Biotechnol J 4:575–603

Streatfield SJ (2007) Approaches to achieve high-level heterologous protein production in plants. Plant Biotechnol J 5:2–15

Svab Z, Hajdukiewicz P, Maliga P (1990) Stable transformation of plastids in higher plants. Proc Natl Acad Sci USA 87:8526–8530

Terpe K (2003) Overview of tag protein fusions: from molecular and biochemical fundamentals to commercial systems. Appl Microbiol Biotechnol 60:523–533

Tian Q, Stepaniants SB, Mao M, Weng L, Feetham MC, Doyle MJ, Yi EC, Dai H, Thorsson V, Eng J, Goodlett D, Berger JP, Gunter B, Linseley PS, Stoughton RB, Aebersold R, Collins SJ, Hanlon WA, Hood LE (2004) Integrated genomic and proteomic analyses of gene expression in Mammalian cells. Mol Cell Proteomic 3:960–969

Torrent M, Llompart B, Lasserre-Ramassamy S, Llop-Tous I, Bastida M, Marzabal P, Westerholm-Parvinen A, Saloheimo M, Heifetz PB, Ludevid M (2009) Eukaryotic protein production in designed storage organelles. BMC Biol 7:5

Tremblay R, Wang X, Jevnikar AM, Ma S (2008) Expression of a fusion protein consisting of cholera toxin B subunit and an anti-diabetic peptide (p277) from human heat shock protein in transgenic tobacco plants. Transgenic Plant J 2:186–191

Tremblay R, Wang D, Jevnikar AM, Ma S (2010) Tobacco, a highly efficient green bioreactor for production of therapeutic proteins. Biotechnol Adv 28:214–221

Tremblay R, Feng M, Menassa R, Huner NP, Jevnikar AM, Ma S (2011a) High-yield expression of recombinant soybean agglutinin in plants using transient and stable systems. Transgenic Res 20:345–356

Tremblay R, Diao H, Huner NP, Jevnikar AM, Ma S (2011b) The development, characterization, and demonstration of a novel strategy for purification of recombinant proteins expressed in plants. Transgenic Res. doi:10.1007/s11248-011-9498-6 (published online)

Twyman RM, Stoger E, Schillberg S, Christou P, Fischer R (2003) Molecular farming in plants: host systems and expression technology. Trends Biotechnol 21:570–578

Urry DW (1992) Free energy transduction in polypeptides and proteins based on inverse temperature transitions. Prog Biophys Mol Biol 57:23–57

Van Molle I, Joensuu JJ, Buts L, Panjikar S, Kotiaho M, Bouckaert J, Wyns L, Niklander-Teeri V, De Greve H (2007) Chloroplasts assemble the major subunit FaeG of Escherichia coli F4 (K88) fimbriae to strand-swapped dimers. J Mol Biol 368:791–799

van Rooijen GJ, Moloney MM (1995) Structural requirements of oleosin domains for subcellular targeting to the oil body. Plant Physiol 109:1353–1361

Vaquero C, Sack M, Schuster F, Finnern R, Drossard J, Schumann D, Reimann A, Fischer R (2002) A carcinoembryonic antigen-specific diabody produced in tobacco. FASEB J 16:408–410

Verma D, Daniell H (2007) Chloroplast vector systems for biotechnology applications. Plant Physiol 145:1129–1143

Vitale A, Ceriotti A (2004) Protein quality control mechanisms and protein storage in the endoplasmic reticulum. A conflict of interests? Plant Physiol 136:3420–3426

Wirth S, Calamante G, Mentaberry A, Bussmann L, Lattanzi M, Barañao L, Bravo-Almonacid F (2004) Expression of active human epidermal growth factor (hEGF) in tobacco plants by integrative and non-integrative systems. Mol Breed 13:23–35

Witte CP, Noël LD, Gielbert J, Parker JE, Romeis T (2004) Rapid one-step protein purification from plant material using the eight-amino acid StrepII epitope. Plant Mol Biol 55:135–147

Witters LA (2001) The blooming of the French lilac. J Clin Invest 108:1105–1107

Xu J, Ge X, Dolan MC (2011) Towards high-yield production of pharmaceutical proteins with plant cell suspension cultures. Biotechnol Adv 29:2782–2799

Yusibov V, Modelska A, Steplewski K, Agadjanyan M, Weiner D, Hooper DC, Koprowski H (1997) Antigens produced in plants by infection with chimeric plant viruses immunize against rabies virus and HIV-1. Proc Natl Acad Sci USA 94:5784–5788

Zhou F, Badillo-Corona JA, Karcher D, Gonzalez-Rabade N, Piepenburg K, Borchers AM, Maloney AP, Kavanagh TA, Gray JC, Bock R (2008) High-level expression of human immunodeficiency virus antigens from the tobacco and tomato plastid genomes. Plant Biotechnol J 6:897–913

Chapter 2
Induction of Oral Tolerance to Treat Autoimmune and Allergic Diseases by Using Transgenic Plants

Shengwu Ma and Anthony M. Jevnikar

Abstract In recent years, the use of plants as a green bioreactor for production of recombinant pharmaceutical proteins, a technology known as plant molecular farming or biofarming, has gained increasing attention. This new technology has the potential to produce large quantities of the required protein at competitive low costs. Moreover, edible tissues or organs offer the possibility of direct oral delivery of pharmaceutical proteins expressed by plants with minimal processing, significantly reducing production costs and accelerating product development. To date, a number of recombinant proteins of pharmaceutical interest have been produced in plants, ranging from monoclonal antibodies, vaccines, hormones to enzymes. Furthermore, many plant-made pharmaceutical proteins have been tested in pre-clinical animal models of disease with promising results, with some plant-made vaccines and monoclonal antibodies advanced to human clinical trials. This chapter highlights the progress made towards the utilization of transgenic plants to express and deliver recombinant autoantigens or allergens to induce oral tolerance for the treatment of autoimmunity and allergy.

S. Ma (✉)
Transplantation Immunology Group, Lawson Health Research Institute, London, ON N6A 4G5, Canada

Department of Biology, University of Western Ontario, London, ON N6A 5B7, Canada
e-mail: sma@uwo.ca

A.M. Jevnikar
Transplantation Immunology Group, Lawson Health Research Institute, London, ON N6A 4G5, Canada

A. Wang and S. Ma (eds.), *Molecular Farming in Plants: Recent Advances and Future Prospects*, DOI 10.1007/978-94-007-2217-0_2,

2.1 Introduction

In recent years there has been a rapidly growing demand for protein-based therapeutics, largely due to a better understanding of the molecular mechanisms underlying the pathogenesis of various human diseases and the identification of new molecular targets. Moreover, protein-based therapeutics have several advantages over small-molecule drugs, including serving highly specific and complex set of functions that cannot be mimicked by chemical compounds, and having highly specific actions that produce no or minimal interference with normal biological processes (Douthwaite and Jermutus 2006). Today most protein pharmaceuticals, including insulin, growth hormones, and most monoclonal antibodies, are produced through recombinant methods. Bacterial, yeast and mammalian cell cultures are the most commonly used hosts for the expression of recombinant proteins. While these conventional expression systems offer certain advantages, they are generally limited by low yields and high production costs, mainly due to the requirement for large fermenters, sterile conditions, and expensive media. It has become more and more apparent that current expression systems are inadequate to meet the ever increasing demand for protein pharmaceuticals. In recent years, plants have emerged as a promising alternative system for the production of heterologous proteins.

There are several potential advantages of using plants as a protein production system. First, a major advantage of plant systems over conventional cell culture-based production systems is the anticipated cost savings, reflecting the large amount of biomass that can be produced in a short time with no need for specialized equipment or expensive media. Plants can be contained, grown easily and inexpensively in large quantities, can be harvested and processed with available agronomic infrastructures, and have simple, unlimited scalability. It is estimated that protein production in transgenic plants can be as much as four orders of magnitude less expensive than production in mammalian cell culture, on a per gram of unpurified protein basis (Dove 2002). Secondly, in contrast to the bacterial expression systems, plants, being higher eukaryotes, are capable of performing many of the complex protein processing steps, such as glycosylation. Thirdly, plants do not harbor infectious agents such as viruses and prions harmful to humans as these agents cannot replicate in plants. Safety is a primary concern when any therapeutic proteins for human use are prepared from animal tissues or bacterial cells (Tremblay et al. 2010). Finally, pharmaceutical proteins synthesized in the edible tissues of transgenic plants can be delivered by ingestion of transgenic plant tissue without tedious and expensive downstream processing.

Indeed, an increasing number of pharmaceutical proteins have been expressed in plants, ranging from monoclonal antibodies, vaccines, hormones to enzymes. Moreover, many plant-made proteins have been tested in preclinical animal models with promising results, with several plant-derived monoclonal antibodies and vaccines advanced into different stages of human clinical trials (Penney et al. 2011; Paul and Ma 2011). Recently, there has been a growing interest in using transgenic plants as a novel therapeutic strategy for the treatment of autoimmune diseases and allergies.

Autoimmune diseases such as diabetes and multiple sclerosis or allergies such as asthma are manifestations of immunological hypersensitivity. They arise when mechanisms controlling responses to host's self proteins or to innocuous environmental antigens break down (Larché and Wraith 2005). Currently available therapies mainly treat the symptoms of autoimmune or allergic diseases through global immunosuppression and are associated with increased risk of infections. Specific suppression of unwanted autoimmune responses or allergic responses without affecting the normal function of the host's immune system is the major goal of treatment for autoimmune disorders or allergies. Antigen-specific immunosuppression would allow for the inhibition of autoimmune or allergic responses without adversely affecting the function of the immune system. Mucosal administration of protein antigens to induce oral tolerance is an attractive therapeutic option for the treatment of autoimmune diseases and allergies. The use of transgenic plants to express and deliver recombinant autoantigens or allergens represents a novel strategy for oral tolerance induction. This chapter highlights the progress made towards the utilization of transgenic plants to express and deliver recombinant autoantigens or allergens to induce oral tolerance for the treatment of autoimmune diseases and allergies.

2.2 Oral Tolerance and Its Therapeutic Potential in Autoimmunity and Allergy

Mucosal tolerance is classically defined as a state of hyporesponsiveness to subsequent parenteral injections of proteins to which an individual or animal has been previously exposed via the mucosal route (Faria and Weiner 2006). It is now well recognized that oral tolerance is an immunoregulatory strategy used by the gut and its associated lymphoid tissues to render the peripheral immune system unresponsive to nonpathogenic proteins, such as food, airborne antigens or the commensal bacterial flora (Faria and Weiner 2006; Weiner et al. 2011). The gut-associated lymphoid tissue (GALT) is a well-developed immune network that has not only developed the inherent property of preventing the host from reacting to ingested proteins, but has also evolved to protect the host from ingested pathogens. It is generally agreed that oral tolerance is established and maintained at the T cell level (Faria and Weiner 2006; Weiner et al. 2011).

Oral tolerance can occur via a number of mechanisms, depending on the amount of an antigen administered. Administration of high doses of an antigen induces tolerance via the mechanism of clonal anergy/deletion of effector T cells, whereas low doses of mucosally administered antigen induce tolerance via the mechanism of an active suppression of effector T cells through the induction of antigen-specific regulatory T cells (Treg), which produce downregulatory cytokines such as IL-4, IL-10 and TGF-b, a Th2/Th3 cytokine pattern. After activation in the GALT, the regulatory T cells migrate to the site of inflammation, and on re-encountering the fed antigen, they display their specific suppressive effect, resulting in an attenuated

T cell-mediated immune response. Low-dose and high-dose tolerance may not be mutually exclusive and may have overlapping functionality (Weiner et al. 2011).

The induction of antigen-specific oral tolerance is an attractive therapeutic approach for treatment of autoimmune, allergic, and inflammatory diseases. Oral tolerance has been shown to be effective in treating animal models of autoimmunity and allergy. Examples include experimental allergic encephalomyelitis (EAE), arthritis, diabetes mellitus, uveitis, airway eosinophilia, allergy and food hypersensitivity (Weiner et al. 2011). Oral tolerance has also been studied in humans. In rheumatoid arthritis (RA), patients with active RA were fed dnaJP1, which is a 15-mer dominant epitope heat-shock protein thought to be involved in RA pathogenesis, though independent from the primary trigger of disease. After 6 months of treatment, patients treated with dnaJP1 showed a significant reduction in T cells producing TNF and a trend toward an increase of T cells producing IL-10, indicating a positive effect on the disease (Park et al. 2009; Koffeman et al. 2009). Another human trial conducted involves oral insulin for prevention of type 1 diabetes [Diabetes Prevention Trial 10 (DPT-10)]. Although there were no differences between the oral insulin and placebo groups in the primary outcome, a subset of individuals in the oral insulin prevention trial with high levels of insulin autoantibodies had an apparent several year delay in progression to diabetes ($P=0.01$), and a follow-up study is planned (Skyler 2008; Weiner et al. 2011). Oral tolerance has also been investigated in food allergy. Clark et al. (2009) reported that oral administration of peanut flour induced clinical tolerance to peanut protein and protected peanut-allergic children from developing severe peanut allergy.

It is likely that the translation of oral tolerance to a realistic therapy for human autoimmunity and allergy will require the use of a mucosal adjuvant to enhance the efficacy of oral tolerance induction (Faria and Weiner 2006; Weiner et al. 2011). Recent studies have shown that oral administration of *Lactacoccus lactis* cells engineered to secrete OVA and/or IL-10 enhances oral tolerance in mice (Huibregtse et al. 2007; Frossard et al. 2007). Oral, nasal, or sublingual administration of antigen coupled to the cholera toxin B subunit (CTB) was also shown to enhance mucosal tolerance (Sun et al. 2010). CTB increases mucosal antigen uptake and presentation to antigen presenting cells by binding to GM1 ganglioside and the induction of regulatory cells (Sun et al. 2010; Weiner et al. 2011).

2.3 Transgenic Plants as a Novel Strategy for Oral Tolerance Induction

The induction of oral tolerance requires repeated ingestion of large amounts of autoantigens or allergens. A major consideration for clinical application of oral tolerance is the cost. It is essential to develop such a heterologous expression system that is able to provide recombinant autoantigens or allergens in sufficient quantities and at an affordable cost. While cell culture-based conventional expression systems may allow the production of sufficient quantities of recombinant proteins, they are unlikely to be cost effective, partly because recombinant therapeutic proteins derived from these systems need to be purified before use. Downstream protein purification and processing are a complex, labour-intensive and expensive process that can eliminate the economic advantage of any production system. The use of transgenic plants for oral tolerance strategy has considerable clinical appeal, not only for efficacy but also for simplicity of production and delivery, advantages of cost, absence of contamination risk with human pathogens, and perhaps, if used in edible plants, increased patient acceptance. Also augmented immune responses to plant produced vaccines may suggest increased stability for plant expressed recombinant proteins to gastrointestinal degradation. There is now increasing evidence that foreign proteins compartmentalized within plant cells are protected in the stomach through bioencapsulation by the plant cell (Verma et al. 2010).

Both nuclear and chloroplast transgenic plants are being exploited to express and deliver recombinant autoantigens and allergens for oral tolerance induction. Transformation of plant nuclear genomes by *Agrobacterium* or biolistic methods has become routine in plant research. The number of plant species amenable to transformation and regeneration is now quite large (Tzfira and Citovsky 2006). The main advantage of using nuclear transgenic plants as an expression platform is its flexibility and the efficiency in scaling up the production of recombinant proteins, which can be achieved simply by planting more acreages of the biotech crops. An additional advantage is the long-term continuous production of recombinant proteins with little to no external input because foreign genes are stably integrated into the nuclear genome of the host plant and are inherited in the next generation and as such, stable and predictable transgene expression can be maintained over many generations. Moreover, the synthesis of foreign proteins in nuclear transgenic plants can be readily targeted to edible plant organs such as leaves, seeds and fruits, allowing for oral delivery of a palatable product. However, a limitation of using nuclear-transformed plants for protein production is the relatively low-level accumulation of recombinant proteins, with typical expression levels within the range of 0.0001 to ~1% of total soluble protein (TSP) (Tremblay et al. 2010).

Chloroplast transformed plants can also be an economic source of autoantigens and allergens. One of the main advantages of chloroplasts as bioreactors is the potential for accumulation of large quantities of recombinant proteins. A typical tobacco leaf cell contains as many as 100 chloroplasts per cell with up to 100 genome copies per chloroplast, and therefore the copy number of any introduced transgene can be amplified by as many as 10,000 per cell, leading to extraordinarily high levels of foreign protein products (Chebolu and Daniell 2009). Other advantages of chloroplast bioreactors include high levels of transgene containment, no transgene silencing and no position effects as often occurred in nuclear transgenic plants, and the ability to express multiple genes in a single transformation event (transgene stacking) (Chebolu and Daniell 2009). A limitation of chloroplast recombinant protein production, however, is that like bacteria they are unable to perform glycosylation essential for proper protein folding and functions.

Recently, the chloroplast of the unicellular green alga *Chlamydomonas reinhardtii* has been gaining increasing attention as a new bioreactor for the production of autoantigens and other therapeutic proteins (Wang et al. 2008). Compared to chloroplast transgenic plants, the use of chloroplast transgenic algae as a bioreactor offers several additional advantages. Microalgae, such as *C. reinhardtii*, grow and reproduce faster than any other terrestrial or aquatic plant, doubling its biomass in approximately 8 h. Microalgae are non-toxic and non-polluting, thus environmentally friendly for mass cultivation and commercial exploitation. Also, there will be a significant reduction in the time required to generate transgenic algae as compared to the time required to generate transplastomic plants. In general, stable transplastomic algal lines can be obtained in as little as 3 weeks, with the potential to scale up to mass production in an additional 4–6 weeks (Franklin and Mayfield 2004; Rasala and Mayfield 2011). Moreover, *C. reinhardtii* is rich in essential amino acids and protein, with the protein content comprising up to 25% of its dry weight (Franklin and Mayfield 2004). These attributes, and the fact that green algae are generally regarded as safe (GRAS) by the U.S. Food and Drug Administration (FDA) for human consumption, make *C. reinhardtii* a particularly attractive system for oral tolerance induction.

2.4 Treatment of Type 1 Diabetes with Transgenic Plants Expressing Diabetes-Associated Autoantigens

Type 1 diabetes (T1D) is a chronic autoimmune disease in which a loss of self-tolerance to insulin-producing β cells in the pancreatic islets results in impaired glucose homeostasis (Eisenbarth 2004). While insulin replacement therapy has transformed T1D from a fatal disease to a chronic one, it is still associated with significant morbidity. Complications of diabetes lead to kidney failure, blindness, amputations, and increased risk of macrovascular disease. Intensive insulin therapy is associated with significant adverse events related to hypoglycemia (Arabi et al. 2009). Restoration of self-tolerance is considered as the most satisfactory solution to the prevention and cure of T1D. A number of beta-cell-specific autoantigens have been identified including glutamic acid decarboxylase (GAD), insulin, insulinoma-associated antigen (IA-2) and heat shock protein 60 (Hsp60) (Eisenbarth 2004; Jasinski and Eisenbarth 2005). The identification of islet cell autoantigens holds promise as targets for antigen-specific immunotherapy in T1D.

2.4.1 Induction of Oral Tolerance by Mucosal Administration of Transgenic Plants Expressing GAD

GAD is recognized as the major and early islet autoantigen in T1D. There are two isomers of GAD, GAD65 and 67, both of which are implicated in T1D (Bu et al. 1992; Elliott et al. 1994). Rat and human islets express GAD65 predominantly, whereas mouse islets majorly express GAD67 (Elliott et al. 1994). We are the first group to demonstrate that transgenic plants can be used as a novel strategy to induce antigen-specific oral tolerance for the treatment of type 1 diabetes in non-obese diabetic (NOD) mice. Initially we generated transgenic tobacco and potato plants expressing murine GAD67, with expression levels up to 0.4% of TSP (Ma et al. 1997). To demonstrate that GAD-specific oral tolerance can be induced by administration of GAD67 transgenic plants, young pre-diabetic female NOD mice were fed GAD67 transgenic potato tuber or tobacco leaf tissue as dietary supplementation for a period of 7 months starting at 5 weeks of age. Control mice received an equivalent amount of vector-minus insert transformed tobacco or potato tissue. As expected, GAD67 plant fed mice were protected from diabetes, whereas those fed control plant tissue were not protected from diabetes. The protection was associated with the inhibition of proliferation of GAD-reactive splenic T cells as well as a Th1 to Th2 cytokine profile shift.

We subsequently produced transgenic tobacco plants expressing human GAD65 (Ma et al. 2004). Compared to GAD67 expression in plants, human GAD65 accumulated to a lower level (0.04% of TSP). Oral administration of GAD65 plant tissue delivering approximately 8–10 μg GAD65/per mouse daily was shown not to provide protection against diabetes in NOD mice. To enhance oral tolerance to GAD65, we additionally produced transgenic plants expressing the anti-inflammatory Th2 cytokine IL-4 for use as a mucosal adjuvant. Co-administration of IL-4 plant tissue delivering approximately 1–2 μg IL-4/per mouse daily resulted in suppression of diabetes in NOD mice, suggesting the effectiveness of IL-4 in potentiating induction of oral tolerance, especially in cases where only low expression levels of recombinant autoantigens can be achieved in transgenic plant systems. Frequency analysis of cytokine-secreting cells in spleens showed that the ratio of IL-4/IFN-γ-secreting cells was increased in IL-4 plus GAD65 plant fed mice but was unchanged in control plant treated mice, suggesting a shift in the Th1/Th2 balance toward Th2 dominance. Furthermore, the results of adoptive transfer experiments indicated that the protection from diabetes in IL-4+GAD65 plant fed mice was associated with the induction of regulatory T cells.

2.4.2 Induction of Oral Tolerance by Mucosal Administration of Transgenic Plants Expressing Insulin

Insulin is another major early autoantigen in T1D (Jasinski and Eisenbarth 2005; Zhang and Eisenbarth 2011). Arakawa et al. (1998) generated transgenic potato plants synthesizing human proinsulin fused to cholera toxin B subunit (CTB-INS). Fusion to CTB was intended to enhance oral tolerance induction to insulin. The plant-derived fusion protein retained the GM1 ganglioside-binding activity of CTB and the antigenicity of insulin. NOD mice fed transgenic potato tubers delivering μg quantities of CTB-INS fusion protein had a reduction in insulitis, and a delay in the progression of clinical diabetes, while control mice fed transgenic potato tubers expressing insulin or CTB protein alone were not protected from diabetes.

Recently, Ruhlman et al. (2007) expressed human proinsulin fused to CTB (CTB-Pins) in chloroplasts of both lettuce and tobacco, with accumulation levels up to ~16% of TSP in tobacco and up to ~2.5% of TSP in lettuce. In a short-term feeding study, 5-week-old female NOD mice were fed 8 mg of powdered tobacco leaf material expressing CTB-Pins or, as negative controls, CTB–green fluorescent protein (CTB-GFP) or interferon–GFP (IFN-GFP), or untransformed leaf, each week for 7 weeks. Histological analysis of the pancreatic islets from CTB-Pins treated mice showed decreased lymphocytic infiltration in the islets (insulitis) compared with the controls. Moreover, increased expression of immunosuppressive cytokines, such as IL-4 and IL-10, was observed in the pancreas of CTB-Pins-treated NOD mice. Serum levels of immunoglobulin G1 (IgG1), but not IgG2a, were also elevated in CTB-Pins-treated mice. Taken together, the prevention of pancreatic insulitis in CTB-Pins treated mice may be likely due to the induction of Th2 lymphocyte-mediated oral tolerance.

2.5 Treatment of Arthritis with Transgenic Plants Expressing Arthritis-Associated Autoantigens

Rheumatoid arthritis (RA) is an autoimmune disease that causes chronic inflammation of the joints. Rheumatoid arthritis can also cause inflammation of the tissue around the joints, as well as in other organs in the body. Numerous antigenic targets have been identified in RA patients including type II collagen (CII), human chondrocyte glycoprotein 39, and various members of the heat-shock protein family (Ichim et al. 2008). To date, treatment interventions have all been associated with non antigen-specific inhibition of inflammatory processes, causing concerns regarding increased susceptibility to infections. Induction of oral tolerance offers the promise of antigen-specific immunotherapy for the treatment of RA.

2.5.1 Induction of Oral Tolerance by Mucosal Administration of Transgenic Plants Expressing Type II-Collagen Peptides

Hashizume et al. (2008) produced a fusion protein, consisting of glutelin and four tandem repeats of a $CII_{250-270}$ peptide (GluA-4X$CII_{250-270}$) containing a human T cell epitope, in transgenic rice and conducted feeding experiments to determine whether oral administration of transgenic rice seeds can provide protection against RA in a mouse model of RA. In these experiments, DBA/1 mice were fed either transgenic rice or wild-type rice for 2 weeks before immunization with type II collagen, and an average daily intake of the $CII_{250-270}$ peptide as a diet was estimated to be about 25 μg per mouse. Mice treated with GluA-4X$CII_{250-270}$ transgenic rice showed lower delayed serum specific-IgG2a response against challenge with type II collagen.

2.5.2 *Induction of Oral Tolerance by Mucosal Administration of Transgenic Plants Expressing Mycobacterial Hsp65*

Heat shock proteins (Hsp) are a group of highly conserved proteins that serve as intracellular chaperons that are induced under stress conditions. Recent research has implicated Hsp65 as an autoantigen in RA (Massa et al. 2007). To develop a novel oral antigen therapy for RA using plant-based systems, Rodriguez-Narciso et al. (2011) produced transgenic tobacco plants expressing recombinant mycobacterial Hsp65 protein and examined the potential of orally administered plant-made Hsp65 protein for treating adjuvant-induced arthritis in Lewis rats. The results showed that oral feeding of Lewis rats with transgenic tobacco leaf tissue delivering 10 μg of Hsp65 per rat daily, which was started 10 days post induction of arthritis with a M. tuberculosis strain, significantly promoted recovery of the body weight and reduced joint inflammation.

2.6 Induction of Oral Tolerance to Treat Allergic Diseases with Transgenic Plants Expressing Allergens

Allergic diseases have become one of the most important health problems throughout the world. Allergic diseases are caused by inappropriate immunological responses to harmless antigens driven by a Th2-mediated immune response. Currently, the major therapies used to control allergic diseases are glucocorticoids. These steroid hormones act nonselectively and their extended use can lead to numerous side effects such as increased risk of infection (Moghadam-Kia and Werth 2010). Allergen-specific immunotherapy is considered the preferred approach for treating allergy. A major problem with specific allergen immunotherapy is that the allergen extracts that are currently used for vaccination are unfractionated preparations (Niederberger and Valenta 2004). The inclusion of non-allergenic materials may elicit new allergenicity. Recombinant allergen–based immunotherapy will improve current immunotherapy practice and may open possibilities for prophylactic vaccination. Induction of oral tolerance by mucosal administration of transgenic plants expressing specific recombinant allergens may provide a novel and effective therapeutic strategy against allergy.

2.6.1 *Induction of Oral Tolerance by Mucosal Administration of Transgenic Plants Expressing a Full-Length Antigen*

The sunflower seed methionine-rich 2S albumin (SSA) is an IgE-binding protein and has been identified as the major allergen in sunflower seed (Asero et al. 2002). Smart et al. (2003) produced transgenic lupin (Lupinus angustifolius L.) synthesizing

SSA and investigated whether oral administration of SSA-lupin could protect against the development of experimental asthma. They showed that oral consumption of SSA-lupin seed meal attenuated mucus hypersecretion, pulmonary eosinophilic inflammation, and airway hyperreactivity following subsequent allergen exposure in a mouse model of allergic airway disease. The suppression of experimental asthma was associated with the production of $CD4^{+}$ T cell-derived IFN-γ and IL-10, and the induction of $CD4^{+}$ $CD45^{low}$ regulatory T cells. This work provides a proof of concept for a plant-based new approach to treating allergy.

Recently, Lee et al. (2011) expressed Der p2 protein in transgenic tobacco plants for oral tolerance induction as a strategy to treat house dust mite-induced airway allergy. Der p2 is a major allergen from Dermatophagoides pteronyssinus, the main species of house dust mite and a major inducer of asthma by activating cells in the respiratory tract (Smith et al. 2001). To determine the effect, mice were fed total protein extracted from Der p2 transgenic plants once per day over 6 days. Der p2-treated mice showed a decrease in serum Der p2-specific IgE and IgG1 titers, a decrease in IL-5 and eotaxin levels in bronchoalveolar lavage fluid, and a decrease in eosinophil infiltration into the airway. Hyper-responsiveness was also decreased in mice treated with Der p2 containing total protein, but the number of CD4 + CD25 + Foxp3+ regulatory T cells were significantly increased in mediastinal and mesenteric lymph nodes. Furthermore, splenocytes isolated from Der p2-treated mice exhibited decreased proliferation and increased IL-10 secretion following stimulation with yeast-derived recombinant Der p2.

2.6.2 Induction of Oral Tolerance by Mucosal Administration of Transgenic Rice Expressing Allergen-Specific T Cell Epitopes

Takagi et al. (2005) produced transgenic rice expressing mouse dominant T cell epitope peptides of Japanese cedar (Cryptomeria japonica) pollen allergens (Cry j I and Cry j II), fused to soybean seed storage protein glycinin. Animal studies showed that oral administration of transgenic rice seeds inhibited the proliferative activity of allergen-specific CD4+ T cells, decreased serum IgE levels and reduced clinical allergy symptoms such as sneezing in mice. Moreover, the levels of Th2 cytokines such as IL-4, IL-5 and IL-13 as well as IgE-mediated histamine release were also significantly decreased in treated mice. Transgenic rice synthesizing human dominant T cell epitopes of Cry j 1 and Cry j 2 allergens was also produced (Takaiwa 2007).

To further enhance the efficiency of rice seed-based vaccines in inducing oral tolerance against allergy, this group have recently expressed T-cell epitopes of Cry j 1 and Cry j 2 as a fusion protein with the mucosal adjuvant CTB in rice seed (Takagi et al. 2008). They showed that feeding mice with rice seeds containing CTB-fused T-cell epitopes suppressed allergen-specific IgE responses and pollen-induced clinical symptoms at 50-fold lower doses of T-cell epitopes compared with the same results obtained when control rice seeds expressing T-cell epitopes fused to rice glutelin acidic subunit were used as feed.

Recently, Suzuki et al. (2011) reported the production of a subunit vaccine consisting of a fragment (p45–145) of mite allergen (Der p 1) containing immunodominant human and mouse T cell epitopes in transgenic rice for oral tolerance induction to treat asthma. They showed that prophylactic oral vaccination of mice with transgenic rice seeds reduced the serum levels of allergen-specific IgE and IgG. Allergen-induced CD4+ T cell proliferation and production of Th2 cytokines in vitro, infiltration of eosinophils, neutrophils and mononuclear cells into the airways and bronchial hyperresponsiveness were also inhibited by oral vaccination. The immune response induced by the rice vaccine was antigen-specific, because the levels of specific IgE and IgG in mice immunized with control antigen Der f 2 or ovalbumin were not significantly suppressed by oral vaccination with the Der p 1 expressing transgenic rice.

2.7 Treatment of Inflammatory Bowel Disease by Oral Administration of Transgenic Plants Expressing Interleukin-10

Inflammatory bowel disease (IBD), which includes Crohn's disease (CD) and ulcerative colitis (UC), is a relapsing and remitting condition characterized by chronic inflammation at various sites in the gastrointestinal (GI) tract, which results in diarrhea and abdominal pain (Kozuch and Hanauer 2008). Inflammation results from a T cell-mediated immune response in the GI mucosa. The precise etiology is unknown, but evidence suggests that IBD may result from a breakdown of immunological tolerance towards the gut flora in genetically susceptible hosts, resulting in an inappropriate immune response responsible for the chronic inflammatory process that characterizes CD and UC (Bouma and Strober 2003). Both UC and CD are characterised by the activation of macrophages and dendritic cells with production of pro-inflammatory cytokines such as IL-1, IL-6 and tumour necrosis factor-alpha (TNF-a) (Bouma and Strober 2003; Balding et al. 2004). At present, there is no cure, and therapeutics are primarily aimed at suppressing inflammatory responses. Treatment with anti-TNF-α antibodies has met with some success, however, its therapeutic limitations, elevated cost, and the side effects of other conventional immunosuppressive drugs underscore the need for novel therapeutic strategies (Kalischuk and Buret 2010).

Menassa et al. (2007) expressed human IL-10 in transgenic tobacco plants and investigated the potential of the plant-made cytokine as a luminal therapy for IBD in animal models of colitis. IL-10 is a potent anti-inflammatory cytokine with therapeutic applications in several autoimmune and inflammatory diseases (Beebe et al. 2002). IL-10 −/− mice, which spontaneously develop colitis, were used for feeding experiments. Mice were fed IL-10 tobacco leaf tissue delivering up to 9 μg of human

IL-10 per mouse daily for 4 weeks. Mice treated with IL-10 had reduced severity of colitis. The TNF-α expression at the sites of inflammation was also down- regulated in IL-10 treated mice. Gut histology was significantly improved relative to controls (P=0.002) and was correlated with a decrease in small bowel TNF-α mRNA levels and an increase in IL-2 and IL-1β mRNA levels.

Recently, Bortesi et al. (2009) expressed viral and murine IL-10 in transgenic tobacco. *In vitro* characterization showed that both plant-derived molecules formed stable dimers, were able to activate the IL-10 signaling pathway and to induce specific anti-inflammatory responses in mouse macrophage cells. Their long-term goals are to treat type 1 diabetes and other inflammatory diseases through oral delivery of plant-made IL-10.

2.8 Conclusions

There is growing evidence to support the therapeutic value of oral tolerance induction in the treatment of autoimmune disease, allergy, organ transplant rejection and many other inflammatory diseases. The use of transgenic plants as an oral tolerance induction strategy offers a number of advantages, including simplicity of production and delivery and low-costs. The number of recombinant autoantigens or allergens delivered to intestinal mucosa by transgenic plants for successful oral tolerance induction is increasing. There are a number of approaches that one could use to further increase the efficiency of transgenic plant-based approach for oral tolerance induction. One is to improve the yield of recombinant autoantigens or allergens in transgenic plants, as this will lead to increased delivery of a target protein without increasing the amount of transgenic plant tissue consumed. Recently, Morandini et al. (2011) reported the expression of GAD65 at a significantly higher level in transgenic plants, accounting for 7.7% of TSP when expressed in Arabidopsis seeds. Another approach is to express autoantigens or allergens as fusion proteins containing a mucosal delivery-enhancing molecule, such as human serum transferrin, to improve plant-based mucosal delivery of target proteins (Brandsma et al. 2010). Transgenic plants are believed to hold great promise for oral tolerance therapy against autoimmune disease, allergy, and many other inflammatory diseases.

References

Arabi YM, Tamim HM, Rishu AH (2009) Hypoglycemia with intensive insulin therapy in critically ill patients: predisposing factors and association with mortality. Crit Care Med 37(9): 2536–2544

Arakawa T, Yu J, Chong DK, Hough J, Engen PC, Langridge WH (1998) A plant-based cholera toxin B subunit-insulin fusion protein protects against the development of autoimmune diabetes. Nat Biotechnol 16(10):934–938

Asero R, Mistrello G, Roncarolo D, Amato S (2002) Allergenic similarity of 2S albumins. Allergy 57(1):62–63

Balding J, Livingstone WJ, Conroy J, Mynett-Johnson L, Weir DG, Mahmud N, Smith OP (2004) Inflammatory bowel disease: the role of inflammatory cytokine gene polymorphisms. Mediators Inflamm 13(3):181–187
Beebe AM, Cua DJ, de Waal MR (2002) The role of interleukin-10 in autoimmune disease: systemic lupus erythematosus (SLE) and multiple sclerosis (MS). Cytokine Growth Factor Rev 13(4–5):403–412
Bortesi L, Rossato M, Schuster F, Raven N, Stadlmann J, Avesani L, Falorni A, Bazzoni F, Bock R, Schillberg S, Pezzotti M (2009) Viral and murine interleukin-10 are correctly processed and retain their biological activity when produced in tobacco. BMC Biotechnol 9:22
Bouma G, Strober W (2003) The immunological and genetic basis of inflammatory bowel disease. Nat Rev Immunol 3(7):521–533
Brandsma ME, Diao H, Wang X, Kohalmi SE, Jevnikar AM, Ma S (2010) Plant-derived recombinant human serum transferrin demonstrates multiple functions. Plant Biotechnol J 8(4):489–505
Bu DF, Erlander MG, Hitz BC, Tillakaratne NJ, Kaufman DL, Wagner-McPherson CB, Evans GA, Tobin AJ (1992) Two human glutamate decarboxylases, 65-kDa GAD and 67-kDa GAD, are each encoded by a single gene. Proc Natl Acad Sci USA 89(6):2115–2119
Chebolu S, Daniell H (2009) Chloroplast-derived vaccine antigens and biopharmaceuticals: expression, folding, assembly and functionality. Curr Top Microbiol Immunol 332:33–54
Clark AT, Islam S, King Y, Deighton J, Anagnos- tou K, Ewan PW (2009) Successful oral tolerance induction in severe peanut allergy. Allergy 64(8):1218–1220
Douthwaite J, Jermutus L (2006) Exploiting directed evolution for the discovery of biologicals. Curr Opin Drug Discov Dev 9(2):269–275
Dove A (2002) Uncorking the biomanufacturing bottleneck. Nat Biotechnol 20(8):777–779
Eisenbarth GS (2004) Type 1 diabetes: molecular, cellular and clinical immunology. Adv Exp Med Biol 552:306–310
Elliott JF, Qin HY, Bhatti S, Smith DK, Singh RK, Dillon T, Lauzon J, Singh B (1994) Immunization with the larger isoform of mouse glutamic acid decarboxylase (GAD67) prevents autoimmune diabetes in NOD mice. Diabetes 43(12):1494–1499
Faria AM, Weiner HL (2006) Oral tolerance: therapeutic implications for autoimmune diseases. Clin Dev Immunol 13(2–4):143–157
Franklin SE, Mayfield SP (2004) Prospects for molecular farming in the green alga Chlamydomonas. Curr Opin Plant Biol 7(2):159–165
Frossard CP, Steidler L, Eigenmann PA (2007) Oral administration of an IL-10-secreting Lactococcus lactis strain prevents food-induced IgE sensitization. J Allergy Clin Immunol 119(4):952–959
Hashizume F, Hino S, Kakehashi M, Okajima T, Nadano D, Aoki N, Matsuda T (2008) Development and evaluation of transgenic rice seeds accumulating a type II-collagen tolerogenic peptide. Transgenic Res 17(6):1117–1129
Huibregtse IL, Snoeck V, de Creus A, Braat H, De Jong EC, Van Deventer SJ, Rottiers P (2007) Induction of ovalbumin-specific tolerance by oral administration of Lactococcus lactis secreting ovalbumin. Gastroenterology 133(2):517–528
Ichim TE, Zheng X, Suzuki M, Kubo N, Zhang X, Min LR, Beduhn ME, Riordan NH, Inman RD, Min WP (2008) Antigen-specific therapy of rheumatoid arthritis. Expert Opin Biol Ther 8(2):191–199
Jasinski JM, Eisenbarth GS (2005) Insulin as a primary autoantigen for type 1A diabetes. Clin Dev Immunol 12(3):181–186
Kalischuk LD, Buret AG (2010) A role for Campylobacter jejuni-induced enteritis in inflammatory bowel disease? Am J Physiol Gastrointest Liver Physiol 298(1):G1–9
Koffeman EC, Genovese M, Amox D, Keogh E, Santana E, Matteson EL, Kavanaugh A, Molitor JA, Schiff MH, Posever JO, Bathon JM, Kivitz AJ, Samodal R, Belardi F, Dennehey C, van den Broek T, van Wijk F, Zhang X, Zieseniss P, Le T, Prakken BA, Cutter GC, Albani S (2009) Epitope-specific immunotherapy of rheumatoid arthritis: clinical responsiveness occurs with immune deviation and relies on the expression of a cluster of molecules associated with T cell tolerance in a double-blind, placebo-controlled, pilot phase II trial. Arthritis Rheum 60(11):3207–3216

Kozuch PL, Hanauer SB (2008) Treatment of inflammatory bowel disease: a review of medical therapy. World J Gastroenterol 14(3):354–377

Larché M, Wraith DC (2005) Peptide-based therapeutic vaccines for allergic and autoimmune diseases. Nat Med 11(4 Suppl):S69–S76

Lee C, Ho H, Lee K, Jeng S, Chiang B (2011) Construction of a Der p2-transgenic plant for the alleviation of airway inflammation. Cell Mol Immunol 8(5):404–414

Ma S, Zhao DL, Yin ZQ, Mukherjee R, Singh B, Qin HY, Stiller CR, Jevnikar AM (1997) Transgenic plants expressing autoantigens fed to mice to induce oral immune tolerance. Nat Med 3(7):793–796

Ma S, Huang Y, Yin Z, Menassa R, Brandle JE, Jevnikar AM (2004) Induction of oral tolerance to prevent diabetes with transgenic plants requires glutamic acid decarboxylase (GAD) and IL-4. Proc Natl Acad Sci USA 101(15):5680–5685

Massa M, Passalia M, Manzoni SM, Campanelli R, Ciardelli L, Yung GP, Kamphuis S, Pistorio A, Meli V, Sette A, Prakken B, Martini A, Albani S (2007) Differential recognition of heat-shock protein DNAJ-derived epitopes by effector and Treg cells leads to modulation of infl ammation in juvenile idiopathic arthritis. Arthritis Rheum 56(5):1648–1657

Menassa R, Du C, Yin ZQ, Ma S, Poussier P, Brandle J, Jevnikar AM (2007) Therapeutic effectiveness of orally administered transgenic low-alkaloid tobacco expressing human interleukin-10 in a mouse model of colitis. Plant Biotechnol J 5(1):50–59

Moghadam-Kia S, Werth VP (2010) Prevention and treatment of systemic glucocorticoid side effects. Int J Dermatol 49(3):239–248

Morandini F, Avesani L, Bortesi L, Van Droogenbroeck B, De Wilde K, Arcalis E, Bazzoni F, Santi L, Brozzetti A, Falorni A, Stoger E, Depicker A, Pezzotti M (2011) Non-food/feed seeds as biofactories for the high-yield production of recombinant pharmaceuticals. Plant Biotechnol J. doi:10.1111/j.1467-7652.2011

Niederberger V, Valenta R (2004) Recombinant allergens for immunotherapy. Where do we stand? Curr Opin Allergy Clin Immunol 4(6):549–554

Park KS, Park MJ, Cho ML, Kwok SK, Ju JH, Ko HJ, Park SH, Kim HY (2009) Type II collagen oral tolerance: mechanism and role in ollagen-induced arthritis and rheumatoid arthritis. Mod Rheumatol 19(6):581–589

Paul M, Ma JK (2011) Plant-made pharmaceuticals: leading products and production platforms. Biotechnol Appl Biochem 58(1):58–67

Penney CA, Thomas DR, Deen SS, Walmsley AM (2011) Plant-made vaccines in support of the Millennium Development Goals. Plant Cell Rep 30(5):789–798

Rasala BA, Mayfield SP (2011) The microalga Chlamydomonas reinhardtii as a platform for the production of human protein therapeutics. Bioeng Bugs 2(1):50–54

Rodríguez-Narciso C, Pérez-Tapia M, Rangel-Cano RM, Silva CL, Meckes-Fisher M, Salgado-Garciglia R, Estrada-Parra S, López-Gómez R, Estrada-García I (2011) Expression of Mycobacterium leprae HSP65 in tobacco and its effectiveness as an oral treatment in adjuvant-induced arthritis. Transgenic Res 20(2):221–229

Ruhlman T, Ahangari R, Devine A, Samsam M, Daniell H (2007) Expression of cholera toxin B-proinsulin fusion protein in lettuce and tobacco chloroplasts – oral administration protects against development of insulitis in non-obese diabetic mice. Plant Biotechnol J 5(4):495–510

Skyler JS (2008) Update on worldwide efforts to prevent type 1 diabetes. Ann N Y Acad Sci 1150: 190–196

Smart V, Foster PS, Rothenberg ME, Higgins TJ, Hogan SP (2003) A plant-based allergy vaccine suppresses experimental asthma via an IFN-gamma and CD4+CD45RBlow T cell-dependent mechanism. J Immunol 171(4):2116–2126

Smith AM, Benjamin DC, Hozic N, Derewenda U, Smith WA, Thomas WR, Gafvelin G, van Hage-Hamsten M, Chapman MD (2001) The molecular basis of antigenic cross-reactivity between the group 2 mite allergens. J Allergy Clin Immunol 107(6):977–984

Sun JB, Czerkinsky C, Holmgren J (2010) Mucosally induced immunological tolerance, regulatory T cells and the adjuvant effect by cholera toxin B subunit. Scand J Immunol 71(1):1–11

Suzuki K, Kaminuma O, Yang L, Takai T, Mori A, Umezu-Goto M, Ohtomo T, Ohmachi Y, Noda Y, Hirose S, Okumura K, Ogawa H, Takada K, Hirasawa M, Hiroi T, Takaiwa F (2011) Prevention of allergic asthma by vaccination with transgenic rice seed expressing mite allergen: induction of allergen-specific oral tolerance without bystander suppression. Plant Biotechnol J. doi:10.1111/j.1467-7652.2011

Takagi H, Hiroi T, Yang L, Tada Y, Yuki Y, Takamura K, Ishimitsu R, Kawauchi H, Kiyono H, Takaiwa F (2005) A rice-based edible vaccine expressing multiple T cell epitopes induces oral tolerance for inhibition of Th2-mediated IgE responses. Proc Natl Acad Sci USA 102(48): 17525–17530

Takagi H et al (2008) Efficient induction of oral tolerance by fusing cholera toxin B subunit with allergen-specific T-cell epitopes accumulated in rice seed. Vaccine 26:6027–6030

Takaiwa F (2007) A rice-based edible vaccine expressing multiple T-cell epitopes to induce oral tolerance and inhibit allergy. Immunol Allergy Clin North Am 27(1):129–139

Tremblay R, Wang D, Jevnikar AM, Ma S (2010) Tobacco, a highly efficient green bioreactor for production of therapeutic proteins. Biotechnol Adv 28:214–221

Tzfira T, Citovsky V (2006) Agrobacterium-mediated genetic transformation of plants: biology and biotechnology. Curr Opin Biotechnol 17:147–154

Verma D, Moghimi B, LoDuca PA, Singh HD, Hoffman BE, Herzog RW, Daniell H (2010) Oral delivery of bioencapsulated coagulation factor IX prevents inhibitor formation and fatal anaphylaxis in hemophilia B mice. Proc Natl Acad Sci USA 107(15):7101–7106

Wang X, Brandsma M, Tremblay R, Maxwell D, Jevnikar AM, Huner N, Ma S (2008) A novel expression platform for the production of diabetes-associated autoantigen human glutamic acid decarboxylase (hGAD65). BMC Biotechnol 8:87

Weiner HL, da Cunha AP, Quintana F, Wu H (2011) Oral tolerance. Immunol Rev 241(1): 241–259

Zhang L, Eisenbarth GS (2011) Prediction and prevention of Type 1 diabetes mellitus. J Diabetes 3(1):48–57

Chapter 3
Molecular Farming Using Bioreactor-Based Plant Cell Suspension Cultures for Recombinant Protein Production

Ting-Kuo Huang and Karen A. McDonald

Abstract The need for biomanufacturing capacity for recombinant protein production to meet the expanding pharmaceutical and industrial market demands has gained increasing importance, leading to the development of new protein expression platforms capable of addressing requirements in terms of protein yield, product quality, and production cost. In the past few decades, molecular farming using plant cell-based expression systems (whole plants and *in vitro* plant cells, organ and tissue cultures) have been investigated as an alternative for the large-scale bioproduction of recombinant proteins. Molecular farming using bioreactor-based plant cell suspension cultures provides attractive features over recombinant microbial fermentation and mammalian cell cultures in terms of intrinsic safety, cost-effective biomanufacturing, and the capability for post-translation modifications. The current research and development, emerging techniques, commercialization and future prospects of molecular farming using bioreactor-based plant cell suspension cultures for production of recombinant proteins will be discussed in this chapter.

3.1 Introduction

The global market of recombinant protein products for pharmaceutical/therapeutic applications reached $86 billion in 2007 and is expected to increase to an estimated $160 billion in 2013, representing an annual growth rate of 11%. Large-scale biomanufacturing of recombinant protein products has received important considerations due to the expanding market demand with the additional applications of

T.-K. Huang • K.A. McDonald (✉)
Department of Chemical Engineering and Materials Science, University of California – Davis, 1031B Kemper Hall, 1 Shields Avenue, Davis, CA 95616, USA
e-mail: kamcdonald@ucdavis.edu

A. Wang and S. Ma (eds.), *Molecular Farming in Plants: Recent Advances and Future Prospects*, DOI 10.1007/978-94-007-2217-0_3,

the industrial/technical recombinant proteins, resulting in new protein expression platforms and emerging technologies development such as molecular farming bioproduction platforms (transgenic animal and plant-based) in addition to mammalian cell cultures and microbial fermentation (Demain and Vaishnav 2009; Hacker et al. 2009).

Plant molecular farming refers to the large scale production of recombinant proteins, including plant-made pharmaceuticals (PMP) and plant-made industrial proteins (PMIP), in transformed plant cells or, transgenic plants. Plant-based molecular farming, including transgenic plants, transplastomic plants, transient expression by agroinfiltration in plants and plant tissues, virus-infected plants, plant tissue and organ cultures, and plant cell suspension cultures, have been investigated as an economical alternative bioproduction platform for recombinant protein production in the past 20 years (Obembe et al. 2011). Plant cell-based molecular farming offers attractive features over traditional microbe- and mammalian cell-based expression systems, including (1) their intrinsic safety (plant cells do not propagate mammalian viruses and pathogens and can be cultivated in animal-derived component free medium which are important considerations for therapeutics production), (2) cost-effective biomanufacturing leading to lower production costs (Shadwick and Doran 2005), and (3) the capability for protein post-translation modifications (such as the ability to produce glycoproteins with similarity to their native counterparts in terms of N-glycan structure compared to mammalian cells) (Gomord et al. 2010).

Plant cell suspension cultures grown in a fully contained bioreactor system offer additional features, compared to molecular farming using whole plants, for economical, sustained recombinant protein production, especially for biopharmaceuticals production, (Franconi et al. 2010; Hellwig et al. 2004) including (1) shorter biomanufacturing timescale of only few weeks required in plant cell culture process; (2) more consistency in product quality and homogeneity of the target protein N-glycan structures in controlled bioreactor operations (De Muynck et al. 2010; Lienard et al. 2007); (3) cost effective purification operations especially for secreted products (Rawel et al. 2007); (4) less contamination from endotoxin and mycotoxin; (5) safer production platform in a contained system, avoiding issues such as gene flow in the environment and potential contamination to the food chains (Franconi et al. 2010); and (6) ease of compliance with cGMP requirements (Shih and Doran 2009), etc. In this chapter, recent developments and future prospects of molecular farming using bioreactor-based plant cell suspension cultures (dedifferentiated plant cells such as tobacco, rice and carrot cell cultures, etc.) for recombinant protein production will be discussed.

3.2 Recombinant Protein Production Using Plant Cell Suspension Cultures

In vitro suspension cultured plant cells under controlled environmental conditions have been developed primarily for the production of valuable medicinal metabolites such as shikonin and paclitaxel (Taxol) in the past 50 years (Hellwig et al. 2004).

Currently various recombinant proteins for therapeutic (monoclonal antibodies, antigens, vaccines, hormones, growth factors and blood proteins, etc.), medical (e.g. gelatin and collagen for drug capsules), and industrial (enzymes such as cellulases, lignases and lipases for biofuel) applications can be expressed using plant cell cultures transformed with appropriate gene expression systems (Huang and McDonald 2009). In this section, the methods to establish transgenic plant cell suspension cultures will be presented.

3.2.1 Features of Plant Cell Suspension Cultures for Recombinant Protein Production

Molecular farming using bioreactor-based plant cell culture should exhibit the following features: (1) ease of genetic manipulation by stable transformation or transient expression, (2) high protein expression level, (3) low endogenous proteolytic activity, (4) high product stability (inside and outside of the cells), (5) low concentration of secondary metabolites, which may cause changes in expressed protein structural and biological properties and complicate downstream processes, (6) post-translational modification capability, uniform glycosylation pattern and proper protein folding, (7) small cell aggregates and good homogeneous dispersion in a bioreactor, (8) high specific growth rate, and (9) long-term cell line genetic and production stability, etc.

3.2.2 Method to Establish Transgenic Plant Cell Suspension Cultures

Transgenic plant cell suspension cultures developed for recombinant protein production are usually derived from stably transformed plant or plant tissues generated by mainly using *Agrobacterium*-mediated transformation methods (Offringa et al. 1990), although they can also be derived from initial transformed callus from an independent transformation event. Leaves, roots, shoots or petioles from grown plants can be used as initial explants. Callus cultures grown on agar containing appropriate plant growth regulating hormones initiated from explants from transgenic plants can be transferred to grow in a chemically defined media with continuous shaking to establish suspended cell cultures (Rao et al. 2009). After 5–14 days, the force from hydrodynamic/mechanical stress in continuous shaken or agitated liquid media results in a population of cell suspensions with healthy cells, dead and/or decaying cell material. The population of an initial suspended cells usually consists roughly of three fractions including (1) free single cells of various shapes, (2) clusters of 3–5 cells, small (up to 20 cells) and big cell aggregates (more than 20 cells), and (3) cell groups with a threadlike morphology, released from the callus. The outermost cell layers of the cell aggregates are removed due to the hydrodynamic force by the agitation of the shaker and then represent the fractions of single cells.

A dilution ratio of 1:1–1:5 (volume ratio of mother culture to fresh medium) is typically used to maintain the cell suspension culture. Fine cell fractions can be obtained and transferred using sterile pipette (2–10 mL). Sterile meshes with different sizes (such as Sigma cell dissociation kit) can be implemented to separate large aggregates and established a suspension culture only with single cells, clusters or small aggregates. A mesh size of 60 is used to initiate a suspension culture composed of big and small aggregates, clusters and single cells, size 80 is used to separate small aggregates, clusters and single cells, size 100 is used to obtain clusters and single cells, and size 150 is to isolate almost single cells. Observations using an inverted microscope can be used to confirm the formation of structures, single cell, clusters or cell divisions.

The preculture stage is carried out to propagate the single cells by inoculating the original cell population containing single cells and small cell aggregates in fresh nutrient media. The addition of plant growth regulators is usually required in the culture medium to promote rapid cell growth and maintain the cell morphology. After 1–2 weeks, part of the preculture (cells and nutrient medium) is transferred to other vessels or a bioreactor. Usually the cell suspension with some nutrient medium can be transferred from the smaller to the larger bioreactor to produce increasing amounts of cell suspension. The optimization of bioreactor operation for recombinant protein production will be discussed in later session. For some plant species, starting with a population of active single cells in a liquid medium, cell aggregates of various sizes will develop soon after initiation of growth, coexisting with some free suspended cells. The size distribution of cell aggregates and single cell morphology are important consideration for bioreactor optimization.

Subcultures are frequently conducted by a dilution of 1:5–1:10 with fresh nutrient medium at 1–2 week intervals to maintain suspended plant cells in a healthy and active condition for the establishment of long-term cell suspension cultures. The optimal dilution ratio (or inoculation ratio) and subculture frequency have to be experimentally determined for each individual species.

3.2.3 *Example of Transgenic Nicotiana benthamiana Cell Suspension Cultures*

3.2.3.1 Plant Cell Transformation and Media

Figure 3.1 shows the schematic presentation of the production of a recombinant protein of interest in transgenic tobacco plants, transgenic tobacco suspension cell cultures and transiently expressed in tobacco suspension cells. The flowchart of how to obtain a cell suspension in Fig. 3.1 can be easily adapted to other plant cell systems.

The gene expression system is stably transformed into *Nicotiana benthamiana* cells using *Agrobacterium*-mediated gene transformation by the recombinant *Agrobacterium tumefaciens* carrying appropriate binary vector containing the gene

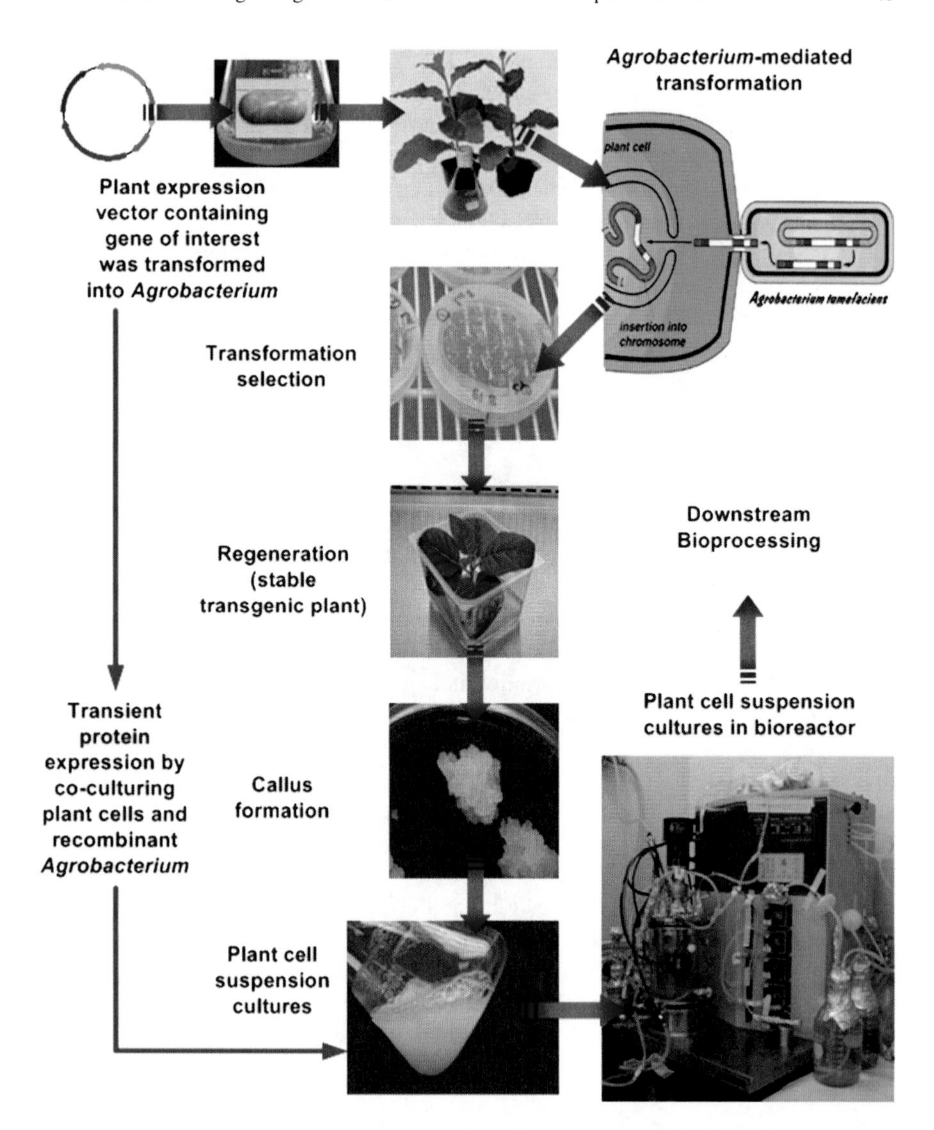

Fig. 3.1 A schematic presentation of the production of a recombinant protein of interest in transgenic plants, transgenic plant suspension cell cultures and transiently expressed in plant suspension cells

of interest. Newly expanded leaves from *N. benthamiana* plants are cut into 1 cm square sections soaked in an *Agrobacterium* solution adjusted to 0.1 OD_{600} for 10 min and incubated on co-cultivation medium consisting of Murashige and Skoog minimal organics (MSO) medium modified with 30 g/L sucrose, 2 mg/L 6-benzylaminopurine (BA), and 200 μM acetosyringone, pH 5.8, at 23–25°C in the dark for 2–3 days. Leaves are transferred to agar-solidified selection medium consisting of MSO medium modified with 30 g/L sucrose, 2 mg/L BA, 400 mg/L carbenicillin, 250 mg/L cefotaxime, and 250 mg/L kanamycin and incubated at 23–25°C for 10 days. Plant tissues are subcultured until shoots formed. Shoots are harvested and transferred to agar-solidified rooting medium consisting of half strength MSO medium modified with 15 g/L sucrose, 2 mg/L BA, 1.3 g/L calcium gluconate, 400 mg/L carbenicillin, 250 mg/L cefotaxime, and 100 mg/L kanamycin. Leaves are removed from each rooted shoot with a portion of the petiole attached and placed on callus-generating medium consisting of MSO medium modified with 30 g/L sucrose, 0.4 mg/L 2,4-dichlorophenoxyacetic acid (2,4-D), 0.1 mg/L kinetin, 400 mg/L carbenicillin, 150 mg/L timentin, and 100 mg/L kanamycin for developing transgenic callus. Transgenic callus are subcultured every 3–4 weeks on agar-solidified KCMS medium consisting of 30 g/L sucrose, 4.3 g/L MS salt mixture, 0.1 g/L myo-inositol, 0.204 g/L KH_2PO_4, 0.5 mg/L nicotinic acid, 0.5 mg/L thiamine-HCl, 0.5 mg/L pyridoxine-HCl, 0.2 mg/L 2,4-D, 0.1 mg/L kinetin, and appropriate antibiotics as selection pressure, pH 5.8.

3.2.3.2 Transgenic Plant Cell Suspension Cultures and Media

Transgenic *N. benthamiana* cell lines are selected from transgenic callus cultures and screened by appropriate protein analysis methods such as ELISA and Western blots. There can often be a significant variation in expression levels between independent transformed lines due to the fact the that the expression cassette is inserted randomly at one or more locations in the nuclear genome, so screening of the independent transformation events, either at the whole plant or callus stage, is an important step. Transgenic plant cell suspension cultures established from callus cultures are subcultured weekly by transferring 20 mL of established suspension cells into 200 mL KCMS medium (pH 5.8), consisting of 30 g/L sucrose, 4.3 g/L MS salt mixture, 0.1 g/L myo-inositol, 0.204 g/L KH_2PO_4, 0.5 mg/L nicotinic acid, 0.5 mg/L thiamine-HCl, 0.5 mg/L pyridoxine-HCl, 0.2 mg/L 2,4-D, and 0.1 mg/L kinetin, and appropriate antibiotics as selection pressure, in a 1 L flask maintained on an orbital shaker at 140 rpm and 25°C. Scale-up of cell biomass expansion can be conducted in larger vessels such as stirred-tank bioreactor.

3.3 Optimization of Recombinant Protein Production in Plant Cell Cultures

Current limitations of plant cell-based expression systems for recombinant protein production are the low product yields and the effects of non-human glycosylation pattern of glycosylated plant-made recombinant proteins on the activity, immunogenicity and allergenicity for therapeutic applications. To achieve higher protein yields and desired quality, recombinant protein expressed in plant cells can be optimized from transcription to protein stability including the selections of promoters, enhancers, integration sites, codon usage/synthetic gene design, gene silencing suppressors, and product compartmentalization. Table 3.1 shows examples of

Table 3.1 Selected example of foreign protein productions by plant cell culture bioreactor systems (Huang and McDonald 2009)

Plant species	Culture type	Protein product	Promoter system	Localization	Bioreactor system	Operation conditions	Production level	Reference
Nicotiana benthamiana (tobacco)	Suspension cells	AAT	CaMV 35S	Secreted, extracellular	STR, pitched blade impeller	25°C, 50 rpm, 40% DO, pH 6.4, batch	<1 μg-FAAT/L (without pH control), 100 μg-FAAT/L (pH control)	Huang et al. (2009)
Nicotiana benthamiana (tobacco)	Suspension cells	AAT	Estradiol inducible XVE system	Secreted, extracellular	STR, pitched blade impeller	25°C, 50 rpm, 40% DO, pH 6.4, batch, 10 μM estradiol as inducer	<1 μg-FAAT/L (without pH control), 60 μg-FAAT/L (pH control)	Huang et al. (2009)
Nicotiana benthamiana (tobacco)	Suspension cells	AAT	Cucumber mosaic virus inducible viral vector (CMViva)	Secreted, extracellular	STR, pitched blade impeller	25°C, 50 rpm, 40% DO, pH 6.4, batch, 10 μM estradiol as inducer	25 μg-FAAT/L (without pH control), 100 μg-FAAT/L (pH control)	Huang et al. (2009)
Nicotiana benthamiana (tobacco)	Suspension cells	AAT	Cucumber mosaic virus inducible viral vector (CMViva)	Secreted, extracellular	STR, pitched blade impeller	25°C, 50–100 rpm, 40% DO, pH 6.4, SCC, 1 μM estradiol as inducer	600 μg-FAAT/L (pH control)	Huang et al. (2010)
O. sativa (rice)	Suspension cells	AAT	RAmy3D (inducible)	Secreted, extracellular	STR, pitched blade impeller	STR, 75 rpm, 27°C, 70% DO, 0.1–0.2 vvm, SCC (multiple growth and production phases)	40–110 mg/L, 3–12 mg/(L-day), 4–8 mg/(g-DCW-day)	Trexler et al. (2005)

(continued)

Table 3.1 (continued)

Plant species	Culture type	Protein product	Promoter system	Localization	Bioreactor system	Operation conditions	Production level	Reference
O. sativa (rice)	Suspension cells	AAT	RAmy3D (inducible)	Secreted, extracellular	Membrane bioreactor	25°C, 130 rpm, two stage culture (sugar starvation)	100–247 mg/L,4–10% of TSP	McDonald et al. (2005)
O. sativa (rice)	Suspension cells	hGM-CSF	RAmy3D (inducible)	Secreted, extracellular	Flask	110 rpm, 13 day, 27°C, batch	129 mg/L, 25% of TSP	Shin et al. (2003)
N. tabacum L. (tobacco)	Suspension cells	hGM-CSF	CaMV 35S	Secreted, extracellular	Flask	110 rpm, 12 day, 25°C, batch	105 μg/L	Hong et al. (2002)

AAT human alpha-1-antitrypsin; *FAAT* functional human alpha-1-antitrypsin; *DO* dissolved oxygen; *DCW* dried cell weight; *SCC* semicontinous culture; *D* dilution rate (1/day); *STR* stirred-tank bioreactor; *TSP* total soluble protein; *VVM* volume of gas per volume of culture per minute recombinant protein production by stably transformed plant cell suspension culture in bioreactors, highlighting the importance of selection of host, expression vector and bioreactor operational strategy. Recent development to improve recombinant protein expression in plant cell cultures will be discussed.

3.3.1 Host Systems

Although tobacco, maize, rice and alfalfa are commonly utilized as hosts for recombinant proteins production, the emerging plant expression systems, like *Lemna minor*, *Physcomitrella patens*, *Chlamydomonas reinhardtii* or higher plant cell suspension cultures are offering new opportunities for molecular farming.

3.3.2 Genetic Transformation Methods

Several methods can be implemented for a recombinant protein expressed stably or transiently in plant cells including (1) stable transformation by incorporation of foreign DNA into nuclear, plastid or chloroplast genomes using *Agrobacterium*-mediated transformation, and by particle bombardment or electroporation of protoplasts (Sharma et al. 2005), and (2) transient expression by *Agrobacterium* agroinfiltration of plant tissue or infection with plant viral vectors (Pogue et al. 2010).

3.3.2.1 Stably Transgene Expression

Stable expression of a foreign gene in plant cell culture requires that the gene of interest is integrated into the nuclear or chloroplast genome of host cells and will be passed to subsequent generations. Transgenic plant-based molecular farming may be a better choice for long-term commercial scale production. The methods of plant transformation will depend on the plant host species and the procedure of stable transformation is well established and usually time consuming, which normally takes at least few months (3–9 months) for development and optimization of transformation, selection of a best transformed clone, and plant regeneration (Twyman et al. 2003).

However, genetic instability of the transgenic dedifferentiated cells has been shown to result in somaclonal variation (gene drift), a potential limiting factor in developing plant cell cultures for long term recombinant protein production (Offringa et al. 1990). Reduction in the product yield is often attributed to genetic instability, transgene loss, variation in growth rate, or other undesirable genetic or epigenic changes and has been observed for dedifferentiated plant cell cultures when maintained by liquid subculturing (Lambe et al. 1995; Vandermaas et al. 1994). Thus, cryopreservation of transgenic plant cell lines for plant cell suspension culture has been discussed to allow for the long-term preservation and maintenance of a stable production system (Cho et al. 2007; Schmale et al. 2006). On the other hand, recombinant protein expression using transient expression in plant cell culture by either recombinant *Agrobacterium*-mediated transfection or plant viruses, common methods applied to whole plants and intact plant tissues (Komarova et al. 2010), is under development as an alternative for recombinant protein expression.

3.3.2.2 Transient Transgene Expression

Transient expression for rapid transgene gene expression has potential for large-scale recombinant protein production in whole plants or plant tissue. Transient expression by the agroinfiltration method (Kapila et al. 1997), in which recombinant *Agrobacterium tumefaciens* are infiltrated into plant tissue, or using plant viral vectors (Gleba et al. 2007) are commonly used. In agroinfiltration the T-DNA is transferred to the nucleus in a large number of plant cells resulting in the expression of milligram amounts of recombinant protein within a short time period (2–14 days depending on the protein of interest, the host and the expression system). Viral vectors have also attracted interest because viral infections are rapid and systemic, and infected cells yield large amounts of virus and viral gene products (Streatfield 2007). Since plant viruses do not integrate into the genome, there is no stable transformation and the transgene is not passed through the germ line.

Transient expression has given higher productivity than stably transformed plants mainly due to the relative timing of the onset of PTGS (post-transcriptional gene silencing) compared to transgene expression (Wroblewski et al. 2005) and position effect of transgene integration into the plant cell genome (Kumar and Fladung 2001). In transient expression, high levels of gene product may accumulate prior to the initiation of PTGS. Currently, only few examples of using transient expression for recombinant protein production in plant cell suspension cultures have been reported. Boivin et al. utilized transient co-expression of a transgene and the p19 viral suppressor by agroinfiltration to enhance the mouse IgG1 antibody production reaching 148 mg IgG1/kg-FW in *N. benthamiana* cell culture, implying the feasibility of using transient expression in a plant cell culture bioreactor system (Boivin et al. 2010).

3.3.3 Gene Expression Systems

The type and characteristics of the promoter system significantly impact the gene expression level by affecting the transcription rate of the target gene, and further determines the bioreactor optimal operation mode. Constitutive or inducible promoters are commonly utilized for gene expression in plant cell cultures. Table 3.2 compares the features of various promoters used in transgenic plants cells (Huang and McDonald 2009).

Table 3.2 Features of various promoters for expressing recombinant protein in plant cells (Huang and McDonald 2009)

Types	Examples	Features
Constitutive promoter	CaMV 35S, Maize ubiquitin	Commonly used in plant-based molecular farming
		Recombinant protein production is growth-associated
		Potential problem of PTGS
Inducible promoter (chemically)	Steroid-regulated	Easy for induction by simply adding inducer
		Low dose required in plant cell cultures
		Inducer-dependent toxic effects
	Ethanol-regulated	Inducer is simple with low toxicity
		Volatility may be a problem
	Tetracycline-regulated	Short half-life of antibiotics
		High dose required for induction, resulting in toxic effect to host cells
		Continuous addition required
Inducible promoter (metabolic)	Metabolite-regulated: rice alpha-amylase RAamy3D (sugar starvation)	Need to conduct media exchange
		Two stage cultures required
		Affect cellular metabolism of host cells
Inducible promoter (physical)	Temperature-regulated	Trigger heat shock response related proteins
		Influence cellular metabolism
		Additional energy required
	Light-regulated	Influence cell growth and cellular metabolism
		Additional energy required

Constitutive promoters, such as the CaMV 35S promoter, directly drive the expression of the target gene which is directly related to cell growth. The target protein expressed by a constitutive promoter is considered to be a growth-associated product and is continuously expressed until host cells reach the stationary phase, resulting in an additional metabolic burden and hence impacting the plant cellular physiology and growth rate due to the potential toxic properties of the product on the host cells or it may interfere with host cell metabolism.

Inducible gene expression systems controlled by the strength of specific external factors or compounds, such as light, temperature, metal ions, alcohols, steroids and herbicides, etc., have been developed to achieve high-level recombinant protein production (Murphy 2007; Padidam 2003; Zuo and Chua 2000), allowing the cell growth and protein production phases to be optimized independently. The inducible promoter systems are attractive particularly when product synthesis is deleterious to host cell growth. The expression of a foreign gene linked to an inducible promoter can be induced at a specific stage during the cell growth cycle, and there is less potential for PTGS found in transgenic plants driven by constitutive promoters (Vaucheret and Fagard 2001). Chemically inducible promoter systems induced by estradiol (Zuo et al. 2000), ethanol (Zhang and Mason 2006) and dexamethasone

(Samalova et al. 2005) have mainly been developed for recombinant protein production in plant systems. Selection of the highest expressing cell lines with an inducible promoter system is complicated by the fact that the optimal timing and level of the inducer is often not known *a priori* and must be determined experimentally.

Plant viral vectors, designed to increase the transgene copy number by the action of the viral replicase, are alternatives for foreign gene expression (Lico et al. 2008). Several genetically modified viral vectors, such as *Tobacco mosaic virus* (TMV), *Cucumber mosaic virus* (CMV) and *Potato virus* X (PVX), have been developed for recombinant protein production in plant molecular farming systems (Lico et al. 2008; Lindbo 2007; Wagner et al. 2004). However, plant viral vectors are prone to vector instability, leading to reduced replication of the viral vector due to methylation, PTGS, or loss of transgene (Angell and Baulcombe 1997; Atkinson et al. 1998).

Therefore, inducible plant viral vectors, which combine the features of inducible promoters and plant viral vectors (Gleba et al. 2007; Lico et al. 2008), are proposed as an alternative (Gleba et al. 2007). Our group has developed a *Cucumber mosaic virus* (CMV) inducible viral amplicon (CMViva) expression system encoding a viral replicase controlled by an estradiol-activated XVE promoter (Zuo et al. 2000), and hence the recombinant viral amplicons are only produced intracellularly under induction conditions. The CMViva system has been demonstrated to allow tightly regulated foreign gene expression and good production of functional recombinant human proteins in non-transgenic plant tissues using transient agroinfiltration and in transgenic tobacco cell suspension culture in bioreactor (Huang et al. 2009, 2010; Plesha et al. 2007, 2009; Sudarshana et al. 2006). Other examples of inducible plant viral vectors in plant cell cultures include the estradiol-inducible *Tomato Mosaic Virus* (ToMV) amplicon system expressing GFP in tobacco BY-2 cells (Dohi et al. 2006) and the ethanol-inducible *Bean Yellow Dwarf Virus* (BeYDV) amplicon expressing Norwalk virus capsid protein (NVCP) in tobacco NT1 cells (Zhang and Mason 2006).

3.3.4 *Stability of Recombinant Proteins*

Recombinant proteins expressed in plant cell cultures can be secreted to the extracellular culture medium or retained in an intracellular compartment such as ER, cytoplasm, or vacuole. Secreted products offer advantages of bioprocessing over intracellular retained products such as simpler downstream processing and flexibility in the bioreactor operational mode allowing continuous culture or multiple production cycles by reusing plant cells, leading to increased the overall productivity. However, secreted recombinant proteins are commonly degraded by proteolytic enzymes produced during plant cell cultivation or resulting from cell death/lysis (Doran 2006a) and/or may be unstable in the simple cell culture medium composition (James and Lee 2001; Tsoi and Doran 2002). Previous studies have proposed a variety of approaches to improve the stability of secreted recombinant proteins and to prevent the recombinant protein loss from proteolytic degradation, including the supplementation of protease inhibitors or protein stabilization agents (such as gelatin, BSA and other low-value proteins), mannitol (to regulate the osmotic pressure of the medium to minimize cell lysis), PVP, PEG, Pluronic F-68 and other polymers (as stabilizing agents for protection of the protein product from denaturing agents produced from the cell cultures) (Doran 2006a, b). Other molecular-level approaches have also been proposed for reducing proteolytic effects on recombinant proteins including (1) co-expression of protease inhibitors hindering endogenous protease activities along the cell secretory pathway or released into the culture medium (Komarnytsky et al. 2006), (2) suppression of protease gene expression using RNA interference (RNAi) (Kim et al. 2008), and (3) development of specific protease-deficient host cells (Schiermeyer et al. 2005).

Optimization of bioreactor operation is an alternative to enhance the stability of secreted recombinant proteins. Huang et al. (2009) proposed a bioreactor strategy involving pH control for improving functional recombinant human protein production in transgenic tobacco cell culture, resulting in enhanced recombinant protein stability and reduced protease activity in cell cultures, as an effective alternative to adding protease inhibitors or protein stabilizing agents in plant cell culture.

3.3.5 Post-translational Modification of Plant-Made Recombinant Proteins

Post-translational modification (PTM) of the plant-made protein is a critical quality attribute and in certain cases may limit the pharmaceutical and industrial applications of plant-made proteins. Although plants and mammalian cells share similar PTM mechanisms such as expressed proteins entering the secretory pathways where N-linked glycosylation occurs in the ER and Golgi apparatus, O-linked glycosylation occurs in the Golgi apparatus, and molecular chaperones in the ER help to fold the protein (Faye et al. 2005), minor differences in PTM have been observed. Plant cells tend to attach α-(1,3)-fucose and β-(1,2)-xylose in the glycan of the plant-made glycoprotein, which are absent in mammalian cells (Sethuraman and Stadheim 2006). In addition, plant-made glycoprotein lacks the terminal galactose and sialic acid residues, which have been observed on human glycoproteins (Gomord and Faye 2004). These differences in the glycan structures of plant-made glycoproteins may affect the expressed protein inherent stability, biological activity and immunogenicity, limiting their applications.

Strategies have been demonstrated to alleviate this issue by expressing "humanized" glycans in N-linked plant-made glycoproteins (Schahs et al. 2007), such as applying RNAi against to the appropriate fucose and xylose transferases in plant cells. Strasser et al. (2008) reported the generation of glyco-engineered *N. benthamiana* lines with downregulated endogenous xylosyltransferase and fucosyltransferase genes by using RNAi. In addition, double "knock-out" of the α-(1,3)-fucose and β-(1,2)-xylose transferases has also been investigated for the production of human vascular endothelial growth factor (hVEGF) in moss cell culture (Koprivova et al. 2004). Another example is to delete the fucose residues in plant N-glycans by repression of the GDP-mannose 4,6-dehydratase gene using virus-induced gene silencing and RNAi (Gomord et al. 2010). Castilho et al. (2010) has introduced the entire biosynthetic pathway for sialyation and terminal galactose in *N. benthamiana* using transient expression for producing MAb with a human-like glycosylation pattern. Other strategies to control the glycosylated protein are to understand how variations of bioreactor process parameters affect protein glycosylation patterns in host cells (del Val et al. 2010).

3.4 Technological Progresses in Plant Cell Culture Bioreactor Systems

In this section, the technological progress in bioreactor systems including bioreactor selection, bioreactor operation considerations and optimization for recombinant protein production in plant cell cultures will be discussed.

3.4.1 Bioreactor Systems

Table 3.3 summarizes important features of various types of bioreactors for *in vitro* plant cell cultures. General criteria for choosing a suitable bioreactor design should consider adequate oxygen mass transfer to cells, low shear stress to cells and adequate mass transfer (nutrient supply to cells and product, and by-product and metabolite removal from cells). The scalability of the bioreactor system for large scale production of a recombinant protein in plant cell culture also needs to be addressed.

3.4.1.1 Stirred-Tank Bioreactor

Stirred-tank bioreactors with suitable impellers can provide adequate volumetric mass transfer coefficients and a homogeneous environment enabling suspended plant cell growth and product production to be controlled. Although Rushton turbines (resulting in a radial flow pattern) can provide complete solids and gas dispersion, they also induce high turbulence around the impeller region and have higher specific power input and energy dissipation rate than other impellers with axial flow patterns (such as marine, paddle, and pitched-blade impellers), resulting in higher shear damage to suspended plant cells. The pitched-blade impeller with the upward-pumping mode provides similar capabilities, compared to Rushton turbines, for cell aggregate dispersion while reducing shear stress to plant cells when the power input was restricted by cell damage considerations (Doran 1999). However, the oxygen mass transfer performance of an upward-pumping pitch-blade impeller was poor in highly cell density cultures (Kieran 2001), compared with that of the same impeller operated in the downward-pumping mode (Junker et al. 1998). Generally, impeller systems exhibiting axial flow patterns with low impeller tip speed (less than 2.5 m/s) are considered acceptable for plant cell cultures (Amanullah et al. 2004).

Approaches have been proposed to reduce cell damage from hydrodynamic shear stress by agitation and from gas bubble bursting by aeration including developing new impeller systems to provide more efficient mixing at lower impeller tip speeds (such as the curved-blade disk turbine, hydrofoil impeller, helical ribbon, centrifugal impellers and cell-lift) (Doran 1999) and designing new aeration systems (such as bubble-free aeration, gas basket, and cage-aeration) for shear-sensitive cell culture processes. In addition, the low-power-number impellers such as Intermig, Prochem Maxflow and Scaba designed for animal cell culture can be implemented for plant cell culture (Varley and Birch 1999). The bioreactor geometric specifications such as impeller diameter, spacing between impellers, impeller off-bottom clearance, the baffles and their width, the sparger type and position, the ratio of liquid height to tank diameter, and the number of impellers are also critical considerations for large-scale bioreactor performance in order to provide sufficient mixing and adequate mass transfer.

3.4.1.2 Pneumatic Bioreactor

The pneumatic bioreactor (such as a bubble column or air-lift), consisting of a cylindrical vessel in which air or a gas mixture is introduced through a sparger at the bottom of the vessel for aeration, mixing and fluid circulation, exhibits features of low capital and operational cost, and ease of scale up for large scale operation of plant cell cultures. In addition, the low shear stress in pneumatic bioreactors is desirable for shear-sensitive plant cells (Eibl and Eibl 2008). However, bubble column bioreactors are less applicable to highly viscous liquid and high cell density cultures due to the lower gas-liquid interfacial area resulting from bubble coalescence in the viscous liquids and lack of mechanical break-up of bubbles.

Air-lift or modified air-lift bioreactors containing a draft tube (internal or external loop) exhibit the following features: (1) preventing bubble coalescence by directing them in one direction; (2) enhancing oxygen mass transfer by increasing the number of bubbles or gas-liquid interfacial area for enhancing mass transfer; (3) distributing shear stress more evenly; and (4) promoting the cyclical movement of fluid resulting in shorter mixing times (Huang et al. 2001, Wang et al. 2002). However, the inadequate oxygen mass transfer and poor fluid mixing in a high cell density culture, leading to inhomogeneities in biomass, nutrient, oxygen and pH, and extensive foaming (resulting from extracellular polysaccharides, proteins, fatty acids, and high superficial gas velocity) may become limiting factors in pneumatic bioreactor operation (Tanaka 2000).

Table 3.3 Comparisons of bioreactor systems for foreign protein production by plant cell cultures

Bioreactor	Stirred-tank	Bubble column	Air-lift bioreactor
	Air	Air	Air Air Air
Features	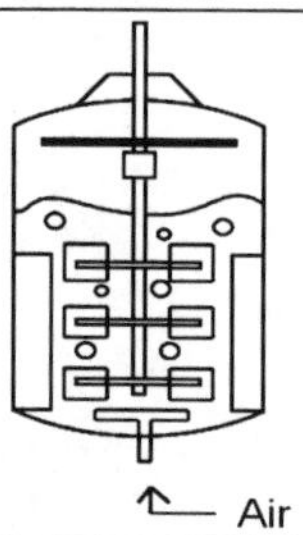	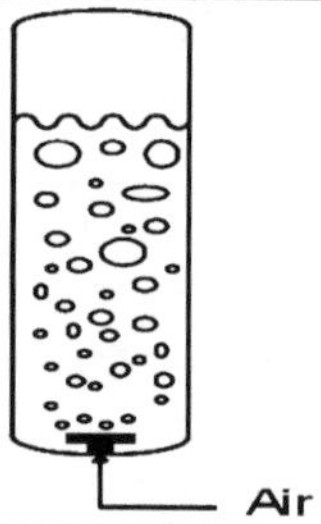	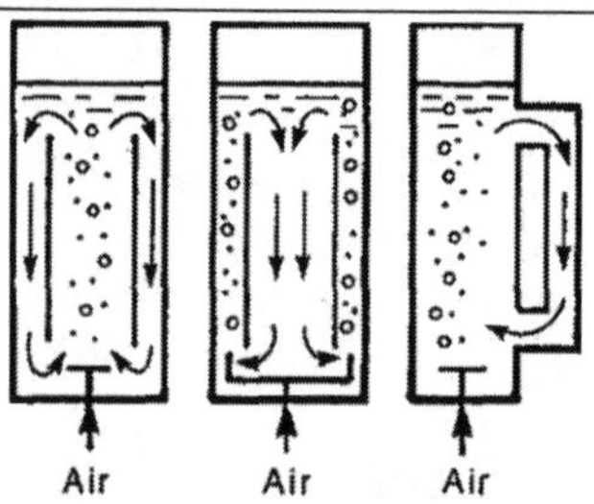
k_La (OTR)	High	Low	Medium
Cell damage by agitation	High	Low	Low
Cell damage by aeration	Medium	High	High
Mixing time	Short	Long	Medium
Operation	Medium	Simple	Simple
Flexibility	High	High	High
CIP/SIP	Yes	Yes	Yes
Scale size	Commercial	Commercial	Commercial
Scale-up	Medium	Easy	Easy
Power consumption	High	Low	Low
Culture type	Suspension; microcarrier	Suspension; microcarrier	Suspension; microcarrier
Cell density	Low/medium	Medium	Medium
Productivity	Medium	Medium	Medium
Monitoring and control	Direct, easy, multiple	Direct, easy, multiple	Direct, easy, multiple
Operation cost	High	Medium	Medium
Equipment cost	High	Medium	Medium
Ease of GMP compliance	Yes	Yes	Yes

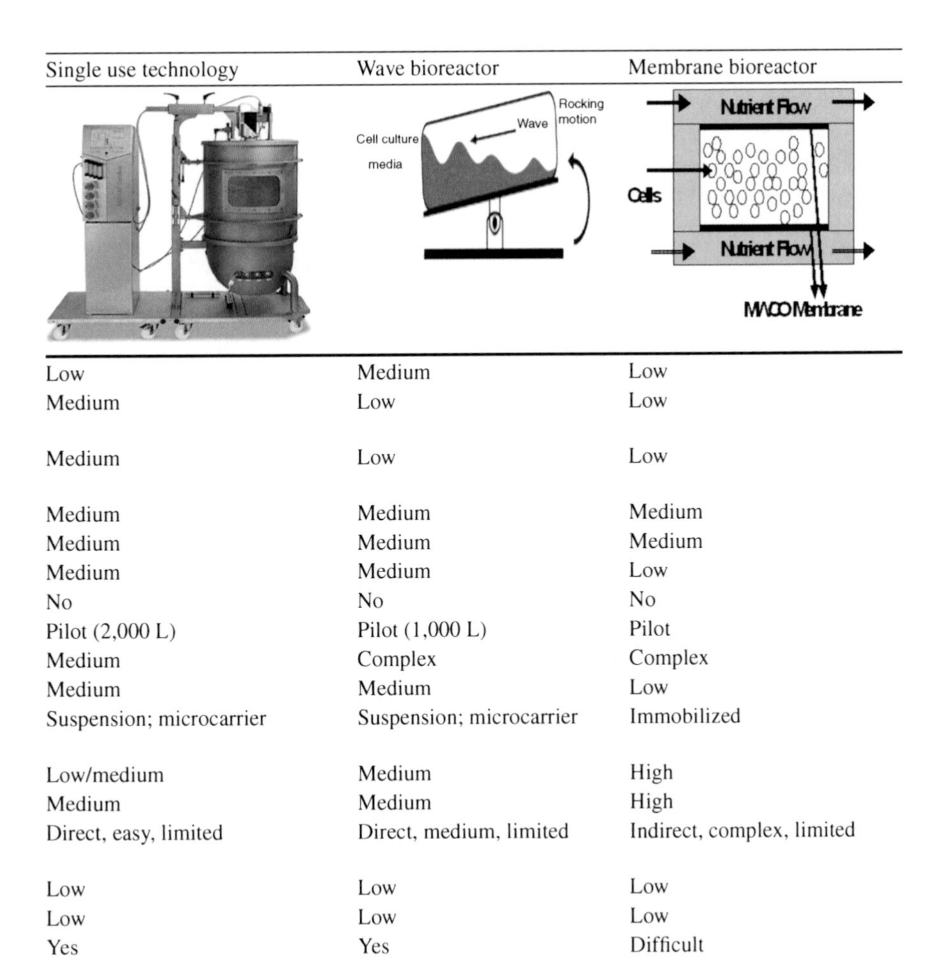

Single use technology	Wave bioreactor	Membrane bioreactor
Low	Medium	Low
Medium	Low	Low
Medium	Low	Low
Medium	Medium	Medium
Medium	Medium	Medium
Medium	Medium	Low
No	No	No
Pilot (2,000 L)	Pilot (1,000 L)	Pilot
Medium	Complex	Complex
Medium	Medium	Low
Suspension; microcarrier	Suspension; microcarrier	Immobilized
Low/medium	Medium	High
Medium	Medium	High
Direct, easy, limited	Direct, medium, limited	Indirect, complex, limited
Low	Low	Low
Low	Low	Low
Yes	Yes	Difficult

3.4.1.3 Disposable Bioreactor (Single-Use Bioreactor)

Disposable bioreactors, in which the cultivation vessel is made of a single-use plastic bag, provide attractive features for the biomanufacturing production of recombinant proteins in terms of time- and cost-savings for clean-in-place, sterilization-in-place, facility design and set-up, validation, capital investment on stainless steel vessels, elimination of cross-contamination and reduction of turnover time between each run, and shorter development time and increased throughput (Eibl et al. 2009a, b; Hacker et al. 2009). Currently disposable bioreactors have been successfully implemented into preclinical, clinical, and production-scale biomanufacturing facilities (Eibl et al. 2010). Although disposable bioreactors with standard stirred-tank configurations up to 2,000 L are available from companies such as HyClone and Xcellerex, they are limited in volume and mostly used for seed expansion and inoculation of the large conventional bioreactors. The disposable shaken bioreactor is another type of disposable bioreactor, which consists of a cylindrical vessel mounted on a circular moving shaker platform and contains a disposable, sterile plastic bag with appropriate connection tubes for seeding, feeding, gas supply and harvesting of these cultures (Micheletti et al. 2006). Protalix utilizes a disposable, bubble column-type bioreactor, which consists of a sterilizable polyethylene bag filled with plant cells and medium, for the prGCD production in transgenic carrot cell culture (Shaaltiel et al. 2008).

3.4.1.4 Wave Bioreactor

Wave bioreactors, another type of disposable bioreactor, possessing the advantages of low-cost and providing a low shear environment, utilize a non-gas permeable sterile bag comprised of plastic film. The mixing, mass and heat transfer in the wave bioreactor are regulated by rocking rate, rocking angle and medium filling volume. The bags can be filled with cell suspension up to 50–60% of their total capacity (up to 1,000 L) (Eibl et al. 2009a). Oxygen is supplied from the air or gas mixture continuously through headspace aeration. While the wave bioreactor is rocking, the liquid surface of the medium in the bag is continuously renewed and bubble-free surface aeration takes place resulting in oxygenation and bulk mixing with less shear stress to suspended cells. Additional advantages include time and cost savings, reduced foaming, easy operation and low risk of contamination. The wave bioreactor can be operated in different culture modes, including batch, fed-batch and perfusion when combined with different cell-retention devices. On-line measurements of pH, DO and other sensing technology (Read et al. 2009) make the wave bioreactor and other disposable bioreactors highly attractive for plant cell cultures (Eibl et al. 2009a). Investigations have demonstrated the application of the wave bioreactor for cultivating tobacco, grape and apple suspension cells up to 100 L working volume (Terrier et al. 2007). Eibl et al. achieved high plant cell (*V. vinifera*) biomass productivities of 40 g-FCW/(L-day) with a doubling time of 2 days and observed that there was no significant change in cell morphology when compared to cultivations in stirred tank bioreactors (Eibl and Eibl 2006).

3.4.1.5 Membrane Bioreactor

Membrane bioreactors are designed to retain host cells and also possibly recombinant protein product in a cell compartment by utilizing specialized membranes with a specific molecular weight cut-off (MWCO) for *in situ* aeration, nutrient supply, and product separation (Qi et al. 2003). The culture medium flow is circulated in the membrane bioreactor for bringing oxygen and nutrients to the cell and removing the waste metabolites. Membranes can be packed into different geometries including plate-and-sheet, tubular, spiral-wound, and hollow-fiber modules. The hollow fiber membrane bioreactor is the most commonly used geometry.

The main features of using membrane bioreactors in plant cell cultures are high cell density and high protein volumetric productivity, resulting from the use of membranes that retain the secreted foreign protein in the cell compartment, concentrating the product before harvest. Another feature is that the shear stress-induced cell damage found in stirred-tank bioreactors can be minimized in a membrane bioreactor because the cells are retained in a relatively quiescent region in which cells are protected from mechanical damage and are not in direct contact with gas bubbles. McDonald et al. (McDonald et al. 2005) applied a membrane bioreactor for the production of recombinant human alpha-1-antitrypsin (AAT) using transgenic rice cell culture with a rice alpha amylase promoter, Ramy3D, which is activated under sugar starvation conditions, resulting in extracellular product titer up to 250 mg/L (equivalent to 4–10% of the extracellular total soluble protein). However, large scale process in membrane bioreactors may cause operation problems resulting in poor cell viability, poor process stability, product heterogeneity, membrane fouling, and diffusion gradients (heat and mass transfer). Therefore the membrane bioreactor is mainly implemented for small scale processes and is difficult to scale up for large scale applications, although it may be useful for smaller scale applications such as patient-specific therapeutics.

3.4.2 Bioreactor Operation Consideration

Plant cell cultures exhibiting features of biological, morphological properties and cellular physiology characteristics, which are distinctive from bacterial and mammalian cells, need to be considered for their scalability in recombinant protein production. Important bioreactor operation considerations for plant cell cultures include hydrodynamics, mass and heat transfer, mixing, cell growth, viability and oxygen demand, cell aggregation, rheological properties and shear sensitivity of plant cell cultures, etc.

3.4.2.1 Plant Cell Growth and Oxygen Demand

Plant cells show a longer doubling time (20–100 h) than that of bacterial (30 min to 1 h), yeast (2–3 h) and mammalian cells (24–48 h). Different species show varied cell growth kinetics in bioreactors. A typical OUR (oxygen update rate) value for plant cell cultures is about 2–10 mmol-O_2/(L-h), compared with 10–90 mmol-O_2/(L-h) for microbial cells and 0.05–10 mmol-O_2/(L-h) or 0.02–0.1 × 10–9 mmol/(cell-h) for mammalian cells (depending on the cell density and cell line type). The specific OUR is reported as 0.8 mmol-O_2/(g-DCW-h) for transgenic rice cell cultures expressing human AAT (Trexler et al. 2002) and 0.3–0.5 mmol-O_2/(g-DCW-h) for transgenic NT-1 cells expressing recombinant GUS. Inadequate oxygen mass transfer has been demonstrated to inhibit transgenic tobacco cell growth and reduce recombinant antibody heavy chain production (Sharp and Doran 2001b). Gao and Lee (1992) found that an increase in the oxygen supply enhanced tobacco cell growth rate, biomass concentration, and GUS production yield in shake flask, stirred-tank and air-lift bioreactors.

To meet the OUR of 2–10 mmol-O_2/(L-h) for plant cell cultures, a typical volumetric oxygen mass transfer coefficient (kLa) required in a bioreactor operation is between 10 and 50 h^{-1} (Curtis and Tuerk 2006), which is lower than that for microbial fermentation (100–1,000 h^{-1}) and slightly higher than that for mammalian cell culture (0.25–10 h^{-1}). The critical dissolved oxygen concentration for plant cell cultures in a bioreactor is reported as 1.3–1.6 g/m^3, corresponding to 20% of air saturation (Doran 1993). The optimal settings of kLa and dissolved oxygen concentration in bioreactor for plant cells expressing recombinant protein need to be investigated individually.

3.4.2.2 Aggregation and Rheological Properties of Plant Cell Cultures

Plant cell aggregate formation and aggregate size distribution are critical for plant cell culture bioreactor operation and are dependent on plant species, method of inoculum preparation, cell growth stage, medium composition, bioreactor types and culture conditions. Formation of plant cell aggregates promotes cellular organization and differentiation resulting in enhancing secondary metabolite production, and also impacts mass transfer, leading to oxygen, nutrient or chemical inducer inhomogeneities inside large cell aggregates (Kieran 2001). Therefore, the inner cells of the aggregates may become oxygen and nutrient deficient, resulting in adverse effects on cell growth and foreign protein production. Although moderate cell aggregation (200–500 μm) is advantageous in some cases since it enhances sedimentation rates, facilitating media exchange as well as *in-situ* recovery of culture broth during downstream processing, generation of large cell aggregates (~1–2 mm) is undesirable since this complicates the bioreactor operation, exacerbates mass transfer limitations and makes cell aggregates more susceptible to shear stress, resulting in cell damage, attributed to aggregate surface attrition (Kieran et al. 2000) and aggregate shattering (Namdev and Dunlop 1995).

Rheological properties of plant cells *in vitro* culture are dependent on cell aggregate size and morphology, biomass concentration, cell growth stage and culture conditions. Plant cells tend to transition from spherical to elongated shapes when cell division is terminated (Cosgrove 1997). Curtis (Curtis and Emery 1993) studied the rheological properties of 10 different plant cells in flasks and found that elongated plant cell morphology in *N. tabacum* batch culture exhibited a power-law type fluid rheological property (with a power law index of 0.6), resulting in higher apparent viscosity, compared to spherical cells. Curtis (Curtis and Emery 1993) also observed that tobacco cell culture displayed Newtonian rheological properties and did not elongate when grown in semicontinuous culture (cells remained nearly spherical in shape), confirming the dependence of rheology on cell morphology. Elongated plant cell morphology may lead to higher packed cell volume (PCV) at a given dried cell weight (DCW), attributed to a more loose cellular network under packed conditions (Su and Arias 2003).

Kato et al. (1978) and Curtis (Curtis and Emery 1993) both found that culture spent media was not responsible for the overall broth viscosity and the viscous and non-Newtonian fluid character of the culture was mainly due to the plant cell morphology (elongated and filamentous cells) and high biomass concentration (with DCW over 10 g/L and PCV over 50–60%). Kato et al. (1978) found that the apparent viscosity of *N. tabacum* cell culture broth was increased by a factor of 27 throughout the batch culture period and the filtrate (cell-free broth) was only increased from 0.9 to 2.2 cP. A typical apparent viscosity of plant cell culture broth is 4–150 cP.

3.4.2.3 Shear Sensitivity of Plant Cell Cultures

Two main shear stresses leading to cell damage in plant cell culture bioreactor operation are hydrodynamic shear force induced by agitation and air bubble bursting caused by gas sparging. A single plant cell (100–500 μm in length and 20–50 μm in diameter) is about 10–100 times larger in size than bacterial (<1 μm in diameter), fungal (<100 μm in length and 5–10 μm in diameter) and mammalian cells (10–100 μm in diameter) and thus are capable of withstanding tensile strain, however, suspended plant cells are considered sensitive to shear stress due to their large volume of intracellular vacuoles (up to 90% of cell volume) and a rigid, inflexible cellulose-based cell wall (Dunlop et al. 1994). Thus, plant cells are more susceptible to shear stress during the late exponential growth and early stationary phases when the cells are of relatively large size and contain large vacuoles (Wagner and Vogelmann 1977), inducing the cellular response changes including cell viability, cell growth rate, membrane integrity, release of intracellular components (proteins or metabolites), metabolism (OUR, mitochondrial activity, ATP concentration, cell wall composition, increase of calcium ions in cytoplasm), cell morphology and aggregation sizes (Kieran et al. 1997, 2000), influenced by the intensity and the exposure duration of the cells to the shear force.

For a stirred-tank bioreactor, shear stress generated by the impeller system reduces the average aggregate size of *Catharanthus roseus* cell culture (80–100 μm in the shake flask versus 64–80 μm in the stirred-tank bioreactor) and has adverse effects on plant cell growth and viability (Tanaka et al. 1988), partially due to the fact that plant cells are subject to the higher shear stress region of the impeller and thus more shear-induced damage on cells is generated (Doran 1999). For shear-sensitive plant cell cultures, therefore, reducing the shear stress intensity by decreasing the agitation speed of the impeller is a general solution, while maintaining adequate mixing, oxygen and heat transfer rates in a high apparent viscosity plant cell culture broth. At high biomass concentration, low agitation rates can also enhance the clumping of cells into cell aggregates of varying sizes. Thus, a concept of critical shear stress (using regrowth of cells as an indicator) above which cell viability may be lost can be applied for impeller design and a critical shear stress between 50 and 200 N/m^2 has been reported (Kieran et al. 1997).

3.4.3 Bioreactor Process Optimization

The optimization of bioreactor operation needs to consider the characteristics of plant cell expression system such as cell growth, viability and oxygen demand, cell aggregation, rheological properties and shear sensitivity of plant cell cultures, foaming and wall growth, interactions between host cell, gene expression system and product formation including the type of gene expression system, expressed product location, impact of expressed products on host cells, etc.

For the growth-associated recombinant protein production (secreted or intracellular product) driven by a constitutive promoter, target protein productivity can be enhanced by reaching high cell density culture and prolonging the exponential active cell growth phase. Fed-batch cultures have been applied to achieve high cell density culture when utilizing an effective substrate feeding strategy (Suehara et al. 1996). However, the accumulation of inhibitory metabolites in fed-batch cultures might limit recombinant protein productivity. Therefore, perfusion culture with a cell retention device can be an alternative to obtain high cell density culture and continuously withdraw cultured medium (De Dobbeleer et al. 2006; Lucumi and Posten 2006; Su and Arias 2003). The cell growth rate in a batch or fed-batch plant cell culture is usually reduced when the PCV reaches about 60–70%, resulting in a reduction of cellular metabolic activity (Maccarthy et al. 1980). Therefore, semi-continuous culture or perfusion culture with a bleed stream has been demonstrated to be more appropriate for high cell density culture compared with fed-batch and perfusion cultures (De Dobbeleer et al. 2006).

For recombinant protein production driven by an inducible promoter, two-stage cultures are typically implemented to allow the cell growth and protein production phases to be independently optimized. The bioreactor operation and conditions for the induction phase (protein production phase) are highly dependent on the type of inducible promoter used (Huang and McDonald 2009). The sugar-starvation Ramy3D promoter system (metabolically regulated) has been investigated to express human proteins including human AAT (Huang et al. 2001) and human granulocyte-macrophage colony stimulating factor (hGM-CSF) (Shin et al. 2003) in transgenic rice cell cultures. In these studies, a media exchange to a sugar-free medium or nutrient medium containing an alternative carbon source (Terashima et al. 2001) for inducing foreign protein production was applied at an appropriate time in the growth phase. During the induction phase without carbon source supplementation, however, the rice cell viability was significantly decreased, resulting in increased protease activity. Therefore, a cyclical semi-continuous process, which alternates between growth and production phases, has been developed to reuse the transgenic rice cells for long-term operation (Trexler et al. 2005). For a chemically inducible system, the timing of induction and concentration and manner of application of inducer (single, multiple or continuous induction) applied to plant cell cultures are important for optimizing inducible plant cell culture bioreactor operation. This will need to be investigated based on the nature of the inducer (inducer stability and toxicity to host cells) and plant species (cell growth rate and aggregates) for enhancing foreign protein expression. Higher inducer concentrations and multiple or continuous application may benefit high cell density operational modes. Semi-continuous/continuous or perfusion bioreactor operation at high cell density with slower specific growth rate for a prolonged protein production phase can be beneficial for inducible production of the recombinant protein and the secreted recombinant protein can be continuously harvested from the cell culture broth. Huang et al. showed that OUR is an important parameter to determine the optimal timing of induction (TOI) (Huang et al. 2010). In the case of a chemically inducible, estrogen receptor-based promoter (XVE) system in tobacco cell culture, the optimal TOI occurs at the maximum OUR which occurs at the end of the exponential phase (Huang et al. 2010). We developed the semicontinuous culture production of human AAT using a chemically inducible plant viral vector in transgenic tobacco cell culture, resulting in fivefold increase in volumetric productivity of biologically functional AAT compared with batch operation (Huang et al. 2010).

Additionally, the productivity of secreted recombinant proteins, which could be rapidly degraded in the culture medium, can be improved through (1) addition of medium additives to enhance product stability and prevent the product from proteolysis derived from proteases generated by host cells (Benchabane et al. 2008; Doran 2006a, b), (2) *in-situ* protein recovery (either by adding resins into the medium or by circulating the culture broth through a chromatography column external to the bioreactor) (James et al. 2002; Sharp and Doran 2001a), and (3) immobilization of plant cells, in which cells are immobilized into a suitable microcarrier or support matrix in a bioreactor (Gilleta et al. 2000; Osuna et al. 2008), to facilitate recovery of secreted protein from the culture broth.

3.4.4 Scale Up Considerations

The development of the large-scale bioreactor operation of plant cell suspension culture has been well established for the production of plant secondary metabolites such as paclitaxel (taxol), ginseng and shikonin (Hellwig et al. 2004). Phyton Biotech (www.phytonbiotech.com) commercially produces paclitaxel compound using plant cell suspension cultures up to 75,000 L and has been a long-term supplier of this small molecule API (active pharmaceutical ingredient) of Bristol-Myers Squibb's Taxol® oncology product. Therefore, the bioreactor technologies available for large scale plant secondary metabolites production could be implemented for the large scale production of recombinant proteins in transgenic plant cell suspension cultures. Currently, the most challenging problem of the scale-up of bioreactor-based plant cell culture is to providing a low-shear environment while maintaining adequate mixing and oxygen transfer in high cell density culture and/or in long-term perfusion culture operation. Although some investigations have tried to meet market demands by increasing bioreactor capacity, the optimization of cell culture productivity in bioreactors appears as a better strategy.

3.5 Current Status of Commercialization

Although molecular farming using bioreactor-based plant cell suspension cultures provides an alternative biomanufacturing platform for large scale production of recombinant proteins for pharmaceutical and industrial applications, only few available commercial examples have been demonstrated.

In February 2006, USDA approved the first transgenic tobacco cell culture-produced recombinant glycoprotein by Dow AgroSciences (www.dowagro.com), as a veterinary vaccine based on the HN antigen derived from immunoprotective particles of Newcastle Disease Virus (NDV) for preventing avian NDV disease (Travis 2008). Other successful examples from Dow AgroSciences for the production of immunogenic proteins include the HA antigen of Avian Influenza Virus (AIV) and the VP2 structural protein of Infectious Bursal Diseases Virus (IBDV), which are driven by CaMV 35S or CsVMV (Cassava Vein Mosaic Virus) constitutive promoter in transgenic tobacco, potato or tomato cell cultures (Mihaliak et al. 2007). The recombinant immunoprotective proteins are expressed and accumulated in the stationary phase of plant cell growth in cytoplasmic, cell wall or membrane structure, with the product titer up to 4–30 mg/L (Mihaliak et al. 2007; Miller et al. 2006).

The most recent case in the plant cell culture made pharmaceuticals is the transgenic carrot suspension cell culture in bioreactors developed by Protalix Biotherapeutics for recombinant glucocerebrosidase (GCD) production dedicated for patients with the genetic disorder Gaucher disease (Shaaltiel et al. 2007). Protalix Biotherapeutics (http://www.protalix.com/) in Israel and Pfizer in the US (http://www.pfizer.com/) have announced a collaboration to market the recombinant glucocerebrosidase enzyme produced by transgenic carrot cell cultures as a therapeutic protein drug, currently in phase III clinical trial, for the treatment of Gaucher's disease in EU and USA (Ratner 2010). This represents an exciting milestone for recognizing plant cell culture-based biomanufacturing as a bio-equivalent and economical alternative to mammalian production of human biopharmaceuticals, further suggesting the possibility of biosimilar products for existing protein drugs. Currently patients are treated with either Ceredase® by Genzyme purified from human placental tissue or Cerezyme® by Genzyme produced in recombinant CHO cell cultures (Ratner 2010). However, the recombinant GCD by CHO cell cultures requires an additional *in vitro* enzymatic reaction to expose the terminal mannose residues of its N-glycan chains to facilitate uptake of the GCD into macrophages, making it one of the most expensive therapeutic proteins to date with an annual treatment cost of nearly US $200,000 per patient (Kaiser 2008). The plant-made recombinant GCD (prGCD) expressed in transgenic carrot cell culture by Protalix is fused to the N-terminal signal peptide from *Arabidopsis thaliana* basic endochitinase and to a C-terminal vacuole targeting sequence from tobacco chitinase A. The prGCD expressed by transgenic carrot cells is retained in ER (endoplasmic reticulum) for glycosylation and then targeted to the vacuole. Therefore, the N-glycan structures of the prGCD are trimmed to expose mannose residues, leading to the correct mannose glycosylation pattern. As a result, the *in vitro* trimming of the glycans for *in vitro* protein modification is eliminated during downstream processing, resulting in significant cost reduction (Shaaltiel et al. 2007). Currently, the prGCD by Protalix is undergoing a Phase III clinical trial to evaluate its safety and efficacy in Gaucher patients.

3.6 Future Prospects

Though cheaper, safer, easier to manipulate and more rapid than most established molecular farming using transgenic plants, plant cell suspension culture is still not the best production platform the plant system can offer, as the overall product yield and usability is often limited by the loss of recombinant protein during the late stationary phase due to increased proteolytic activity (Corrado and Karali 2009). Although the adoption of advanced cell culture technology has been implemented during the past years, the system is still limited to a small number of well-characterized plant cell lines (such as tobacco, rice, carrot, or *Arabidopsis*), which are amenable to develop suspension cell cultures and need improvement before they can be adopted commercially.

Therefore, future directions for enabling the molecular farming using bioreactor-based plant cell suspension cultures as a recombinant protein production biomanufacturing platform include: (1) selection of plant hosts with lower endogenous protease activity to the target protein, (2) development of algorithms for synthetic gene design for optimal expression in plant hosts, (3) generation and selection of the most productive cell lines by automatic high throughput system and/or development of site-specific integration strategies, (4) medium formulation design and optimization, (5) enhancing recombinant protein stability and preventing proteolytic degradation, (6) selection of bioreactor systems according to the interactions between host cells, product formation and bioreactor design, (7) optimization of bioreactor operation strategy, (8) incorporation of gene silencing suppressors, (9) development of large scale transient expression in plant cell culture, (10) adaptation of disposable bioreactor technology, and (11) engineering humanized plant-made glycosylated proteins.

3.7 Conclusions

The promising cases described above, including Protalix's transgenic carrot suspension culture platform for prGCD production, indicate the opportunities for molecular farming using plant cell culture bioreactor system for large scale biomanufacturing of specialty proteins, orphan drugs or personalized medicine, rare genetic diseases, and biosimilars or even biobetter therapeutics, to lower the cost of goods while maintaining or improving the plant-made protein quality. Plant cell culture bioreactor systems exhibit the advantage of plant-made therapeutics in a similar method capable of meeting the EMEA and FDA regulatory requirements that have been set for the past 20 years for microbial and mammalian cells-made proteins. Strategies for the selection of plant species, cell culture types, gene expression systems, bioreactor systems and operation modes, and product of interest for application need to be carefully investigated for making the plant cell bioreactor processes a practical and economical platform for foreign protein production in next phase of development and commercial application.

References

Amanullah A, Buckland BC, Nienow AW (2004) Mixing in the fermentation and cell culture industries. In: Paul EL, Atiemo-Obeng VA, Kresta SM (eds) Handbook of industrial mixing. John Wiley & Sons, Inc., pp 1071–1170

Angell SM, Baulcombe DC (1997) Consistent gene silencing in transgenic plants expressing a replicating potato virus X RNA. EMBO J 16(12):3675–3684

Atkinson RG, Bieleski LRF, Gleave AP, Janssen BJ, Morris BAM (1998) Post-transcriptional silencing of chalcone synthase in petunia using a geminivirus-based episomal vector. Plant J 15(5):593–604

Benchabane M, Goulet C, Rivard D, Faye L, Gomord V, Michaud D (2008) Preventing unintended proteolysis in plant protein biofactories. Plant Biotechnol J 6(7):633–648

Boivin EB, Lepage É, Matton DP, De Crescenzo G, Jolicoeur M (2010) Transient expression of antibodies in suspension plant cell suspension cultures is enhanced when co-transformed with the tomato bushy stunt virus p19 viral suppressor of gene silencing. Biotechnol Prog 26(6):1534–1543

Castilho A, Strasser R, Stadlmann J, Grass J, Jez J, Gattinger P, Kunert R, Quendler H, Pabst M, Leonard R, Altmann F, Steinkellner H (2010) In planta protein sialylation through overexpression of the respective mammalian pathway. J Biol Chem 285:15923–15930. doi:10.1074/jbc.M109.088401

Cho JS, Hong SM, Joo SY, Yoo JS, Kim DI (2007) Cryopreservation of transgenic rice suspension cells producing recombinant hCTLA4Ig. Appl Microbiol Biotechnol 73(6):1470–1476

Corrado G, Karali M (2009) Inducible gene expression systems and plant biotechnology. Biotechnol Adv 27(6):733–743

Cosgrove DJ (1997) Relaxation in a high-stress environment: the molecular bases of extensible cell walls and cell enlargement. Plant Cell 9(7):1031–1041

Curtis WR, Emery AH (1993) Plant-cell suspension-culture rheology. Biotechnol Bioeng 42(4):520–526

Curtis WR, Tuerk AL (2006) Oxygen transport in plant tissue culture systems. Plant Tissue Cult Eng 6:173–186

De Dobbeleer C, Cloutier M, Fouilland M, Legros R, Jolicoeur M (2006) A high-rate perfusion bioreactor for plant cells. Biotechnol Bioeng 95(6):1126–1137
De Muynck B, Navarre C, Boutry M (2010) Production of antibodies in plants: status after twenty years. Plant Biotechnol J 8(5):529–563
del Val IJ, Kontoravdi C, Nagy JM (2010) Towards the implementation of quality by design to the production of therapeutic monoclonal antibodies with desired glycosylation patterns. Biotechnol Prog 26(6):1505–1527
Demain AL, Vaishnav P (2009) Production of recombinant proteins by microbes and higher organisms. Biotechnol Adv 27(3):297–306
Dohi K, Nishikiori M, Tamai A, Ishikawa M, Meshi T, Mori M (2006) Inducible virus-mediated expression of a foreign protein in suspension-cultured plant cells. Arch Virol 151(6):1075–1084
Doran P (1993) Design of reactors for plant cells and organs. Bioprocess Des Control 48:115–168
Doran PM (1999) Design of mixing systems for plant cell suspensions in stirred reactors. Biotechnol Prog 15(3):319–335
Doran PM (2006a) Foreign protein degradation and instability in plants and plant tissue cultures. Trends Biotechnol 24(9):426–432
Doran PM (2006b) Loss of secreted antibody from transgenic plant tissue cultures due to surface adsorption. J Biotechnol 122(1):39–54
Dunlop EH, Namdev PK, Rosenberg MZ (1994) Effect of fluid shear forces on plant-cell suspensions. Chem Eng Sci 49(14):2263–2276
Eibl R, Eibl D (2006) Design and use of the wave bioreactor for plant cell culture. Plant Tissue Cult Eng 6:203–227
Eibl R, Eibl D (2008) Design of bioreactors suitable for plant cell and tissue cultures. Phytochem Rev 7(3):593–598
Eibl R, Eibl D, Eibl R, Werner S, Eibl D (2009a) Bag bioreactor based on wave-induced motion: characteristics and applications. In: Disposable bioreactors, vol 115, Advances in biochemical engineering/biotechnology. Springer, Berlin/Heidelberg, pp 55–87
Eibl R, Werner S, Eibl D (2009b) Disposable bioreactors for plant liquid cultures at Litre-scale. Eng Life Sci 9(3):156–164
Eibl R, Kaiser S, Lombriser R, Eibl D (2010) Disposable bioreactors: the current state-of-the-art and recommended applications in biotechnology. Appl Microbiol Biotechnol 86(1):41–49
Faye L, Boulaflous A, Benchabane M, Gomord W, Michaud D (2005) Protein modifications in the plant secretory pathway: current status and practical implications in molecular pharming. Vaccine 23(15):1770–1778
Franconi R, Demurtas OC, Massa S (2010) Plant-derived vaccines and other therapeutics produced in contained systems. Expert Rev Vaccines 9:877–892. doi:10.1586/erv.10.91
Gao J, Lee JM (1992) Effect of oxygen supply on the suspension culture of genetically modified tobacco cells. Biotechnol Prog 8(4):285–290
Gilleta F, Roisin C, Fliniaux MA, Jacquin-Dubreuil A, Barbotin JN, Nava-Saucedo JE (2000) Immobilization of Nicotiana tabacum plant cell suspensions within calcium alginate gel beads for the production of enhanced amounts of scopolin. Enzyme Microb Technol 26(2–4):229–234
Gleba Y, Klimyuk V, Marillonnet S (2007) Viral vectors for the expression of proteins in plants. Curr Opin Biotechnol 18(2):134–141
Gomord V, Faye L (2004) Posttranslational modification of therapeutic proteins in plants. Curr Opin Plant Biol 7(2):171–181
Gomord V, Fitchette A-C, Menu-Bouaouiche L, Saint-Jore-Dupas C, Plasson C, Michaud D, Faye L (2010) Plant-specific glycosylation patterns in the context of therapeutic protein production. Plant Biotechnol J 8(5):564–587
Hacker DL, De Jesus M, Wurm FM (2009) 25 years of recombinant proteins from reactor-grown cells – Where do we go from here? Biotechnol Adv 27(6):1023–1027
Hellwig S, Drossard J, Twyman RM, Fischer R (2004) Plant cell cultures for the production of recombinant proteins. Nat Biotechnol 22(11):1415–1422

Hong SY, Kwon TH, Lee JH, Jang YS, Yang MS (2002) Production of biologically active hG-CSF by transgenic plant cell suspension culture. Enzyme Microb Technol 30(6):763–767

Huang T-K, McDonald KA (2009) Bioreactor engineering for recombinant protein production in plant cell suspension cultures. Biochem Eng J 45(3):168–184

Huang TK, Wang PM, Wu WT (2001) Cultivation of Bacillus thuringiensis in an airlift reactor with wire mesh draft tubes. Biochemical Engineering Journal 7(1):35–39

Huang JM, Sutliff TD, Wu LY, Nandi S, Benge K, Terashima M, Ralston AH, Drohan W, Huang N, Rodriguez RL (2001) Expression and purification of functional human alpha-1-antitrypsin from cultured plant cells. Biotechnol Prog 17(1):126–133

Huang TK, Plesha MA, Falk BW, Dandekar AM, McDonald KA (2009) Bioreactor strategies for improving production yield and functionality of a recombinant human protein in transgenic tobacco cell cultures. Biotechnol Bioeng 102(2):508–520

Huang T-K, Plesha MA, McDonald KA (2010) Semicontinuous bioreactor production of a recombinant human therapeutic protein using a chemically inducible viral amplicon expression system in transgenic plant cell suspension cultures. Biotechnol Bioeng 106(3):408–421

James E, Lee JM (2001) The production of foreign proteins from genetically modified plant cells. Adv Biochem Eng Biotechnol 72:127–156

James E, Mills DR, Lee JM (2002) Increased production and recovery of secreted foreign proteins from plant cell cultures using an affinity chromatography bioreactor. Biochem Eng J 12(3):205–213

Junker BH, Stanik M, Barna C, Salmon P, Buckland BC (1998) Influence of impeller type on mass transfer in fermentation vessels. Bioprocess Eng 19(6):403–413

Kaiser J (2008) Is the drought over for pharming? Science 320(5875):473–475. doi:10.1126/science.320.5875.473

Kapila J, DeRycke R, VanMontagu M, Angenon G (1997) An Agrobacterium-mediated transient gene expression system for intact leaves (vol 122, p 101, 1997). Plant Sci 124(2):227–227

Kato A, Kawazoe S, Soh Y (1978) Biomass production of tobacco cells .4. Viscosity of broth of tobacco cells in suspension culture. J Ferment Technol 56(3):224–228

Kieran PM (2001) Bioreactor design for plant cell suspension cultures. In: Tramper J, Cabral JMS, Mota M (eds) Multiphase bioreactor design. Taylor & Francis Ltd, Routledge, pp 391–426

Kieran PM, MacLoughlin PF, Malone DM (1997) Plant cell suspension cultures: some engineering considerations. J Biotechnol 59(1–2):39–52

Kieran P, Malone D, MacLoughlin P (2000) Effects of hydrodynamic and interfacial forces on plant cell suspension systems. Influ Stress Cell Growth Prod Form 67:139–177

Kim NS, Kim TG, Kim OH, Ko EM, Jang YS, Jung ES, Kwon TH, Yang MS (2008) Improvement of recombinant hGM-CSF production by suppression of cysteine proteinase gene expression using RNA interference in a transgenic rice culture. Plant Mol Biol 68(3):263–275

Komarnytsky S, Borisjuk N, Yakoby N, Garvey A, Raskin I (2006) Cosecretion of protease inhibitor stabilizes antibodies produced by plant roots. Plant Physiol 141(4):1185–1193

Komarova TV, Baschieri S, Donini M, Marusic C, Benvenuto E, Dorokhov YL (2010) Transient expression systems for plant-derived biopharmaceuticals. Expert Rev Vaccines 9:859–876

Koprivova A, Stemmer C, Altmann F, Hoffmann A, Kopriva S, Gorr G, Reski R, Decker EL (2004) Targeted knockouts of Physcomitrella lacking plant-specific immunogenic N-glycans. Plant Biotechnol J 2(6):517–523

Kumar S, Fladung M (2001) Controlling transgene integration in plants. Trends Plant Sci 6(4):155–159

Lambe P, Dinant M, Matagne RF (1995) Differential long-term expression and methylation of the hygromycin phosphotransferase (Hph) and beta-glucuronidase (Gus) genes in transgenic pearl-millet (Pennisetum-Glaucum) Callus. Plant Sci 108(1):51–62

Lico C, Chen Q, Santi L (2008) Viral vectors for production of recombinant proteins in plants. J Cell Physiol 216(2):366–377

Lienard D, Dinh OT, van Oort E, Van Overtvelt L, Bonneau C, Wambre E, Bardor M, Cosette P, Didier-Laurent A, de Borne FD, Delon R, van Ree R, Moingeon P, Faye L, Gomord V (2007) Suspension-cultured BY-2 tobacco cells produce and mature immunologically active house dust mite allergens. Plant Biotechnol J 5(1):93–108

Lindbo JA (2007) High-efficiency protein expression in plants from agroinfection-compatible Tobacco mosaic virus expression vectors. BMC Biotechnol 7:52–62

Lucumi A, Posten C (2006) Establishment of long-term perfusion cultures of recombinant moss in a pilot tubular photobioreactor. Process Biochem 41(10):2180–2187

Maccarthy JJ, Ratcliffe D, Street HE (1980) The effect of nutrient medium composition on the growth-cycle of catharanthus-roseus G. Don cells grown in batch culture. J Exp Bot 31(124):1315–1326

McDonald KA, Hong LM, Trombly DM, Xie Q, Jackman AP (2005) Production of human alpha-1-antitrypsin from transgenic rice cell culture in a membrane bioreactor. Biotechnol Prog 21(3):728–734

Micheletti M, Barrett T, Doig SD, Baganz F, Levy MS, Woodley JM, Lye GJ (2006) Fluid mixing in shaken bioreactors: implications for scale-up predictions from microlitre-scale microbial and mammalian cell cultures. Chem Eng Sci 61(9):2939–2949

Mihaliak CA, Fanton MJ, Mcmillen JK (2007) Preparation of vaccine master cell lines using recombinant plant suspension cultures United States Patent 20070107086

Miller TJ, Fanton MJ, Webb SR (2006) Stable immunoprophylactic and therapeutic compositions derived from transgenic plant cells and methods for production United States Patent 20060222664

Murphy DJ (2007) Improving containment strategies in biopharming. Plant Biotechnol J 5(5):555–569

Namdev PK, Dunlop EH (1995) Shear sensitivity of plant-cells in suspensions – present and future. Appl Biochem Biotechnol 54(1–3):109–131

Obembe OO, Popoola JO, Leelavathi S, Reddy SV (2011) Advances in plant molecular farming. Biotechnol Adv 29(2):210–222

Offringa R, de Groot MJ, Haagsman HJ, Does MP, van den Elzen PJ, Hooykaas PJ (1990) Extrachromosomal homologous recombination and gene targeting in plant cells after Agrobacterium mediated transformation. EMBO J 9(10):3077–3084

Osuna L, Moyano E, Mangas S, Bonfill M, Cusido RM, Pinol MT, Zamilpa A, Tortoriello J, Palazon J (2008) Immobilization of Galphimia glauca plant cell suspensions for the production of enhanced amounts of Galphimine-B. Planta Med 74(1):94–99

Padidam M (2003) Chemically regulated gene expression in plants. Curr Opin Plant Biol 6(2):169–177

Plesha MA, Huang TK, Dandekar AM, Falk BW, McDonald KA (2007) High-level transient production of a heterologous protein in plants by optimizing induction of a chemically inducible viral amplicon expression system. Biotechnol Prog 23(6):1277–1285

Plesha MA, Huang TK, Dandekar AM, Falk BW, McDonald KA (2009) Optimization of the bioprocessing conditions for scale-up of transient production of a heterologous protein in plants using a chemically inducible viral amplicon expression system. Biotechnol Prog 25(3):722–734

Pogue GP, Vojdani F, Palmer KE, Hiatt E, Hume S, Phelps J, Long L, Bohorova N, Kim D, Pauly M, Velasco J, Whaley K, Zeitlin L, Garger SJ, White E, Bai Y, Haydon H, Bratcher B (2010) Production of pharmaceutical-grade recombinant aprotinin and a monoclonal antibody product using plant-based transient expression systems. Plant Biotechnol J 8(5):638–654

Qi HN, Goudar CT, Michaels JD, Henzler HJ, Jovanovic GN, Konstantinov KB (2003) Experimental and theoretical analysis of tubular membrane aeration for mammalian cell bioreactors. Biotechnol Prog 19(4):1183–1189

Rao AQ, Bakhsh A, Kiani S, Shahzad K, Shahid AA, Husnain T, Riazuddin S (2009) The myth of plant transformation. Biotechnol Adv 27(6):753–763

Ratner M (2010) Pfizer stakes a claim in plant cell-made biopharmaceuticals. Nat Biotechnol 28(2):107–108

Rawel HM, Kroll J, Kulling S (2007) Effect of non-protein components on the degradability of proteins. Biotechnol Adv 25(6):611–613

Read EK, Park JT, Shah RB, Riley BS, Brorson KA, Rathore AS (2009) Process analytical technology (PAT) for biopharmaceutical products: Part I. concepts and applications. Biotechnol Bioeng 105(2):276–284

Samalova M, Brzobohaty B, Moore I (2005) pOp6/LhGR: a stringently regulated and highly responsive dexamethasone-inducible gene expression system for tobacco. Plant J 41(6): 919–935

Schahs M, Strasser R, Stadlmann J, Kunert R, Rademacher T, Steinkellner H (2007) Production of a monoclonal antibody in plants with a humanized N-glycosylation pattern. Plant Biotechnol J 5(5):657–663

Schiermeyer A, Schinkel H, Apel S, Fischer R, Schillberg S (2005) Production of Desmodus rotundas salivary plasminogen activator alpha 1 (DSPA alpha 1) in tobacco is hampered by proteolysis. Biotechnol Bioeng 89(7):848–858

Schmale K, Rademacher T, Fischer R, Hellwig S (2006) Towards industrial usefulness – cryo-cell-banking of transgenic BY-2 cell cultures. J Biotechnol 124(1):302–311

Sethuraman N, Stadheim TA (2006) Challenges in therapeutic glycoprotein production. Curr Opin Biotechnol 17(4):341–346

Shaaltiel Y, Bartfeld D, Hashmueli S, Baum G, Brill-Almon E, Galili G, Dym O, Boldin-Adamsky SA, Silman I, Sussman JL, Futerman AH, Aviezer D (2007) Production of glucocerebrosidase with terminal mannose glycans for enzyme replacement therapy of Gaucher's disease using a plant cell system. Plant Biotechnol J 5(5):579–590

Shaaltiel Y, Baum G, Bartfeld D, Hashmueli S, Lewkowicz A (2008) Production of high mannose proteins in plant culture. United States Patent 20080038232

Shadwick FS, Doran PM (2005) Foreign Protein Expression Using Plant Cell Suspension and Hairy Root Cultures. In: Fischer R, Schillberg S (eds) Molecular Farming: Plant-Made Pharmaceuticals and Technical Proteins, Wiley-VCH Verlag GmbH & Co. KGaA, Weinheim, FRG, pp 13–36. doi: 10.1002/3527603638.ch2

Sharma KK, Bhatnagar-Mathur P, Thorpe TA (2005) Genetic transformation technology: status and problems. Vitro Cell Dev BiolPlant 41(2):102–112

Sharp JM, Doran PM (2001a) Characterization of monoclonal antibody fragments produced by plant cells. Biotechnol Bioeng 73(5):338–346

Sharp JM, Doran PM (2001b) Strategies for enhancing monoclonal antibody accumulation in plant cell and organ cultures. Biotechnol Prog 17(6):979–992

Shih SMH, Doran PM (2009) Foreign protein production using plant cell and organ cultures: advantages and limitations. Biotechnol Adv 27(6):1036–1042

Shin YJ, Hong SY, Kwon TH, Jang YS, Yang MS (2003) High level of expression of recombinant human granulocyte-macrophage colony stimulating factor in transgenic rice cell suspension culture. Biotechnol Bioeng 82(7):778–783

Strasser R, Stadlmann J, Schähs M, Stiegler G, Quendler H, Mach L, Glössl J, Weterings K, Pabst M, Steinkellner H (2008) Generation of glyco-engineered Nicotiana benthamiana for the production of monoclonal antibodies with a homogeneous human-like N-glycan structure. Plant Biotechnol J 6(4):392–402

Streatfield SJ (2007) Approaches to achieve high-level heterologous protein production in plants. Plant Biotechnol J 5(1):2–15

Su WW, Arias R (2003) Continuous plant cell perfusion culture: bioreactor characterization and secreted enzyme production. J Biosci Bioeng 95(1):13–20

Sudarshana MR, Plesha MA, Uratsu SL, Falk BW, Dandekar AM, Huang TK, McDonald KA (2006) A chemically inducible cucumber mosaic virus amplicon system for expression of heterologous proteins in plant tissues. Plant Biotechnol J 4(5):551–559

Suehara KI, Takao S, Nakamura K, Uozumi N, Kobayashi T (1996) Optimal expression of GUS gene from methyl jasmonate-inducible promoter in high density culture of transformed tobacco cell line BY-2. J Ferment Bioeng 82(1):51–55

Tanaka H (2000) Technological problems in cultivation of plant cells at high density (Reprinted from Biotechnol Bioeng 23:1203–1218, 1981). Biotechnol Bioeng 67(6):775–790

Tanaka H, Semba H, Jitsufuchi T, Harada H (1988) The effect of physical stress on plant-cells in suspension-cultures. Biotechnol Lett 10(7):485–490

Terashima M, Ejiri Y, Hashikawa N, Yoshida H (2001) Utilization of an alternative carbon source for efficient production of human alpha(1)-Antitrypsin by genetically engineered rice cell culture. Biotechnol Prog 17(3):403–406

Terrier B, Courtois D, Henault N, Cuvier A, Bastin M, Aknin A, Dubreuil J, Petiard V (2007) Two new disposable bioreactors for plant cell culture: the wave and undertow bioreactor and the slug bubble bioreactor. Biotechnol Bioeng 96(5):914–923

Travis J (2008) Is the drought over for pharming? Science 320(5875):473–477

Trexler MM, McDonald KA, Jackman AP (2002) Bioreactor production of human α_1-antitrypsin using metabolically regulated plant cell cultures. Biotechnol Prog 18(3):501–508

Trexler MM, McDonald KA, Jackman AP (2005) A cyclical semicontinuous process for production of human alpha(1)-antitrypsin using metabolically induced plant cell suspension cultures. Biotechnol Prog 21(2):321–328

Tsoi BMY, Doran PM (2002) Effect of medium properties and additives on antibody stability and accumulation in suspended plant cell cultures. Biotechnol Appl Biochem 35:171–180

Twyman RM, Stoger E, Schillberg S, Christou P, Fischer R (2003) Molecular farming in plants: host systems and expression technology. Trends Biotechnol 21(12):570–578

Vandermaas HM, Dejong ER, Rueb S, Hensgens LAM, Krens FA (1994) Stable transformation and long-term expression of the Gusa reporter gene in callus lines of perennial ryegrass (Lolium-Perenne L). Plant Mol Biol 24(2):401–405

Varley J, Birch J (1999) Reactor design for large scale suspension animal cell culture. Cytotechnology 29(3):177–205

Vaucheret H, Fagard M (2001) Transcriptional gene silencing in plants: targets, inducers and regulators. Trends Genet 17(1):29–35

Wagner F, Vogelmann H (1977) Cultivation of plant tissue culture in bioreactors and formation of secondary metabolites. In: Barz W, Reinhard E, Zenk MH (eds) Plant tissue culture and its biotechnological application. Springer, Berlin, pp 245–252

Wagner B, Fuchs H, Adhami F, Ma Y, Scheiner O, Breiteneder H (2004) Plant virus expression systems for transient production of recombinant allergens in Nicotiana benthamiana. Methods 32(3):227–234

Wang PM, Huang TK, Cheng HP, Chien YH, Wu WT (2002) A modified airlift reactor with high capabilities of liquid mixing and mass transfer. J Chem Eng Japan 35(4):354–35

Wroblewski T, Tomczak A, Michelmore R (2005) Optimization of Agrobacterium-mediated transient assays of gene expression in lettuce, tomato and Arabidopsis. Plant Biotechnol J 3(2):259–273

Zhang XR, Mason H (2006) Bean Yellow Dwarf Virus replicons for high-level transgene expression in transgenic plants and cell cultures. Biotechnol Bioeng 93(2):271–279

Zuo JR, Chua NH (2000) Chemical-inducible systems for regulated expression of plant genes. Curr Opin Biotechnol 11(2):146–151

Zuo JR, Niu QW, Chua NH (2000) An estrogen receptor-based transactivator XVE mediates highly inducible gene expression in transgenic plants. Plant J 24(2):265–273

Chapter 4
Chloroplast-Derived Therapeutic and Prophylactic Vaccines

James S. New*, **Donevan Westerveld***, **and Henry Daniell**

Abstract Despite the development of advanced vaccine technology, as many as 15 million deaths occur annually as a result of inadequate prophylactic or therapeutic treatments against infectious diseases (World Health Organization 2008). The emphasis on profitability by the pharmaceutical industry has led to development of high-cost vaccines targeting diseases with high profit margins, resulting in an annual death toll for developing countries that is largely preventable. Daniell and co-investigators published the first expression of a vaccine antigen, cholera toxin subunit B, through transgenic tobacco chloroplasts in 2001. The polyploidy nature of the chloroplast genome enables engineering of a high copy number of target gene, while the translational machinery of the plastid directs the synthesis of bioactive proteins with proper folding, disulfide bond formation, and lipidation. Furthermore, gene integration is site-specific, expression is polycistronic, and natural gene containment occurs due to the maternal inheritance of the chloroplast genome. The chloroplast transformation technology (CTT) has been used to express proteins of bacterial, viral, protozoan, and recently mammalian origins that may be subsequently utilized in immunization strategies to produce protective or therapeutic immunity. Such chloroplast-derived vaccines represent an inexpensive and effective means of producing antigen-subunit vaccines. Furthermore, transformation of edible crops such as lettuce could generate stable orally-deliverable vaccine antigens through inexpensive field production and agricultural drying techniques. This platform could therefore obviate cold-chain logistics and the requirement of sterile injectables, drastically reducing downstream production costs of such biopharmaceuticals. With these proof-of-concept studies, focus within CTT is shifting towards establishing a viable platform for human immunization by demonstrating

*These authors made equal contributions.

J.S. New • D. Westerveld • H. Daniell (✉)
Department of Molecular Biology and Microbiology, College of Medicine, University of Central Florida, 336 Biomolecular Science Building, Orlando, FL 32816, USA
e-mail: daniell@mail.ucf.edu

A. Wang and S. Ma (eds.), *Molecular Farming in Plants: Recent Advances and Future Prospects*, DOI 10.1007/978-94-007-2217-0_4,

the functionality of such chloroplast derived proteins, emphasizing the need to develop edible plant-based alternatives to tobacco, and improving efficacy through additional peptide fusion-domains.

4.1 Introduction

Current commercial systems for recombinant protein production include bacterial, yeast, insect and mammalian cell culture systems. Although each system is endowed with advantages, high production costs related to fermentation and purification measures occlude inexpensive therapeutics from the market. Through chloroplast transformation technology (CTT), fermentation costs may be eliminated and replaced with green houses and even large-scale acreage bringing the concept of molecular farming to fruition. Field production of vaccines would eliminate the necessity of cold chain logistics and the engagement of many sectors. Furthermore, plants carry the promise of long-term protein stabilization through low-cost drying techniques, allowing for cheaper vaccine distribution. Recent reports have demonstrated stability of insoluble protein inclusions even in dried, senescent tissues (Boyhan and Daniell 2011); this observation argues for the utilization of dried-plant matter as a vehicle for oral immunization.

Inherent bioencapsulation by the cellulosic cell walls protects accumulated proteins from enzymatic digestion, enabling release of proteins in the small intestine via oral delivery, and overcoming the cost requirements of sterile-injectables and cold-chain transport (Streatfield 2005). Additionally, this method of delivery promotes the induction of both mucosal and systemic immunity, providing more efficacious protection upon subsequent pathogen exposure (Davoodi-Semiromi et al. 2010). The low storage cost, minimal contamination risks, high product quality, and large scale-up capacity make the transgenic plant an ideal system for the production of recombinant proteins (Ma et al. 2003). Plant-derived biopharmaceuticals therefore offer an affordable route of protein synthesis, abolishing much of the downstream processing that contributes to the high cost of currently available vaccines.

Several approaches for producing recombinant plants exist including stable transformation of the nuclear and chloroplast genomes, and transient viral expression (Daniell et al. 2009b). Transformation of the chloroplast genome is achieved by particle bombardment with vectors that facilitate homologous recombination with the host chloroplast genome. Gene integration is site-specific, consequently normalizing gene expression across transgenic lines by minimizing position effect (Verma and Daniell 2007). The plasmid-based nature of the chloroplast genome provides other advantages such as an inherent gene containment system, owing to the largely maternal inheritance of plastid genomes (Daniell 2007), the absence of epigenetic gene silencing mechanisms, and pleiotropic effects (Daniell et al. 2001; Rigano et al. 2009; Verma et al. 2010a). The polyploid nature of this organelle provides a high transgene copy number (up to 10,000 copies) allowing considerably higher expression than nuclear counterparts (Verma and Daniell 2007); reports have demonstrated as high as a 500-fold increase in antigen yield between nuclear and chloroplast based approaches (Daniell et al. 2001; Rigano et al. 2009; Scotti et al. 2009).

Chloroplasts have been shown to produce fully functional mammalian proteins through proper protein folding and disulfide bond formation (Arlen et al. 2007; Boyhan and Daniell 2011; Daniell et al. 2009a, b; Staub et al. 2000; Verma and Daniell 2007; Lee et al. 2010), as well as lipidation (Glenz et al. 2006). Reported expression levels of transplastomic lines are unchallenged by other recombinant plant systems, with transgene expression approaching 72% of the total protein in several cases (Oey et al. 2009b; Ruhlman et al. 2010), twofold higher than the expression of RuBisCo. Coincidentally, plant systems have proven remarkably capable of dealing with massive accumulation of foreign proteins (Bally et al. 2009), furthermore sub-cellular localization of chloroplast derived antigen limits its escape. This containment likely further promotes the maintenance of homeostasis and the preservation of phenotype, explaining why hyper-accumulation of target protein is rarely shown to hinder growth (Ruhlman et al. 2010).

CTT has begun to develop in applications beyond model organisms. Edible plants facilitate simple and efficacious oral delivery of vaccine antigens without posing any form of health risk. Several reports have opted for more digestible options such as lettuce to produce antigens of bacterial (Davoodi-Semiromi et al. 2010; Ruhlman et al. 2010), viral (Kanagaraj et al. 2011), and protozoan origin (Davoodi-Semiromi et al. 2010) as well as human autoantigens (Ruhlman et al. 2007; Verma et al. 2010b; Boyhan and Daniell 2011). Another chloroplast-based system of growing interest is the eukaryotic algae *Chlamydomonas reinhardtii*, which is capable of expressing bacterial antigens and human therapeutics (Dreesen et al. 2010; Rasala et al. 2010). This system, however, invokes the use of bioreactors to culture this unicellular organism to sufficient culture densities and therefore is associated with a higher production cost. Despite this, algae culturing has demonstrated its commensurate capacity to produce such complex proteins as monoclonal antibodies (Tran et al. 2009). Interest in harnessing the chloroplast's biosynthetic machinery for biotechnological applications is growing rapidly, and with the advancement of platforms that foster oral delivery, the true potential of this technology is being demonstrated through functional evaluation of chloroplast-derived biopharmaceuticals.

4.2 Chloroplast Derived Vaccine Antigens

4.2.1 Bacterial Antigens

In 2001 Daniell et al. published the first account of achieving the expression of a vaccine antigen, cholera toxin subunit B (CTB), through plastid genetic engineering (Daniell et al. 2001), with functionality of this tobacco-derived subunit vaccine demonstrated through GM1 ganglioside binding, and by transcytosis of fusions peptide GFP into the intestinal mucosa (Limaye et al. 2006). Later, a murine immunization model wherein mice treated subcutaneously or orally with tobacco derived

CTB developed strong IgG1 and, in the case of orally immunized mice, IgA titers that were able to resist the intestinal osmotic disruption caused by CT (Davoodi-Semiromi et al. 2010). Since the realization of the chloroplast's ability to produce vaccine antigens, considerable accomplishments regarding the expression of bacterial antigens have been achieved. Vaccines against enterogenic *E. coli* (Kang et al. 2004, 2003; Rosales-Mendoza et al. 2009; Sim et al. 2009), *Clostridium tetani* (Tregoning et al. 2003, 2005), *Bacillus anthracis* (Gorantala et al. 2011; Koya et al. 2005; Watson et al. 2004), *Yersinia pestis* (Arlen et al. 2007, 2008), *Borrelia burdoferi* (Glenz et al. 2006), as well as a multi-epitope vaccine against diphtheria, pertusis, and tetanus (Soria-Guerra et al. 2009) have been produced through the chloroplast. The yield of these vaccines is typically antigen-dependant and therefore contingent on accumulation and susceptibility to proteolysis. Examples ranging 31% total soluble protein (TSP) (Molina et al. 2004) to as low as 0.5% TSP (Sim et al. 2009) have been reported. While a multitude of bacterial antigens have been effectively expressed through CTT, only some of these studies instituted an *in vivo* functional evaluation of the recombinant antigen. Moreover, only 3 of 16 vaccines were produced in oral delivery systems i.e. lettuce and algae (Lossl and Waheed 2011). Detailed functional evaluations in edible plants for human use are critical to advancing CTT to higher-animal models and through clinical trials.

4.2.1.1 Bacillus anthracis

Recent biological terrorism threats underscore the requirement for the development of a new generation of anthrax vaccines; the currently used culture filtrates of live *Bacillus anthracis* produce severe reactogenicity, emphasizing the need for a safer and more widely applicable vaccine (Cybulski et al. 2009). Protective antigen (PA) of the tripartite anthrax exotoxin facilitates the translocatation of the enzymatic components lethal factor (LF) and edema factor (EF) into the cytosol of an invaded phagocyte (Passalacqua and Bergman 2006; Thoren and Krantz 2011). PA, the target of commercially available Anthrax Vaccine Adsorbed (AVA), was first expressed in tobacco chloroplasts by Watson et al. in 2004. Chloroplast-derived PA activity was similar to the *B. anthracis* analogue during translocation of LF in the macrophage lysis assay (Watson et al. 2004). This PA was later shown to drive immunity through a murine immunization model wherein mice immunized subcutaneously by this chloroplast-derived antigen developed titers similar to those in mice immunized by *B. anthracis* PA (Koya et al. 2005). Though serum IgG molecules of mice immunized by the chloroplast derivative provided slightly less toxin neutralization, challenge of these animals with 1.5 LD_{100} of anthrax toxin demonstrated 100% survival of test mice that received chloroplast-derived PA. This indicates the potential for CTT to produce efficacious and toxin-free PA. Daniell et al. later generated a similar protective antigen plant line in *Lactuca sativa*, suggesting a potential future for the oral delivery of PA (Ruhlman et al. 2010). Several constructs were generated with an optimized expression cassette reaching levels of approximately 22% TSP.

Recently, another study evaluated the effects of oral immunization against anthrax using tobacco derived domain IV of PA delivered orally in conjunction with cholera toxin; gastric neutralization followed by feeding of tobacco derived PA(dIV) produced levels of immunity comparable to mice immunized though i.p. injection by the same antigen (Gorantala et al. 2011). However, low levels of titers (>10^4) were induced through oral delivery in the absence of subcutaneous priming when compared to previously cited titers of ~300,000 that were shown to confer 100% survival of test animals (Koya et al. 2005); characterization of IgG isotypes suggested elevated IgG1 levels indicative of polarized Th2 cell response, and low secretory IgA was detected from fecal extracts. The sera of these mice were shown to increase macrophage efficiency against PA and neutralize exotoxin in a protective manner. Upon challenge, mice immunized orally by tobacco-derived PA had only a 60% survival rate. Therefore, further functional evaluations of edible systems for the oral delivery of PA need to be fully investigated.

4.2.1.2 Staphylococcus aureus

The opportunistic pathogen *Staphylococcus aureus* may invade to cause bacteraemia, which often leads to life-threatening infection (Lowy 1998; Que and Moreillon 2004). The development of multiple drug resistant lines of this pathogen highlights the necessity to produce more effective means of treatment (Lowy 2003). *S. aureus* pathogenicity is dependant on fibronectin-binding proteins (FnBPs), which enables adherence to host extracellular matrixes (Patti and Hook 1994). FnBPs therefore represent a target to impede invasion and colonization of *S. aureus*. The expression of the D2 Fibronectin-binding domain of *S. aureus* fused to cholera toxin B subunit was achieved in *C. reinhardtii* (Dreesen et al. 2010). CTB-D2 fusion protein formed functional oligomers that recognized GM1. A simulated gastric fluid assay was used to screen the stability of the CTB-D2 vaccine during digestion and its potential to deliver antigen to the intestinal mucosa. While stability for 20 min in Sodium Acetate (pH 1.7, 37°C) supplemented by 0.5 mg/ml pepsin was observed, the parameters utilized therein differ from standardized examples in the literature. US Pharmacopeia cited gastric conditions include a 3.2 mg/ml concentration of Pepsin within a simulated gastric fluid (0.2% (w/v) NaCl in 0.7% (v/v) HCl (Herman et al. 2005, 2006). As the duration of stomach-based digestion is dependant on particle size, disparate stomach contents could prolong the exposure of target antigen to the conditions of the stomach; a time lapse of digestion is important to demonstrate stability for longer periods of time (Pohle and Domschke 2003). Furthermore, a quantitation of the assay would have clearly demonstrated antigen stability. After 20 months of storage at room temperature, lyophilized algal derived antigen induced protective titers when delivered orally in the absence of priming. A large 160 μg oral prime, followed by 4 weekly feedings resulted in a dose-dependant induction of IgG and IgA. These systemic and mucosal responses significantly reduced the pathogen load of the spleen, conferred protection in up to 80% of test animals challenged intraperitoneally with a lethal dose of *S. aureus*.

4.2.2 *Viral Antigens*

Progress in the field of CTT came about with the expression of the canine parvovirus vaccine antigen 2L21 (Molina et al. 2004). Subsequently, several viral epitopes have been expressed including L1 protein of HPV-16 that formed functional VLPs (Fernandez-San Millan et al. 2008), and a mutated form (L1_2xCysM) that formed more thermostable capsomeres (Waheed et al. 2010). Other viral epitopes targeted by CTT include Vaccinia virus envelope protein (Rigano et al. 2009), Hepatitis E E2 (Zhou et al. 2006), Hepatitis C core protein (Madesis et al. 2009), as well as Epstein-Barr viral capsid antigen complex (Lee et al. 2006). Additionally, viral expression platforms have utilized edible systems including lettuce (Kanagaraj et al. 2011) and tomato leaves demonstrating the potential use for this system in humans, although antigen expression was undetectable in ripe tomato fruits (Zhou et al. 2008). However, the evaluation of viral antigen functionality is critical in advancing this system into higher-animal models, yet few examples exist wherein investigators completed an *in vivo* analysis and demonstrated the induction of antigen specific immunity (Fernandez-San Millan et al. 2008; Zhou et al. 2008; Gonzalez-Rabade et al. 2011).

4.2.2.1 Human Immunodeficiency Virus

Human immunodeficiency virus type 1 (HIV-1) is the causative pathogen of acquired immunodeficiency syndrome (AIDS), a deadly disease that is a global problem. Treatments for this virus are exacerbated by high mutation rates leading to multiple clades of this pathogen (Loemba et al. 2002; Young et al. 2006; Walker and Burton 2008). A multi-valent vaccine against HIV-1 therefore holds the most promise to elicit an efficacious immune response (Gonzalez-Rabade et al. 2011). Two highly conserved structural proteins, P24 and negative regulatory protein (Nef), show 80% and 84% homology across HIV-1 clades respectfully (Geyer and Peterlin 2001). Gonzalez-Rabade et al. engineered transplastomic tobacco lines expressing p24 as well as p24-Nef fusion protein that accumulated to ~40% TSP, and evaluated their induction of immunity in mice utilizing a prime-boost strategy. Subcutaneous priming with *E. coli* derived p24 or Nef proteins followed by oral boosting with tobacco p24 and p24-Nef fusion antigen, in conjunction with CTB, developed measureable immune responses. Evaluation of the sera from immunized animals demonstrated a primarily IgG1 and IgG2a driven response which implicates both Th1 and Th2 subsets of thymocytes, indicative of cell-mediated and humoral immunity respectively. Interestingly it remains to be determined if the form of immunity generated by this immunization approach would provide protection upon pathogen exposure. The lack of available viral-model systems in mice requires the advancements of studies such as this to higher primate models wherein analogous viruses such as Simian immunodeficiency virus could be examined.

4.2.2.2 Dengue Virus

Dengue hemorrhagic fever, caused by Dengue virus, is a potentially lethal infectious tropical disease and a recent global resurgence of Dengue fever cases worldwide emphasizes the need for Dengue vaccine development (Kanagaraj et al. 2011; Matsui et al. 2010). The first report of viral antigen expression in an edible oral delivery system was achieved with the generation of a plant-line expressing a fusion of Dengue antigens. Dengue-3 serotype polyprotein (prM/E), consisting of premembrane protein (prM) and truncated envelop protein (E), was shown to assemble into structures resembling VLPs ~20 nm diameter. The obviated threat of replication within immunized hosts and the potent B cell and CTL mediated immune responses associated with VLPs support the use of prM/E VLPs to will induce protective immunity. This structural domain developed in this system contains type specific and complex reactive antigenic sites, considered the dominant neutralizing determinants to Dengue virus, and its expression in an edible crop is a significant advancement (Kanagaraj et al. 2011).

4.2.3 Protozoan Antigens

Interestingly, while vaccines against protozoan-based infectious disease are perhaps the most under represented in CTT, these pathogens produce high mortality rates in many developing countries. Protozoan pathogens include members of the genus *Giardiinae*, *Haemosporida*, *Leishmania*, and *Trypanasoma* among others. While many of these pathogens rely on insect-driven transmission, other common modes of dissemination include ingestion of contaminated water and foodstuffs, or the inhalation of oocysts. The lack of licensed vaccines for the majority of these often fatal protozoan infections requires the advancement of immunization methods. Indeed many cases of protozoan pathogenesis target the intestinal mucosa, suggesting that the mucosal immunity provided by oral plant-based immunization could be well implicated. A 2007 study published by Chebolu et al. described the first transplastomic expression of an antigen against a protozoan pathogen, *Entamoeba histolytica*, and an evaluation of its immunogenicity (Chebolu and Daniell 2007). A later study by Davoodi-Semiromi et al. achieved expression of malarial antigens apical membrane antigen-1 and merozoite surface protein-1 and functional evaluation by inhibition of proliferation of the malarial parasite (Davoodi-Semiromi et al. 2010). While efforts have been made to produce vaccines targeting protozoan parasites, the lack of available challenge models prevents their effective development. With improved models of human disease, protozoan-based infectious disease could be one of the major targets off vaccines produced through CTT.

4.2.3.1 Amoebiasis

The causative agent of amoebiasis is the enteric pathogen *Entamoeba histolytica.* Infection is characterized by colitis and formation of liver abscesses. Of the 50 million cases worldwide more than 100,000 deaths are reported annually, primarily confined to developing countries. Pathogen adherence and cytolysis of immune-effectors cells are functions of the galactose/N-acetyl-D-galactosamine (Gal/GalNAc) lectin, which has been shown to be a potential target for prophylactic vaccination against this pathogen (Vines et al. 1998; Lotter et al. 2000). The expression of the Gal/GalNAc lectin in tobacco chloroplasts was achieved in 2006 and was the first report of this antigen being expressed in plants (Chebolu and Daniell 2007). This vaccine was evaluated in a murine immunization model, with the subcutaneous delivery of crude transplastomic plant extracts. Induction of higher IgG titers (>10^4 in the presence of adjuvant) than previous examples of immunization against this target was reported, however no pathogen challenge was completed due to the abrogated susceptibility of BALB/c mice to *E. histolytica.* While the successful induction of antibodies was demonstrated, this study should be repeated in a relevant model where pathogen challenge can evaluate the protection conferred by the induced immunity. With expression levels of 6.3% TSP, a yield of over 29 million doses of this vaccine was estimated from an acre of transgenic plants, suggesting the potential for mass immunizations facilitated by large-scale field production of plant-derived vaccine antigens against protozoan pathogens.

4.2.3.2 Malaria

Malaria is perhaps the most globally important protozoan-based infectious disease, impacting tropical and subtropical areas in over 100 countries (Greenwood et al. 2005). While the most virulent form of this pathogen is *Plasmodium falciparum*, several causative agents exist. The lack of a licensed vaccine against malaria is a result of the high-degree of pathogen polymorphisms, the unavailability of animal challenge models to evaluate functionality, and the sheer cost of development and delivery of such a vaccine (Aide et al. 2007). While several vaccines targeting candidate antigens from the blood stage of this parasite are in clinical trials (Maher 2008), the global implication of this disease stipulates the use of a low-cost vaccine production and delivery system. Two targets of malaria immunization, the apical membrane antigen-1 (AMA-1) and the merozoite surface protein-1 (MSP-1), were expressed in the chloroplasts of both tobacco and lettuce (Davoodi-Semiromi et al. 2010). Expressed as CTB-fusion proteins, these antigens accumulated to 10–13% in tobacco, and 6–7% in lettuce. Mice subcutaneously primed with each of these respective antigens were boosted with tobacco-derived subunit vaccines in either subcutaneous or oral delivery fashions. Significant IgG1 titers were detected in both groups, though mice boosted-orally displayed >twofold weaker titers. However, oral boosting with these antigens drove antibody class switching resulting in an additional IgA response. Furthermore, the antibodies generated in mice were shown to be functionally active through their reactivity of cell-free parasite extracts from each of the life cycles of the parasite. Anti-AMA-1 antibodies bound proteins form the schizont phase, whereas anti-MSP-1 antibodies reacted with both the ring and schizont stages. These sera samples were used to visualize *P. falciparum* through immunofluorescence microscopy, ultimately demonstrating the capacity for this chloroplast-derived vaccine to confer the ability to recognize *P. falciparum* to the mammalian immune system.

4.2.4 Autoantigens

Autoimmune diseases occur when activated lymphocytes recognize endogenous proteins as foreign antigens resulting in an immune response targeting self-tissues. The mechanisms that drive the acquisition of such self-polarized lymphocytes are poorly understood, largely due to the variety of multifariousness of these diseases (Goronzy and Weyand 2007). However, the induction of a unique set of thymic lymphocytes, termed regulatory T-cells (T_{Reg}), may serve to attenuate the destructive responses of cytotoxic T-cells and effectively halt disease pathology (Weiner et al. 2011). The first autoimmune disease that became the target of CTT was Type 1 Diabetes (T1D), with the development of chloroplast-derived insulin in tobacco and lettuce systems (Ruhlman et al. 2007). While diabetes is among the most widely studied autoimmune disorders, multiple targets of autoimmunity that drive diabetic pathology have been identified in the pancreas. Another such diabetic autoantigen that has been produced in algal chloroplasts is glutamic acid decarboxylase (Wang et al. 2008). Functionality of chloroplast derived GAD65 was confirmed through conformational-dependant interactions with diabetic patient sera and was further substantiated by its ability to promote β-cell proliferation of NOD mice.

4.2.4.1 Insulin

Insulin dependent diabetes, or Type 1 diabetes (T1D), is characterized by lymphocytic infiltration to the Islet of Langerhans of the pancreas, leading to a decreased β-cell mass and subsequent hyperglycemia, termed insulitis (Nagata et al. 1994). While current patients of T1D rely on self-administration of injectable recombinant insulin to maintain appropriate blood glucose levels, the threat of more serious complications such as nephropathy, high blood pressure, heart disease, and diabetic neuropathy emphasizes the need for a cure. The development of proinsulin producing tobacco and lettuce lines was accomplished in 2007 (Ruhlman et al. 2007). Expression of CTB-Proinsulin fusion protein resulted in accumulation of up to ~24% in lettuce, with activity illustrated through the GM1 binding assay. Oral delivery of chloroplast-derived proinsulin to 5-week old NOD mice significantly reduced observable insulitis, and blood and urine glucose levels, that was accompanied by islet preservation. Furthermore, the increased expression levels of immunosuppressive cytokines interleukin-4 and IL-10 suggested an active attenuation of cytotoxic immunity. Characterization of the immune response induced established a predominantly IgG1 response; the lack of IgG2a involvement suggests that a T-helper 2 cell response was invoked to drive the induction of oral tolerance to CTB-Pins. A later study completed by Boyhan and Daniell instituted a tripartite furin cleavage site that enabled *in vivo* pro-protein processing to occur outside of the pancreas (Boyhan and Daniell 2011). This is the first report of proinsulin gene expression in plants with simultaneous delivery of C-peptide which enhances glucose disposal and metabolism control, and aids in the prevention of diabetic nephropathy and neuropathy (Hills and Brunskill 2009). While expression of complete proinsulin cholera toxin B (CTB) fusion protein in lettuce chloroplast reached up to 53% of TLP, mice orally gavaged with this material displayed a marked decrease in blood glucose levels in comparison to the control mice given wild-type lettuce, demonstrating that correct folding, and pro-peptide processing occurred to deliver active chloroplast-derived insulin to the circulatory system (Boyhan and Daniell 2011). Together, these studies emphasize the potential for chloroplast-derived insulin products to replace commercially available systems in the form of orally deliverable pharmaceuticals.

4.2.4.2 Factor IX

Haemophilia B is a disease characterized by a loss of functionality in blood clotting factor IX. Replacement therapy with intravenous delivery of recombinant human factor IX results in the development of inhibitory antibodies in 9–23% of patients, potentiating life threatening anaphylactic reactions to this therapeutic (DiMichele 2006). Expression of recombinant human Factor IX in transplastomic tobacco led to expression of CTB-FIX or CTB-FFIX (engineered Furin cleavage site) at 3.8% TSP (Verma et al. 2010b). Oral delivery of FIX induced oral tolerance such that the development of fatal anaphylactic reactions upon subsequent intravenous delivery of FIX was prevented. Systemic release of this protein occurred 2–5 h after feeding, and sights of absorption were localized to the M-cells of the ileum. The reduction of inhibitory antibodies and a Th2-dependant IgE response resulted in >90% survival of mice upon intravenous exposure to recombinant human Factor IX, when compared to 20% survival of untreated animals that received far fewer injections. This observed regression of anaphylactic responses, in comparison to unfed animals or those that received untransformed material, was described as the effects of Th3 cells driving IgG2b and IgA production through TGF-β mediated modulation of B-cell activity. This effect was contingent on consistent oral exposure to FIX, as a hiatus in the feeding schedule resulted in elevated levels of inhibitors. However, resumption of feeding diminished these levels and anaphylactic responses were against prevented. Chloroplast-derived autoantigens were previously demonstrated to induced regulatory thymic lymphocytes upon oral delivery that combated self-recognition of insulin (Ruhlman et al. 2007), therefore successful induction of T_{Reg} in haemophilia B patients through induction of oral tolerance would enable proper replacement therapies that avoid neutralizing reactions of intravenously delivered Factor IX.

4.2.5 Veterinary Advancement

Further advancements of the field through veterinary applications help substantiate the promise of chloroplast-derived vaccines, with their efficacy in livestock immunization demonstrating the viability of this platform for human use. Canine Parvovirus vaccines targeting the 2L21 antigen was among the first attempts made at generating viral vaccines in chloroplasts (Molina et al. 2004). CTT has since been used to produce vaccine candidates in algal chloroplast against Swine Fever Virus; the CSFV E2 protein derived from tobacco chloroplast was found to be immunogenic in mice (He et al. 2007). These advancements will enable the study of challenge models in relevant animals, and therefore lend themselves to study the efficacy of prophylactic immunization. However, a deficiency in the evaluations of these vaccines' biological activity has prevented further advancement. Lax regulations over the development of live-stock pharmaceuticals should promote the evaluation of CTT in this field.

4.2.5.1 Canine Parvovirus

Canine parvovirus causes acute gastroenteritis leading to a high mortality rate in unvaccinated dogs. The highly immunogenic peptide 2L21 derived from the canine parvovirus (CPV) protein (Langeveld et al. 1994), has been credited as the first demonstration of functional vaccine expression in transgenic chloroplast (Molina et al. 2004). CTB-2L21 fusion protein accumulated to 31% TSP in tobacco chloroplast and was subsequently analyzed for immunogenicity. Parenteral injections of enriched plant-extracts in mice and rabbits induced high titer antibody production against 2L21 (Molina et al. 2005). Rabbit sera displayed the same capacity for neutralizing 2L21 antigen as the monoclonal antibody 3C9. Furthermore, orally delivered pulverized leafs expressing CTB-2L21 generated systemic IgG and secretory IgA when delivered in a prime-boost strategy following parenteral administration of a prime vaccination. While the priming generated a humoral IgG response, oral boosting was critical to drive iso-type switching to mucosal IgA. However, the antibodies induced by oral immunization were ineffective at neutralizing 2L21 during *in vitro* assays suggesting the breakdown of key epitopic determinants (Molina et al. 2005). The high yield of subunit vaccine generated through this approach vindicates the generation of similar chloroplast-derived vaccines in edible food systems. The use of a rabbit parenteral immunization model in this study exemplifies the requirements for CTT to advance to higher animal models, but such studies should be implicated for the study of oral immunization to observe the possible differences in efficacy that may arise due to the variable intestinal surface area of mammals.

4.2.5.2 Foot and Mouth

Endemic to developing countries, Foot and Mouth Disease holds the potential to cause massive economic breakdowns by affecting live stock and may impede the progression of an impoverished society (Forman et al. 2009; Rweyemamu and Astudillo 2002). Expression of Foot and mouth disease virus (FMDV) vaccines through chloroplasts have been achieved in tobacco chloroplast (Lentz et al. 2009). Recent progress in the development of an effective vaccine via chloroplast transformation against FMDV is evident in the work by Lentz et al. Expression of the FMDV capsid protein VP1, fused to the B-glucuronidase reporter gene (uidA), was accomplished in transplastomic tobacco that accumulated to 51% TSP. Subsequent inoculation of mice with the chloroplast derived antigen demonstrated the induction of VP1 specific antibodies (Lentz et al. 2009). The importance of an effective plant derived vaccine against FMDV is apparent when considering the financial impact this disease may have on the impoverished countries in the world. The ability to transport such a vaccine with disregard to cold chain logistics or even a needle-based system would provide much needed relief in combating global indigence.

4.3 Challenges to Chloroplast-Derived Vaccine Antigens and Conclusions

The major challenges to chloroplast-derived vaccines are obtaining a sufficient overall yield of vaccine-antigen, demonstrating the biological activity of said antigens, and developing systems for oral delivery. Research within the past year has approached these issues to further substantiate the potential of chloroplast-derived vaccines in human immunization. The promise of prophylactic immunization by virtue of chloroplast derived vaccines has led to the development of subunit vaccines against an array of human pathogens, as well as several autoantigens that were previously described in comprehensive plant-made vaccine reviews (Daniell et al. 2009b; Lossl and Waheed 2011). Current research appears to be more concerned with expressing novel antigens, rather than focusing on methods of improving the viability of this platform for human use.

Optimization of gene regulatory elements as well as improving antigen stability *in planta* has been the two major approaches to improving the yield of antigens. Factors that contribute to the accumulation of a target protein include rates of transcription, translation, and susceptibility to degradation. The use of strong promoters has a direct influence on the protein levels in the chloroplast (Verma and Daniell 2007), as well as the 5′ and 3′ UTRs which initiate translation and confer mRNA stability respectively (Eibl et al. 1999). While a detailed list of the potential transcriptional elements has been reviewed (Verma and Daniell 2007), endogenous transcriptional activators provide unchallenged levels of expression (Ruhlman et al. 2010). Additionally, sustained protein hyper accumulation may negatively affect plant metabolism by altering phenotype. Oey et al. reported the expression of bactericidal phage lytic protein to more than 70% TSP with an optimized codon in transplastomic tobacco (Oey et al. 2009a). The marked stability of this protein resulted in translational limitations characterized by depleted levels of such plastid-encoded proteins as RuBisCo, which severely stunted growth. The production of recombinant proteins at the expense of the materials necessary for growth (Bally et al. 2009) exhausted the metabolic machinery and consequently limited plant growth. However, other reports demonstrated 70% TSP expression without observed phenotypical changes, indicating the antigen dependency of this effect (Ruhlman et al. 2007; Ruhlman et al. 2010). A recent 2010 publication produced a synthetic plastid encoded riboswitch which enabled inducible expression of a full grown plant induced by the exogenously applied ligand: theophylline (Verhounig et al. 2010). Fusion domains may also contribute to the stability of recombinant antigens *in planta* (Molina et al. 2009; Ruhlman et al. 2007, 2010; Boyhan and Daniell 2011; Verma et al. 2010b; Daniell et al. 2009b; Lee et al. 2011; Morgenfeld et al. 2009; Ortigosa et al. 2009; Scotti et al. 2009). These fusion domains were effective in improving the yield of small viral peptides such as E7 HPV type 16 protein, 2L21 of Canine Parvovirus VP2, as well as $PR55^{gag}$ of HIV. Such fusion peptides may also improve immunogenicity by promoting particularization of the antigen (Davoodi-Semiromi et al. 2010). Certain chaperones, for example the CRY chaperone (encoded by the orf2 gene), fold nascent proteins into cuboidal crystals, offering heightened protection from the chloroplast proteases and promoting higher levels of accumulation. Future design of the transformation protocol must include the appropriate determinants regulating protein stability in plastids which are at present time relatively unknown (Apel et al. 2010).

At the forefront of challenges in the field of chloroplast transformation technology stands the lack of available genome sequences (Verma and Daniell 2007). Our limitations in this regard govern the repertoire of plants available for genetic modification. With a few exceptions, most chloroplast-based antigen expression systems have been completed in tobacco cultivars. While this system will never be implicated for human use, progression into plant systems such as lettuce, tomato, and algae are critical. Furthermore, current tomato-based systems accumulate the majority of antigen in the leaves with low levels detectable in immature-green fruits and no detectable levels in mature tomatoes (Zhou et al. 2008). This system therefore requires enhanced targeted expression to the fruit that humans actually consume. Sequencing the chloroplast genomes of high-biomass, edible plants would promote the transition of CTT into platforms viable for humans. Naturally, further elucidation in the field will add to our collection of tools and broaden recombinant possibilities. Few active research groups and poor funding from the pharmaceutical sector (Daniell et al. 2009b; Kirk and Webb 2005) underline the callowness of this technology.

Yet another limitation to CTT is the scarcity of animal models or functional evaluation of chloroplast-derived antigens. While differences between the pathogens that affect humans and mice exist, examining the level of protection conferred to mice against a human pathogen via challenge may be impractical. An important concept, however, is an *in vitro* evaluation of the reactivity of the chloroplast derived antigen with clinical serum samples of patients previously sensitized to the pathogen (Rigano et al. 2009). Alternatively, demonstrating the opsonization of chloroplast- or pathogen-derived antigens by chloroplast-derived vaccine immunized-mice sera further substantiates the reactivity of antibodies generated through chloroplast-derived vaccines (Davoodi-Semiromi et al. 2010; Zhou et al. 2006). Demonstrating efficacy in these mouse-models is required to warrant examination in higher animal models. Advancing through trials in rabbits has been accomplished by some groups (Molina et al. 2005); however, this trend must become more widespread and continue to advance through other systems such as gerbils, pigs, dogs, and eventually non-human primates before human clinical evaluation of chloroplast-derived antigens may occur. These models will aid in the development of antigen doses, and immunization strategies that will be effective in humans. Also, this platform must additionally be evaluated on the basis of field production. It is unknown if similar yields of antigen will be achieved, and other than a single report (Arlen et al. 2007), no examples of chloroplast-derived vaccines produced in the field have been evaluated. This is required for the consideration of long-term antigen stability, wherein preservation processes must be applied to transplastomic plant materials to obtain stable subunit vaccines.

Further characterization of the pathways controlling immune-sensitization versus tolerance would provide insight into the rational design of pharmaceuticals expressed in chloroplasts and delivered orally (Weiner et al. 2011). Fusion proteins with adjuvant or effector domains enhance the *in vivo* activity of the proteinaceous drug. Oral adjuvants such as CTB work through their ability to invoke their natural invasive behaviour in conjunction with the vaccine antigen, subsequently promoting immunity to both (Davoodi-Semiromi et al. 2010; Limaye et al. 2006). Recent reports have implicated the effectiveness of CTB adjuvant properties by delivering a mixture of plant-derived antigen and purified CT (Dreesen et al. 2010). Effector domains, such as the Fc regions antibodies, are capable of inducing complex molecular reactions and may be generated as fusion peptides to vaccine antigens. Cross-linking of antibodies will generate a particle infused with antigens to be internalized by APCs, endowing vaccine antigens with elevated immunogenicity. Chloroplasts have successfully expressed both Fc portions of antibodies, in addition to full monoclonal antibodies (Tran et al. 2009); therefore, effector domain potential appeals highly to this platform.

The translational capacity of the chloroplast is such that it has been used to express a diverse repertoire of proteinaceous molecules ranging from small antigens to complex human monoclonal antibodies. This enables for the inexpensive production of both prophylactic and therapeutic vaccines in recombinant chloroplasts. This platform holds great promise for applications in the context of bioterrorist threats or natural disasters due to its rapid implementation of oral vaccines; however, in order for CTT to burgeon into its full potential there must be an increase in studies that evaluate functionality and enable the efficacious oral delivery of chloroplast-derived antigens.

Acknowledgement The contribution here was supported in part by funding from the Arnold and Mabel Beckman Foundation to James S. New, and by grants from USDA-CSREES 2009-39200-19972, USDA-NIFA 2010-39200-21704 and NIH R01 GM 63879–08 to Henry Daniell.

References

Aide P, Bassat Q, Alonso PL (2007) Towards an effective malaria vaccine. Arch Dis Child 92: 476–479

Apel W, Schulze WX, Bock R (2010) Identification of protein stability determinants in chloroplasts. Plant J 63:636–650

Arlen PA, Falconer R, Cherukumilli S, Cole A, Cole AM, Oishi KK, Daniell H (2007) Field production and functional evaluation of chloroplast-derived interferon-alpha2b. Plant Biotechnol J 5:511–525

Arlen PA, Singleton M, Adamovicz JJ, Ding Y, Davoodi-Semiromi A, Daniell H (2008) Effective plague vaccination via oral delivery of plant cells expressing F1-V antigens in chloroplasts. Infect Immun 76:3640–3650

Bally J, Nadai M, Vitel M, Rolland A, Dumain R, Dubald M (2009) Plant physiological adaptations to the massive foreign protein synthesis occurring in recombinant chloroplasts. Plant Physiol 150:1474–1481

Boyhan D, Daniell H (2011) Low-cost production of proinsulin in tobacco and lettuce chloroplasts for injectable or oral delivery of functional insulin and C-peptide. Plant Biotechnol J 9:585–598

Chebolu S, Daniell H (2007) Stable expression of Gal/GalNAc lectin of Entamoeba histolytica in transgenic chloroplasts and immunogenicity in mice towards vaccine development for amoebiasis. Plant Biotechnol J 5:230–239

Cybulski RJ Jr, Sanz P, O'Brien AD (2009) Anthrax vaccination strategies. Mol Aspects Med 30:490–502

Daniell H (2007) Transgene containment by maternal inheritance: effective or elusive? Proc Natl Acad Sci USA 104:6879–6880

Daniell H, Lee SB, Panchal T, Wiebe PO (2001) Expression of the native cholera toxin B subunit gene and assembly as functional oligomers in transgenic tobacco chloroplasts. J Mol Biol 311: 1001–1009

Daniell H, Ruiz G, Denes B, Sandberg L, Langridge W (2009a) Optimization of codon composition and regulatory elements for expression of human insulin like growth factor-1 in transgenic chloroplasts and evaluation of structural identity and function. BMC Biotechnol 9:33

Daniell H, Singh ND, Mason H, Streatfield SJ (2009b) Plant-made vaccine antigens and biopharmaceuticals. Trends Plant Sci 14:669–679

Davoodi-Semiromi A, Schreiber M, Nalapalli S, Verma D, Singh ND, Banks RK, Chakrabarti D, Daniell H (2010) Chloroplast-derived vaccine antigens confer dual immunity against cholera and malaria by oral or injectable delivery. Plant Biotechnol J 8:223–242

DiMichele DM (2006) Inhibitor treatment in haemophilias A and B: inhibitor diagnosis. Haemophilia 12(Suppl 6):37–41, discussion 41–32

Dreesen IA, Charpin-El Hamri G, Fussenegger M (2010) Heat-stable oral alga-based vaccine protects mice from Staphylococcus aureus infection. J Biotechnol 145:273–280

Eibl C, Zou Z, Beck A, Kim M, Mullet J, Koop HU (1999) In vivo analysis of plastid psbA, rbcL and rpl32 UTR elements by chloroplast transformation: tobacco plastid gene expression is controlled by modulation of transcript levels and translation efficiency. Plant J 19:333–345

Fernandez-San Millan A, Ortigosa SM, Hervas-Stubbs S, Corral-Martinez P, Segui-Simarro JM, Gaetan J, Coursaget P, Veramendi J (2008) Human papillomavirus L1 protein expressed in tobacco chloroplasts self-assembles into virus-like particles that are highly immunogenic. Plant Biotechnol J 6:427–441

Forman S, Le Gall F, Belton D, Evans B, Francois JL, Murray G, Sheesley D, Vandersmissen A, Yoshimura S (2009) Moving towards the global control of foot and mouth disease: an opportunity for donors. Rev Sci Tech 28:883–896

Geyer M, Peterlin BM (2001) Domain assembly, surface accessibility and sequence conservation in full length HIV-1 Nef. FEBS Lett 496:91–95

Glenz K, Bouchon B, Stehle T, Wallich R, Simon MM, Warzecha H (2006) Production of a recombinant bacterial lipoprotein in higher plant chloroplasts. Nat Biotechnol 24:76–77
Gonzalez-Rabade N, McGowan EG, Zhou F, McCabe MS, Bock R, Dix PJ, Gray JC, Ma JK (2011) Immunogenicity of chloroplast-derived HIV-1 p24 and a p24-Nef fusion protein following subcutaneous and oral administration in mice. Plant Biotechnol J. doi:10.1111/ j.1467-7652.2011.00609.x
Gorantala J, Grover S, Goel D, Rahi A, Jayadev Magani SK, Chandra S, Bhatnagar R (2011) A plant based protective antigen [PA(dIV)] vaccine expressed in chloroplasts demonstrates protective immunity in mice against anthrax. Vaccine. doi:10.1016/j.vaccine.2011.03.082
Goronzy JJ, Weyand CM (2007) The innate and adaptive immune systems. In: Goldman L, Ausiello D (eds) Cecil medicine, 23rd edn. Saunders Elsevier, Philadelphia, chap 42
Greenwood BM, Bojang K, Whitty CJ, Targett GA (2005) Malaria. Lancet 365:1487–1498
He DM, Qian KX, Shen GF, Zhang ZF, Li YN, Su ZL, Shao HB (2007) Recombination and expression of classical swine fever virus (CSFV) structural protein E2 gene in Chlamydomonas reinhardtii chroloplasts. Colloids Surf B Biointerfaces 55:26–30
Herman RA, Korjagin VA, Schafer BW (2005) Quantitative measurement of protein digestion in simulated gastric fluid. Regul Toxicol Pharmacol 41:175–184
Herman RA, Storer NP, Gao Y (2006) Digestion assays in allergenicity assessment of transgenic proteins. Environ Health Perspect 114:1154–1157
Hills CE, Brunskill NJ (2009) Cellular and physiological effects of C-peptide. Clin Sci (Lond) 116:565–574
Kanagaraj AP, Verma D, Daniell H (2011) Expression of dengue-3 premembrane and envelope polyprotein in lettuce chloroplasts. Plant Mol Biol 76:323–333
Kang TJ, Loc NH, Jang MO, Jang YS, Kim YS, Seo JE, Yang MS (2003) Expression of the B subunit of E. coli heat-labile enterotoxin in the chloroplasts of plants and its characterization. Transgenic Res 12:683–691
Kang TJ, Han SC, Kim MY, Kim YS, Yang MS (2004) Expression of non-toxic mutant of Escherichia coli heat-labile enterotoxin in tobacco chloroplasts. Protein Expr Purif 38: 123–128
Kirk DD, Webb SR (2005) The next 15 years: taking plant-made vaccines beyond proof of concept. Immunol Cell Biol 83:248–256
Koya V, Moayeri M, Leppla SH, Daniell H (2005) Plant-based vaccine: mice immunized with chloroplast-derived anthrax protective antigen survive anthrax lethal toxin challenge. Infect Immun 73:8266–8274
Langeveld JP, Casal JI, Osterhaus AD, Cortes E, de Swart R, Vela C, Dalsgaard K, Puijk WC, Schaaper WM, Meloen RH (1994) First peptide vaccine providing protection against viral infection in the target animal: studies of canine parvovirus in dogs. J Virol 68:4506–4513
Lee MY, Zhou Y, Lung RW, Chye ML, Yip WK, Zee SY, Lam E (2006) Expression of viral capsid protein antigen against Epstein-Barr virus in plastids of Nicotiana tabacum cv. SR1. Biotechnol Bioeng 94:1129–1137
Lee SB, Li B, Jin S, Daniell H (2010) Expression and characterization of antimicrobial peptides Retrocyclin-101 and Protegrin-1 in chloroplasts to control viral and bacterial infections. Plant Biotechnol J 9:100–115
Lentz EM, Segretin ME, Morgenfeld MM, Wirth SA, Dus Santos MJ, Mozgovoj MV, Wigdorovitz A, Bravo-Almonacid FF (2009) High expression level of a foot and mouth disease virus epitope in tobacco transplastomic plants. Planta 231:387–395
Limaye A, Koya V, Samsam M, Daniell H (2006) Receptor-mediated oral delivery of a bioencapsulated green fluorescent protein expressed in transgenic chloroplasts into the mouse circulatory system. FASEB J 20:959–961
Loemba H, Brenner B, Parniak MA, Ma'ayan S, Spira B, Moisi D, Oliveira M, Detorio M, Wainberg MA (2002) Genetic divergence of human immunodeficiency virus type 1 Ethiopian clade C reverse transcriptase (RT) and rapid development of resistance against nonnucleoside inhibitors of RT. Antimicrob Agents Chemother 46:2087–2094

Lossl AG, Waheed MT (2011) Chloroplast-derived vaccines against human diseases: achievements, challenges and scopes. Plant Biotechnol J 9:527–539

Lotter H, Khajawa F, Stanley SL Jr, Tannich E (2000) Protection of gerbils from amebic liver abscess by vaccination with a 25-mer peptide derived from the cysteine-rich region of Entamoeba histolytica galactose-specific adherence lectin. Infect Immun 68:4416–4421

Lowy FD (1998) Staphylococcus aureus infections. N Engl J Med 339:520–532

Lowy FD (2003) Antimicrobial resistance: the example of Staphylococcus aureus. J Clin Invest 111:1265–1273

Ma JK, Drake PM, Christou P (2003) The production of recombinant pharmaceutical proteins in plants. Nat Rev Genet 4:794–805

Madesis P, Osathanunkul M, Georgopoulou U, Gisby MF, Mudd EA, Nianiou I, Tsitoura P, Mavromara P, Tsaftaris A, Day A (2009) A hepatitis C virus core polypeptide expressed in chloroplasts detects anti-core antibodies in infected human sera. J Biotechnol 145: 377–386

Maher B (2008) Malaria: the end of the beginning. Nature 451:1042–1046

Matsui K, Gromowski GD, Li L, Barrett AD (2010) Characterization of a dengue type-specific epitope on dengue 3 virus envelope protein domain III. J Gen Virol 91:2249–2253

Molina A, Hervas-Stubbs S, Daniell H, Mingo-Castel AM, Veramendi J (2004) High-yield expression of a viral peptide animal vaccine in transgenic tobacco chloroplasts. Plant Biotechnol J 2:141–153

Molina A, Veramendi J, Hervas-Stubbs S (2005) Induction of neutralizing antibodies by a tobacco chloroplast-derived vaccine based on a B cell epitope from canine parvovirus. Virology 342: 266–275

Morgenfeld M, Segretin ME, Wirth S, Lentz E, Zelada A, Mentaberry A, Gissmann L, Bravo-Almonacid F (2009) Potato virus X coat protein fusion to human papillomavirus 16 E7 oncoprotein enhance antigen stability and accumulation in tobacco chloroplast. Mol Biotechnol 43:243–249

Nagata M, Santamaria P, Kawamura T, Utsugi T, Yoon JW (1994) Evidence for the role of CD8+ cytotoxic T cells in the destruction of pancreatic beta-cells in nonobese diabetic mice. J Immunol 152:2042–2050

Oey M, Lohse M, Kreikemeyer B, Bock R (2009a) Exhaustion of the chloroplast protein synthesis capacity by massive expression of a highly stable protein antibiotic. Plant J 57:436–445

Oey M, Lohse M, Scharff LB, Kreikemeyer B, Bock R (2009b) Plastid production of protein antibiotics against pneumonia via a new strategy for high-level expression of antimicrobial proteins. Proc Natl Acad Sci USA 106:6579–6584

Ortigosa SM, Fernandez-San Millan A, Veramendi J (2009) Stable production of peptide antigens in transgenic tobacco chloroplasts by fusion to the p53 tetramerisation domain. Transgenic Res 19:703–709

Passalacqua KD, Bergman NH (2006) Bacillus anthracis: interactions with the host and establishment of inhalational anthrax. Future Microbiol 1:397–415

Patti JM, Hook M (1994) Microbial adhesins recognizing extracellular matrix macromolecules. Curr Opin Cell Biol 6:752–758

Pohle T, Domschke W (2003) Gastric function measurements in drug development. Br J Clin Pharmacol 56:156–164

Que YA, Moreillon P (2004) Infective endocarditis. Nat Rev Cardiol 8:322–336

Rasala BA, Muto M, Lee PA, Jager M, Cardoso RM, Behnke CA, Kirk P, Hokanson CA, Crea R, Mendez M, Mayfield SP (2010) Production of therapeutic proteins in algae, analysis of expression of seven human proteins in the chloroplast of Chlamydomonas reinhardtii. Plant Biotechnol J 8:719–733

Rigano MM, Manna C, Giulini A, Pedrazzini E, Capobianchi M, Castilletti C, Di Caro A, Ippolito G, Beggio P, De Giuli MC, Monti L, Vitale A, Cardi T (2009) Transgenic chloroplasts are efficient sites for high-yield production of the vaccinia virus envelope protein A27L in plant cellsdagger. Plant Biotechnol J 7:577–591

Rosales-Mendoza S, Alpuche-Solis AG, Soria-Guerra RE, Moreno-Fierros L, Martinez-Gonzalez L, Herrera-Diaz A, Korban SS (2009) Expression of an Escherichia coli antigenic fusion protein comprising the heat labile toxin B subunit and the heat stable toxin, and its assembly as a functional oligomer in transplastomic tobacco plants. Plant J 57:45–54

Ruhlman T, Ahangari R, Devine A, Samsam M, Daniell H (2007) Expression of cholera toxin B-proinsulin fusion protein in lettuce and tobacco chloroplasts–oral administration protects against development of insulitis in non-obese diabetic mice. Plant Biotechnol J 5:495–510

Ruhlman T, Verma D, Samson N, Daniell H (2010) The role of heterologous chloroplast sequence elements in transgene integration and expression. Plant Physiol 152:2088–2104

Rweyemamu MM, Astudillo VM (2002) Global perspective for foot and mouth disease control. Rev Sci Tech 21:765–773

Scotti N, Alagna F, Ferraiolo E, Formisano G, Sannino L, Buonaguro L, De Stradis A, Vitale A, Monti L, Grillo S, Buonaguro FM, Cardi T (2009) High-level expression of the HIV-1 Pr55gag polyprotein in transgenic tobacco chloroplasts. Planta 229:1109–1122

Sim J-S, Pak H-K, Kim D-S, Lee S-B, Kim Y-H, Hahn B-S (2009) Expression and characterization of synthetic heat-labile enterotoxin B subunit and hemagglutinin–neuraminidase-neutralizing epitope fusion protein in *Escherichia coli* and tobacco chloroplasts. Plant Mol Biol Rep 27: 388–399

Soria-Guerra RE, Alpuche-Solis AG, Rosales-Mendoza S, Moreno-Fierros L, Bendik EM, Martinez-Gonzalez L, Korban SS (2009) Expression of a multi-epitope DPT fusion protein in transplastomic tobacco plants retains both antigenicity and immunogenicity of all three components of the functional oligomer. Planta 229:1293–1302

Staub JM, Garcia B, Graves J, Hajdukiewicz PT, Hunter P, Nehra N, Paradkar V, Schlittler M, Carroll JA, Spatola L, Ward D, Ye G, Russell DA (2000) High-yield production of a human therapeutic protein in tobacco chloroplasts. Nat Biotechnol 18:333–338

Streatfield SJ (2005) Mucosal immunization using recombinant plant-based oral vaccines. Methods Mucosal Immun 38:150–157

Thoren KL, Krantz BA (2011) The unfolding story of anthrax toxin translocation. Mol Microbiol 80:588–595

Tran M, Zhou B, Pettersson PL, Gonzalez MJ, Mayfield SP (2009) Synthesis and assembly of a full-length human monoclonal antibody in algal chloroplasts. Biotechnol Bioeng 104:663–673

Tregoning JS, Nixon P, Kuroda H, Svab Z, Clare S, Bowe F, Fairweather N, Ytterberg J, van Wijk KJ, Dougan G, Maliga P (2003) Expression of tetanus toxin Fragment C in tobacco chloroplasts. Nucleic Acids Res 31:1174–1179

Tregoning JS, Clare S, Bowe F, Edwards L, Fairweather N, Qazi O, Nixon PJ, Maliga P, Dougan G, Hussell T (2005) Protection against tetanus toxin using a plant-based vaccine. Eur J Immunol 35:1320–1326

Verhounig A, Karcher D, Bock R (2010) Inducible gene expression from the plastid genome by a synthetic riboswitch. Proc Natl Acad Sci USA 107:6204–6209

Verma D, Daniell H (2007) Chloroplast vector systems for biotechnology applications. Plant Physiol 145:1129–1143

Verma D, Kanagaraj A, Jin S, Singh ND, Kolattukudy PE, Daniell H (2010a) Chloroplast-derived enzyme cocktails hydrolyse lignocellulosic biomass and release fermentable sugars. Plant Biotechnol J 8:332–350

Verma D, Moghimi B, LoDuca PA, Singh HD, Hoffman BE, Herzog RW, Daniell H (2010b) Oral delivery of bioencapsulated coagulation factor IX prevents inhibitor formation and fatal anaphylaxis in hemophilia B mice. Proc Natl Acad Sci USA 107:7101–7106

Vines RR, Ramakrishnan G, Rogers JB, Lockhart LA, Mann BJ, Petri WA Jr (1998) Regulation of adherence and virulence by the Entamoeba histolytica lectin cytoplasmic domain, which contains a beta2 integrin motif. Mol Biol Cell 9:2069–2079

Waheed MT, Thones N, Muller M, Hassan SW, Razavi NM, Lossl E, Kaul HP, Lossl AG (2010) Transplastomic expression of a modified human papillomavirus L1 protein leading to the assembly of capsomeres in tobacco: a step towards cost-effective second-generation vaccines. Transgenic Res 20:271–282

Walker BD, Burton DR (2008) Toward an AIDS vaccine. Science 320:760–764

Wang X, Brandsma M, Tremblay R, Maxwell D, Jevnikar AM, Huner N, Ma S (2008) A novel expression platform for the production of diabetes-associated autoantigen human glutamic acid decarboxylase (hGAD65). BMC Biotechnol 8:87

Watson J, Koya V, Leppla SH, Daniell H (2004) Expression of Bacillus anthracis protective antigen in transgenic chloroplasts of tobacco, a non-food/feed crop. Vaccine 22:4374–4384

Weiner HL, da Cunha AP, Quintana F, Wu H (2011) Oral tolerance. Immunol Rev 241:241–259

Worlds Health Organization (2008) Global Burden of Diseae (GBD). The global burden of diseases 2004 update. World Health Organization, Geneva. Available from http://www.who.int/healthinfo/global_burden_disease/2004_report_update/en/index.html

Young KR, McBurney SP, Karkhanis LU, Ross TM (2006) Virus-like particles: designing an effective AIDS vaccine. Methods 40:98–117

Zhou YX, Lee MY, Ng JM, Chye ML, Yip WK, Zee SY, Lam E (2006) A truncated hepatitis E virus ORF2 protein expressed in tobacco plastids is immunogenic in mice. World J Gastroenterol 12:306–312

Zhou F, Badillo-Corona JA, Karcher D, Gonzalez-Rabade N, Piepenburg K, Borchers AM, Maloney AP, Kavanagh TA, Gray JC, Bock R (2008) High-level expression of human immunodeficiency virus antigens from the tobacco and tomato plastid genomes. Plant Biotechnol J 6:897–913

Chapter 5
Seed Expression Systems for Molecular Farming

Allison R. Kermode

Abstract Plant seeds are potentially one of the most economical systems for large-scale production of recombinant proteins for industrial and pharmaceutical uses. Plant-based systems in general have several advantages over the current production systems for biopharmaceuticals, such as yeasts, fungi, insect cells, mammalian cell cultures and transgenic animals. These advantages include the ability to post-translationally modify recombinant proteins, cost-effective production of recombinant proteins, due to the ease of scale-up and the availability of established protocols for harvesting, transport, storage and processing, and a minimal possibility of product contamination by animal/human pathogens. Plant seeds have a further significant advantage, that of providing a stable repository for recombinant proteins. Despite significant progress in the use of transgenic plants and seeds for therapeutic protein production, there are still technical challenges that must be surmounted before these systems can be fully embraced as suitable, viable alternatives for large-scale production of biopharmaceuticals and other products. These challenges include: (1) Enhancing the yields of recombinant proteins to maximize economic feasibility; (2) Manipulating the post-translational machinery of the plant or seed so that the recombinant protein is structurally and functionally similar to the native protein; (3) Developing efficient methods for downstream processing of recombinant proteins, including the removal of any foreign amino acid motifs, *in vitro* post-translational processing and protein purification, and (4) Addressing regulatory issues, such as suitable mechanisms for containment of the transgenic materials to prevent inadvertent transgene flow. Considerable strides have been made toward addressing these challenges particularly in the past 5 years. This has culminated in some plant-derived pharmaceutical proteins including antibodies, vaccines, human

A.R. Kermode (✉)
Department of Biological Sciences, Simon Fraser University, 8888 University Dr., Burnaby, BC V5A 1S6, Canada
e-mail: Kermode@sfu.ca

A. Wang and S. Ma (eds.), *Molecular Farming in Plants: Recent Advances and Future Prospects*, DOI 10.1007/978-94-007-2217-0_5,

blood products and growth regulators reaching the stage of preclinical studies or commercial development, and the promise of the first plant-made pharmaceutical glycoprotein intended for human parenteral administration. Below this progress is reviewed, with an emphasis on enhancing protein yields in seeds and controlling the N-glycan status of seed-produced recombinant proteins.

5.1 Introduction

With an increased understanding of the underlying pathophysiology of human diseases, there has been a great surge in the development of recombinant proteins as therapeutics. The past 30 years has witnessed a significant increase in the number of developed and approved protein therapeutics, and an even larger increase in the number of protein therapeutics in clinical trials. During this time, well over 100 new therapeutic proteins, or peptides ("biologics") have been licensed for production, primarily using microbial systems and mammalian cell cultures (e.g. human fibroblasts or Chinese Hamster Ovary cells) as the production hosts. In some years, the rapid growth in the number of protein targets being pursued as disease therapeutics has produced critical shortages of some drugs as a result of limited manufacturing capacity; the industry has responded by increasing the number of facilities devoted to therapeutic protein production, and by improving protein yields and the efficiency of recombinant protein purification/processing utilizing the established production hosts (Karg and Kallio 2009). Companies and academic researchers have also pursued alternative production hosts, including transgenic plants, plant cultures, and seeds, with the added promise of potentially overcoming some of the biochemical, technical and economic concerns with the traditional production systems (reviewed in Gomord and Faye 2004; Faye and Gomord 2010; Gomord et al. 2010).

The choice of a suitable system for production of a given recombinant therapeutic protein relies on the extent to which the heterologous host (e.g. yeast, fungus, plant/seed, animal or cultured cell system) is capable of post-translationally modifying the protein in such a way that it is identical or very similar to the native protein. In this regard, plant cells are generally able to properly fold, post-translationally modify and otherwise process recombinant proteins in a manner that is similar to that of mammalian systems. Indeed the success of select plant- or seed-based systems for production of biopharmaceutical proteins is evident in several recent reviews showing the range of these proteins, some of which are currently in clinical trials, are on the market, or are about to enter the market (Lau and Sun 2009; Karg and Kallio 2009), and by the increase in publication activity over the past 20 years related to biopharmaceutical products generated in plant hosts (Faye and Gomord 2010). The advantages of plant- and seed-based systems for production of biopharmaceuticals include: (1) potential cost-effective production of recombinant proteins; (2) minimal possibility of product contamination by animal/human pathogens and thus the transmission of human and animal viruses and prions; and (3) the availability of natural storage organs such as tubers, fruits and seeds to facilitate the stable accumulation of recombinant proteins (Twyman et al. 2003; Fischer et al. 2004). The use of plant seeds to host recombinant protein production offers further advantages (Giddings et al. 2000; Twyman et al. 2003). Developing seeds are already geared for high rates of protein synthesis and the stable accumulation of proteins; in the mature dry state, seeds are viable for long periods, thus facilitating storage and distribution of transgenic material (Stoger et al. 2005; Ma et al. 2003). The capacity of seeds to remain viable in the dry, quiescent and stored state, allows for the additional advantage of a 'decoupling' of the processing of the materials to obtain the purified recombinant protein from the generation and harvesting of the transgenic seeds; methods for cost-effective manufacturing of seed-based biopharmaceuticals in the realm of GMP (good manufacturing practice) can also be devised (Boothe et al. 2010).

Despite the advantages associated with plant- and seed-based production systems, there are nonetheless significant challenges associated with them; these obstacles must be surmounted before these systems can be fully embraced as suitable, viable alternatives for large-scale production of biopharmaceuticals and other products. These challenges include: (1) Enhancing the yields of recombinant proteins to maximize economic feasibility; (2) Manipulating the post-translational machinery of the plant or seed so that the recombinant protein is structurally and functionally similar to the native protein; (3) Developing efficient methods for downstream processing of recombinant proteins, including purification and *in vitro* post-translational processing; and (4) Addressing regulatory issues, such as suitable mechanisms for containment of the transgenic materials to prevent inadvertent transgene flow. Progress in seed-based production systems is reviewed below, with an emphasis on enhancing protein yields in seeds and controlling the N-glycan status of seed-produced recombinant proteins. Wherever possible, seed-based production of human recombinant lysosomal enzymes for treatment of lysosomal storage diseases is used as an example.

5.2 The Need for Cost-Effective and Safer Recombinant Protein Production Systems

The importance of developing systems that ensure both cost-effective production, and generate safe, high-quality biopharmaceuticals cannot be underestimated. Most recently a major pharmaceutical company – Genzyme Corporation – experienced significant problems in relation to product contamination of their mammalian (*C*hinese *h*amster *o*vary; CHO) cell cultures. This included a viral contamination that both reduced the protein yields of the cultures and caused the shut-down of some of the protein production systems that were being used to generate recombinant therapeutic enzymes for several months in 2009 and 2010 (Allison 2010). The shortage of recombinant proteins led to the interruption of treatment of patients, dose reduction, or the initiation of alternative treatments. These problems are eliminated with plant- and seed-based systems for recombinant protein production.

Another major issue is often the cost of a therapeutic generated by a given heterologous production system, especially those that are mammalian-cell based. Certainly across all production platforms, the major cost is typically associated with the downstream processing/purification of the recombinant protein. An extreme example here is the prices set for approved therapeutics for rare diseases known as lysosomal storage diseases. Lysosomal storage diseases are a broad class of progressive genetic diseases; collectively they represent a diverse class of over 50 disorders that represent a significant proportion of childhood metabolic disease. Most are caused by a deficiency of a single hydrolytic enzyme that resides in the lysosome; because of this there is a block in the stepwise catabolism of certain macromolecules, a process critical for the growth and homeostasis of tissues. These diseases are unique in that they are generally amenable to enzyme therapies (ERT or *E*nzyme *R*eplacement *T*herapy) (Desnick and Schuchman 2002), the intravenous (parenteral) administration of the purified recombinant enzyme. However, the current systems used to produce recombinant lysosomal enzymes for this purpose (most commonly human fibroblasts or Chinese hamster ovary cells, some of which have led to approved, commercialized products for a few of the lysosomal storage diseases) are prohibitively expensive, costing an average of $170,000 US per patient annually (Werber 2004), and often much more. For example, an average 8-year-old child of 40 kg would incur an annual drug cost of over $400,000 US for Aldurazyme™ – recombinant α-L-iduronidase, a treatment for one of the lysosomal storage diseases known as mucopolysaccharidosis I, and generated using Chinese hamster ovary cell cultures. The high prices of the therapeutics place a considerable burden on health care systems. In addition, for those lysosomal storage diseases that are considered to be 'ultra-rare', there is little commercial interest toward pursuing the development of the associated therapeutics.

5.3 Enhancing Yields of Seed-Produced Recombinant Proteins by Improving Gene Expression and Stable Protein Accumulation

Seeds are versatile hosts for the production of recombinant proteins of several types, all the way from short, simpler peptides, to larger complex, multi-subunit proteins. Various antibodies and other immunoglobulins, insulin, human growth hormone, lysozyme, lactoferrin, and the lysosomal enzymes, β-glucocerebrosidase, and α-L-iduronidase, are among the examples of seed-produced biopharmaceuticals (reviewed in Kermode 2006; Lau and Sun 2009; Boothe et al. 2010). Seeds of major cereal crops have been utilized as host production systems for recombinant proteins of clinical interest, with the major target species including maize, rice, barley and wheat. Likewise, seeds of various dicots, such as soybean, safflower, Arabidopsis and tobacco are convenient hosts for seed-based recombinant protein production, each with its associated advantages and disadvantages. Obviously a significant factor that will ultimately constitute part of the economic feasibility of a given seed platform

Table 5.1 Strategies for optimizing expression of transgenes for greater recombinant protein accumulation in transgenic seeds

Level of control	Enhancement strategy
Transcriptional	Promoter: seed-specific; regulated
	Co-expression with transcription factors
	3′UTR
	MAR sequences
	Coordinated (multicistronic) expression
	Altering seed proteome balance
Post-transcriptional	Intron-mediated enhancement
	3′ UTR (mRNA stability)
	Homology-dependent gene silencing
	Avoid mRNA destabilizing sequences
Translational	5′ UTR and AUG context
	Codon usage
Co-translational	Signal peptide replacement
Post-translational (Protein folding and stability)	Targeting: Apoplast
	ER lumen and ER-derived protein bodies
	Protein storage vacuoles
	Oil bodies
	Ubiquitin fusions
	Chemical enhancement of protein transport/stability
	Co-expression of gene encoding protein stabilizer or molecular chaperone

relies on the yields or accumulation levels of the target recombinant protein. For example, this will influence the efficiency of downstream purification. Improving expression levels in seeds has been, and will continue to be, greatly improved by dissecting information from functional genomics studies tailored to elucidate factors that control seed-specific gene expression, such as the transcription factors that activate seed gene promoters, as well as other regulatory control points. However, attempts to improve the accumulation of a recombinant protein require a ‘holistic’ view of gene expression beyond the transcriptional level, and an appreciation of the interdependent nature of the processes in the pathway leading from the gene to the functional protein, in which each step is functionally linked to the next. Improving yields of recombinant proteins in seeds by careful choice of promoters and other transcriptional control elements has been recently reviewed (Stoger et al. 2005; Stoger 2005; Lau and Sun 2009; Boothe et al. 2010) and so will not be covered here. Throughout this section, reference is made to Table 5.1, which illustrates some of the strategies that are being used to achieve high yields of recombinant proteins in seeds and in other plant-based systems.

Beyond this challenge, several proteins, including human enzymes have been shown to be stable in the mature dry and stored seed (Reggi et al. 2005; Downing et al. 2007; Boothe et al. 2010). This is illustrated well for the recombinant lysosomal enzymes, iduronidase and glucocerebrosidase expressed in Arabidopsis seeds (Downing et al. 2007; He et al. unpublished). Following transfer of the mature seeds to cool, dry storage conditions, the activities of these enzymes are largely retained. Indeed this is one of the key advantages of using seeds as the production system, that of providing a stable repository for therapeutic proteins.

5.3.1 Inducible Systems to Switch on Gene Transcription by Co-expression of Transcription Factors or Exposure to Chemicals

One mechanism that has met with some success is to co-express two transgenes in the plant host – one gene encodes a transcription factor that is capable of transactivating the specific promoter driving expression of your gene of interest (Table 5.1). For example, human lysozyme gene expression driven by the rice storage-protein (globulin) gene promoter is enhanced by almost fourfold when the chimeric gene is co-expressed in transgenic rice with a rice bZIP transcription factor, REB (*r*ice *e*ndosperm *b*ZIP) (Yang et al. 2001). The use of the "global" transcriptional factor, *A*bscisic *A*cid *I*nsensitive3 (ABI3), which controls the transcription of several abscisic acid (ABA)-regulated seed developmental genes, can be used to induce the expression of recombinant proteins in the vegetative tissues of the plant. This can be important in some cases as vegetative tissues may have relative advantages over seeds in relation to regulatory issues such as the need for a high level of containment and the feasibility of separating edible food crops destined for consumption from those destined for biopharmaceutical use (reviewed in Fox 2006). The ectopic expression of transcription factors has been used as a mechanism to trigger inducible expression in vegetative tissues of the plant host. Normally the human lysosomal enzyme α-iduronidase is very unstable in leaf tissues of transgenic tobacco, and is subject to proteolysis when synthesized in a constitutive manner. Ectopic expression of an *ABI3* gene in transgenic tobacco leaves is able to switch on high levels of recombinant α-iduronidase expression (driven by seed-specific 5′ and 3′ regulatory elements), particularly upon exposure to ABA (Fig. 5.1) (Kermode et al. 2007). This approach has also been used successfully to enhance the production of the 65-kDa glutamic acid decarboxylase isoform (GAD 65) in leaves, a major autoantigen involved in the pathophysiology of type 1 diabetes mellitus (Jayaraj et al. unpublished). As well as promoting enhanced accumulation of recombinant proteins in leaves, the same strategy is also effective in inducing expression in young seedlings; thus one can capitalize on the advantages of the use of the mature dry stored seeds as a stable repository of recombinant proteins. Once the recombinant protein is required for purification, the seeds co-expressing the transgene of interest and the *ABI3* gene can be germinated and grown in the presence of ABA; the young

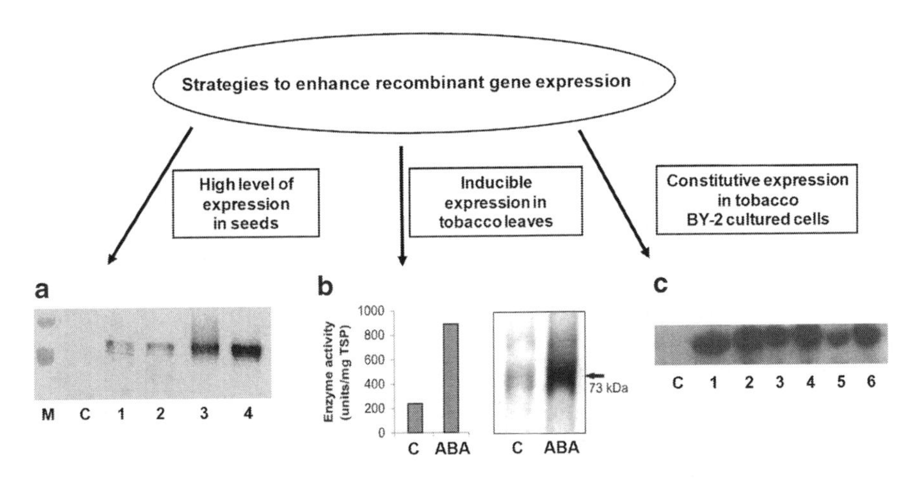

Fig. 5.1 Some strategies used to enhance the synthesis of human recombinant biopharmaceutical proteins in plant and seed hosts using the human lysosomal enzyme, α-L-iduronidase as an example. High-level accumulation is achieved in Arabidopsis (*cgl*) seeds with the use of the arcelin gene regulatory sequences (Downing et al. 2006, 2007), in leaves and other vegetative tissues using co-expression of an ectopically expressed *ABI3* gene (Kermode et al. 2007), and in tobacco BY-2 cultured cells using the CaMV 35S ('constitutive') promoter (Fu et al. 2009)

ABA-treated seedlings can then serve as the materials to purify the recombinant protein whose production has been elicited.

Soybean seeds have been used as a model system to chemically induce transcription of recombinant gene expression (Semenyuk et al. 2010). The underlying mechanism behind the 'genetic switch' involves the intracellular formation of a sensitive, stable ligand-receptor complex. This system was tested utilizing a chimeric synthetic transcription factor *VGE* adapted for use in soybean seed by linking it to the glycinin G1 promoter to drive its expression. The modular transcriptional activator consists of three parts – the Herpes simplex VP16 activation domain (*V*), the yeast Gal4 binding domain (*G*) and the ligand-binding domain of the ecdysone receptor of spruce budworm (*E*). In the absence of ligand the VGE is inactive as a gene transcription activator, but upon interaction with the chemical inducer, a nonsteroidal ecdysone analog, the tripartite transcription factor VGE binds to a specific synthetic promoter upstream of the gene of interest resulting in the expression of that gene. A gene construct encoding an ER-targeted (GFP-KDEL) reporter protein is activated using this chemical induction system in cultured somatic and zygotic embryos of transgenic soybean plants, as well as *in planta* in seeds generated from the transgenic plants grown in the greenhouse. The efficiency of induction of GFP expression by the inducer was strongly influenced by the developmental stage of the seed; the formation of ER-derived GFP-containing protein bodies in the storage parenchyma cells of the seed was correlated with the level of induced expression (Semenyuk et al. 2010).

5.3.2 *Altering the 'Proteome Balance' of Seeds to Improve Recombinant Protein Levels*

An interesting strategy with potential for enhancing the expression of any transgene in seeds is the use of a post-transcriptional gene silencing mechanism to down-regulate a major seed protein gene. This has been attempted in seeds of soybean and Arabidopsis; the suppression of a major seed storage protein leads to a 'rebalancing of the proteome', and can be exploited to produce an enhanced level of a recombinant protein. In Arabidopsis, expression of a transgene (the intact *arcelin5-I* gene of *Phaseolus vulgaris*) is enhanced when an antisense gene encoding an endogenous seed storage protein, 2S albumin, is co-expressed (Goossens et al. 1999b). In control transgenic plants, the Arc5-I protein accumulates to levels of up to 15% TSP; the population of co-transformed plants expressing the antisense 2S gene construct, exhibits higher accumulation of the arcelin protein (up to 24% TSP). Presumably, down-regulation of the 2S albumin storage protein gene made the translational machinery more available to arcelin mRNAs, which competed more efficiently than did other seed transcripts. In soybean seeds, suppression of the synthesis of the α-/α'- subunit of β-conglycinin leads to a corresponding increase in the synthesis and accumulation of glycinin storage protein, some of which is sequestered as proglycinin into *de novo* ER-derived protein bodies. Expression of a chimeric GFP-KDEL construct flanked by 5′ and 3′ regulatory regions of the glycinin gene leads to a fusion protein capable of forming ER-derived protein bodies in soybean seeds; when introgressed into the β-conglycinin-suppression background, there is a four-fold increase in the accumulation of the GFP-KDEL fusion protein which accumulates in ER-derived protein bodies along with glycinin to >7% of the total seed protein (Schmidt and Herman 2008).

5.3.3 *Enhancement of Transcript Stability, Translation, and Co-translational Processes, and the Synergy of the Regulatory Pathway from Gene to Protein*

Among the various strategies at the post-transcriptional level (Table 5.1), these include the use of: (1) an optimal 3′ UTR (3′ flanking region) to enhance mRNA stability, and avoid mRNA destabilizing signals; (2) introns; (3) various 5′ UTRs, either from plant genes or from plant viral genes, that lack secondary structure in the 5′ region of the mRNA and facilitate ribosome scanning for translation initiation; (4) an optimized AUG context, in which the nucleotides surrounding the AUG start codon facilitate translation initiation; (5) Optimized codons that represent host-preferred ones, or those that not limited by availability of the corresponding aminoacyl-tRNAs, for improvement of translational efficiency; and (6) Plant signal peptides (over native signal peptides) to promote efficient signal peptide cleavage during translocation of a recombinant protein into the ER lumen, a process which likely facilitates more efficient protein folding.

The avoidance of mRNA destabilizing sequences to improve the yields of recombinant proteins has been relevant for achieving improved yields of the *E. coli* LT-B antigen (Chikwamba et al. 2002). When the targets are prokaryotic and protozoan genes, a consideration of the high AU contents and presence of RNA destabilizing sequences is important for the stability of transcripts in the plant host (Lau and Sun 2009).

There are several reports of introns enhancing the expression of a reporter gene or a gene encoding a recombinant protein of pharmaceutical interest. Indeed most promoters that are active in dicots need to be modified by the addition of an intron before they will work efficiently in monocots (Stoger et al. 2005); in most cases, the intron is inserted in the untranslated region upstream of the open reading frame. For example, introns from the monocot genes *ADH1*, *SH1*, *UBI1* and *ACT1* genes improve the functioning of the CaMV 35S promoter in transgenic maize and bluegrass, and a dicot intron (*CHS A*) likewise improves expression driven by this promoter by 100-fold (Vain et al. 1996). While the intron-mediated enhancement effect may be as high as 100-fold, generally the effect is often twofold to tenfold, and there tends to be a greater effect in monocots than in dicots, although it is unclear how the enhancing effect is achieved. The most likely mechanism is through enhanced mRNA stability (although there is an exception to this); introns appear to increase steady-state levels of mRNAs without significantly affecting the rate of transcription initiation. It is also evident that while splicing is neither necessary, nor sufficient to achieve the intron-enhancing effect, the mRNAs generated by splicing are more rapidly and efficiently exported from the nucleus to the cytoplasm.

Various 5′ UTRs have been used to boost translation initiation of transgenes in both seed-based systems and transgenic plants. These include the 5′-UTR of the tobacco mosaic virus RNA (omega), the potato virus X RNA and the bean *arcelin5-I* gene of common bean (Gallie et al. 1987; Pooggin and Skryabin 1992; De Jaeger et al. 2002). Toward determining their efficacy for achieving high-level production of recombinant proteins in transgenic rice seeds, the potential of the 5′-UTRs of six seed storage-protein genes to enhance the expression levels of a GUS reporter gene driven by a glutelin (*GluC*) promoter was conducted (Liu et al. 2010). The study included the 5′ UTRs of three glutelin genes (*GluA-1, GluA-2 and GluC*), two prolamine genes (10 and 16 kDa), and the globulin gene, *Glb-1*. All of the 5′ UTRs significantly enhance the GUS expression level without altering the qualitative expression patterns of the glutelin promoter. The 5′-UTRs of *Glb-1* and *GluA-1* genes increase the expression of GUS by about 3.36- and 3.11-fold, respectively, and this is due to an increased translational efficiency of the mRNAs (Liu et al. 2010).

There is a tendency to overlook the connectivity of the regulatory steps involved in the ultimate production of an active mature protein product. Production of the recombinant human lysosomal enzyme, α-L-iduronidase in transgenic Arabidopsis *cgl* seeds serves as a good illustrator of how gene regulatory sequences can function synergistically to promote recombinant protein accumulation (Fig. 5.2) (Downing et al. 2007). This enzyme is deficient in the lysosomal storage disease known as mucopolysaccharidosis (MPS) I (see earlier). Because of the enzyme deficiency, the stepwise degradation of glycosaminoglycans is disrupted; in severely affected

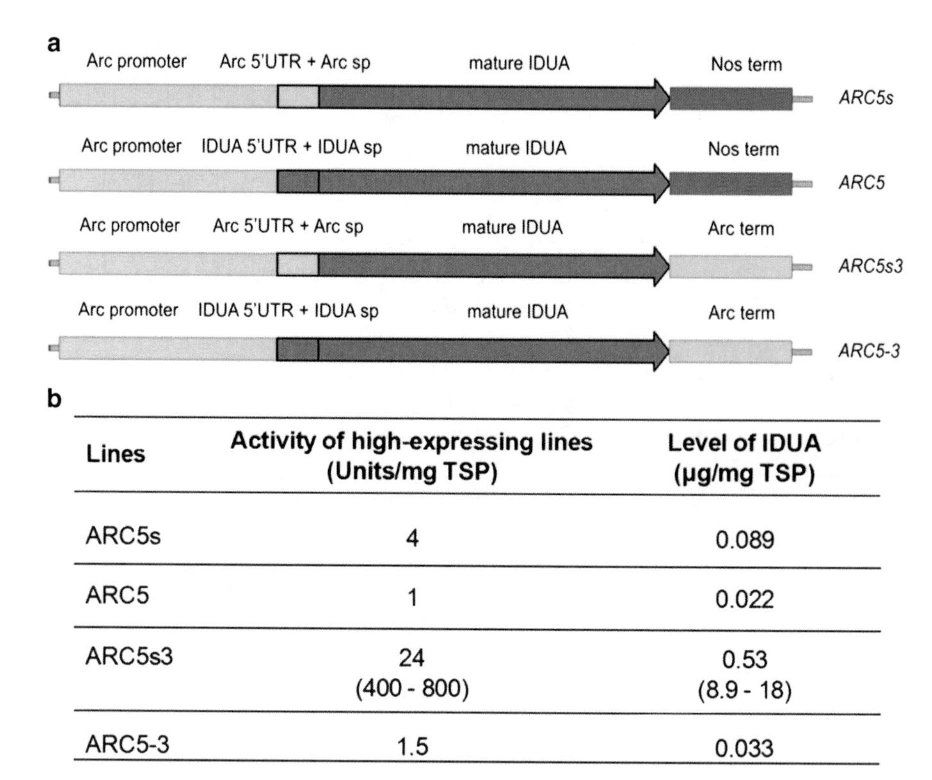

Lines	Activity of high-expressing lines (Units/mg TSP)	Level of IDUA (µg/mg TSP)
ARC5s	4	0.089
ARC5	1	0.022
ARC5s3	24 (400 - 800)	0.53 (8.9 - 18)
ARC5-3	1.5	0.033

Fig. 5.2 (**a**) Schematic diagram of constructs for expression of the gene encoding the human lysosomal enzyme, α-L-iduronidase, in *Arabidopsis cgl* mutant seeds. Gene constructs differ in the 5′-UTR-signal peptide sequences, and in the 3′-UTR-flanking sequences. (**b**) Table of α-L-iduronidase activities (*units per mg total soluble protein*) and α-L-iduronidase protein in extracts of the highest-expressing transformed lines determined from the sceening of at least 30 independent transgenic lines for each construct. The table also shows α-L-iduronidase activities of three atypical *ARC5s3* lines with extremely high levels of α-L-iduronidase gene expression (Reprinted with permission from Kermode (2006) © 2006 Canadian Science Publishing or its licensors)

humans this genetic disease leads to death in early childhood because of profound skeletal, cardiac and neurological disturbances (Clarke 2008). Figure 5.2 shows the constructs geared to express the human α-L-iduronidase gene in Arabidopsis *cgl* seeds. In constructs containing the arcelin gene promoter, exchange of the 5′-UTR and signal peptide sequence of the α-iduronidase gene with those of the arcelin gene results in a considerable increase in α-iduronidase activity (Fig. 5.2). This increase in activity is also reflected in an increased accumulation of protein and in steady-state mRNA levels. A 13-nucleotide-5′-UTR is predicted for *ARC5s* and *ARC5s3* mRNAs while the *ARC5* and *ARC5-3* transcripts, are expected to contain a 20-nucleotide 5′-UTR. Extremely short 5′-UTRs (<20 nucleotides) may inhibit the entry of the ribosomal 43S pre-initiation complex or the recognition of the AUG initiation codon (Kozak 1991; Kawaguchi and Bailey-Serres 2005). However, with respect to the arcelin gene, it is clear that its short 5′-UTR is not an impediment to translation initiation since this gene is abundantly expressed in *Phaseolus vulgaris* seeds and, as a transgene, in seeds of other species (Goossens et al. 1999a; De Jaeger et al. 2002). Furthermore, replacement of the short arcelin 5′-UTR with the 68-nucleotide-long 5′-UTR (translational enhancer) of tobacco mosaic virus (Gallie 2002), has little effect on transgenic expression (De Jaeger et al. 2002). On the other hand, a high GC content (>50%) of the 5′-UTR significantly reduces ribosome loading, while a low GC content (<33%) can promote ribosome loading (Kawaguchi and Bailey-Serres 2005). The much lower predicted GC content (38%) of the arcelin 5′-UTR as compared to the α-iduronidase gene 5′-UTR (about 60%) may improve ribosome loading and hence increase translation initiation.

The recruitment of ribosomes to the initiation codon AUG is the rate-determining step in translation, due to the mechanism by which the small ribosomal subunit scans the mRNA for the first AUG in an appropriate context (Kawaguchi and Bailey-Serres 2002); sequences immediately upstream (A_{-3}) and downstream (G_{+4}) of the initiation codon are important. If one considers an optimal translation AUG context for dicots as: aaA(A/C)aATGGCTNCC(T/A)C (Joshi et al. 1997; Sawant et al. 2001; Niimura et al. 2003), the sequences from the human α-iduronidase gene (TGGCCATGCGTCCCT), especially in the −3 and +4 positions, are far from the ideal as compared to that for the arcelin gene (TGATCATGGCTTCCTC). Indeed, the predicted AUG context of the arcelin gene best matches the canonical sequence found in highly expressed plant genes (Sawant et al. 2001), and is superior to the analogous sequences within other seed protein genes, including those encoding napin and vicilin.

There is a positive correlation between codon-usage bias and the level of gene expression (Duret and Mouchiroud 1999), which likely reflects the composition of the pool of tRNAs available when a gene is expressed. During translation, rare codons could deplete the corresponding tRNAs and lead to ribosome pausing; destabilization of transcripts and ultimately a lowering of translation efficiency may occur (Gustafsson et al. 2004). In some plant hosts, such as tobacco, the presence of rare codons does not appear to affect mRNA stability (van Hoof and Green 1997); however, rare codons may have an effect if they are located just downstream of the start codon or if they occur frequently throughout the transcript. Some investigators have modified codons to optimize the expression of human and microbial genes in plant hosts (e.g. Huang et al. 2002). This is not necessary for high-level expression the human α-iduronidase gene (Downing et al. 2006, 2007). The isolation of three *ARC5s3* lines with extremely high levels of α-iduronidase gene expression implies that the GC-rich sequence encoding the mature α-iduronidase protein does not impede the efficiency of protein synthesis, and may even be an advantage.

The precision and efficiency of processing of the signal peptide during co-translational translocation of the polypeptide into the ER may influence the level of protein accumulation in a plant host (Kermode 1996). The native signal peptides of animal/human proteins can be correctly removed in plant host cells (Gomord and Faye 2004); however in some cases, the efficiency of co-translational cleavage may be greater with plant-gene-derived signal peptides. The arcelin signal peptide is correctly removed from pro-α-iduronidase derived from an *ARC5s3* line (Downing et al. 2006); it was not determined whether this is true for the native α-iduronidase signal peptide in Arabidopsis seeds. Indeed the two constructs that include the arcelin signal peptide (and 5′ UTR) yield greater protein accumulation levels (Downing et al. 2007). Incorrect or inefficient cleavage of the innate (i.e. human-protein) signal peptide could cause improper folding of the nascent protein, and result in protein degradation mediated by the ER quality control system. There is an increased expression of recombinant thrombomodulin in tobacco leaves when the innate signal peptide is replaced by heterologous signal peptides, one of which is plant-derived (Schinkel et al. 2005). Likewise, cleavage of an N-terminal plant-protein-derived signal peptide from human serum albumin occurs in the correct manner in transgenic potato and tobacco plants; the native signal peptide is only partially processed (Sijmons et al. 1990).

Replacement of the nopaline synthase (*nos*) gene 3′ end with the arcelin gene 3′ end (3′ UTR and flanking sequences) results in a significant increase in α-iduronidase activity. However, this increase is highest when other gene regulatory sequences from the arcelin gene are present (5′-UTR and signal peptide), which appear to act synergistically to enhance expression at the protein and activity level. The arcelin gene 3′-end may exert its effects at the transcriptional level due to the presence of a putative MAR (matrix attachment region) motif that would work in concert with the MAR sequence in the 5′ flanking region of the arcelin promoter (Goossens et al. 1999a; De Jaeger et al. 2002). For some plant genes, the 3′-UTR region appears to be important in controlling transcript stability; however, effects on translation due to an interaction between the 3′ poly A tail and 5′ cap of the transcript may also be operative. In addition, some examples of quantitative regulation of transgenic expression by 3′-UTR sequences have been reported (Ali and Taylor 2001; Richter et al. 2000). The extremely high accumulation of α-iduronidase (activity and protein) in three *ARC5s3* lines may be due to several post-transcriptional factors as well as fortuitous transgene integration into transcriptionally active sites of the Arabidopsis genome (Downing et al. 2007).

The results overall suggest that post-transcriptional events have a considerable impact on the level of accumulation of human α-iduronidase. In this system, one of the bottlenecks is likely at the translation initiation step; changes to enhance this also increase the protein accumulation levels. Post-translational modulation of expression, e.g. enhanced protein stability, because of a greater efficiency of signal peptide cleavage, and therefore protein folding, cannot be ruled out. The identification of some very high expressing lines implies that the endomembrane system of developing seeds has a high capacity to properly process, fold and secrete active α-iduronidase.

5.3.4 Subcellular Targeting and Other Post-Translational Approaches to Enhance the Stable Accumulation of Recombinant Proteins in Seeds

Various post-translational processes have been used and can be used to attempt to enhance processing and stability of biopharmaceutical proteins in transgenic plants (Table 5.1). The specific targeting of recombinant proteins within plant hosts can not only determine their function, stability and correct post-translational modification (e.g. proper folding and assembly, disulphide bond formation, and protection from proteases), but can play a multitude of functions including minimizing deleterious effects of the foreign protein on the transgenic plant host, and facilitating downstream processing, including purification. Although exceptions exist, the types of post-translational modifications that many human recombinant proteins undergo in the native host (e.g. N-linked glycosylation), require that they are likewise synthesized and transported along the secretory pathway of the heterologous host. There are several examples of recombinant proteins that are stably accumulated in transgenic seeds when targeted to an apoplastic location, i.e. secreted by adding an N-terminal signal peptide. However, in some instances, manipulating the subcellular localization and deposition site of the recombinant protein has a significant impact on yield. Thus positive outcomes have come from incorporating specific additional targeting motifs, or by fusing the recombinant protein to a very stable partner protein.

5.3.4.1 Targeting of Recombinant Proteins to the ER Membrane

Although not attempted in seeds, the use of a tail anchor domain has been used to enhance the yields of a human immunodeficiency virus protein Nef (negative factor), a promising target for the development of an antiviral vaccine. This antiviral protein is a cytosolic protein; it accumulates to only low levels in the cytosol of transgenic tobacco plants and is very unstable when introduced into the secretory pathway (Barbante et al. 2008). Due to an enhancement of the stability of this protein, its accumulation is very much improved by targeting it to the cytosolic face of the ER membrane using a C-terminal domain of a mammalian ER cytochrome b5, a long-lived, tail-anchored protein. An engineered thrombin cleavage site allows for the Nef protein to be conveniently removed *in vitro* from its tail-anchor (Barbante et al. 2008).

5.3.4.2 Targeting of Recombinant Proteins to the ER Lumen Using KDEL or Similar Motifs

There are several instances, particularly when the expression system is a vegetative tissue of the transgenic plant, in which the addition of a C-terminal ER retention motif (KDEL/HDEL) improves the level of accumulation of the protein of interest. This is the case for human interleukin-4 synthesized in transgenic tobacco leaves and potato tubers (Ma et al. 2005) and for intact immunoglobulins and single chain Fv fragments (reviewed in Twyman et al. 2003; De Muynck et al. 2010). Two forms of the human lysosomal enzyme α-L-iduronidase – one with, and one without a C-terminal ER-retention sequence SEKDEL, have been produced in seeds of both *Brassica napus* and tobacco (*Nicotiana tabacum*) using regulatory sequences from the *arcelin 5-I* gene (Galpin et al. 2010). The C-terminal modification of the human enzyme has little effect on its stability or activity; the specific activities of the purified human enzymes from seeds of both plant species are consistently between 32,000 and 40,000 nmol product min^{-1} mg $protein^{-1}$. Thus, importantly, the addition of this targeting motif does not abolish the biological activity of the recombinant enzyme. The yield-enhancing effects of the KDEL sequence are not specific to vegetative tissues; this targeting motif also has positive effects on improving yields of recombinant proteins in seeds, including vaccine antigens (Moravec et al. 2007), and a human chimeric GAD-67/65 mutant in Arabidopsis seeds, which reaches 7.7% of total soluble protein (Morandini et al. 2011).

It is noteworthy that confinement of proteins within the ER lumen is not guaranteed by the addition of the C-terminal HDEL/KDEL motif, and localization can depend on the status of protein assembly and can even differ in different host species and tissues (reviewed in Stoger et al. 2005). In some cases the predicted localization patterns are particularly unusual in cereals, indicating that more basic knowledge concerning pathways of protein trafficking in cereals and other hosts is required. KDEL-tagged recombinant human serum albumin synthesized with an N-terminal mammalian signal peptide, is targeted to the ER lumen in leaf cells, but is deposited in storage protein aggregates within the vacuole of wheat endosperm cells (Arcalis et al. 2004). It was surmised that the recombinant protein was transported along the same route as the bulk of the endogenous glutelins and prolamins, which in wheat endosperm cells aggregate and bud off from the ER, but later become incorporated into the vacuole by an autophagy-like process.

A KDEL-tagged recombinant antibody fragment, synthesized in rice endoperm cells, with a signal peptide ends up mainly in protein bodies and to some extent in protein storage vacuoles (Torres et al. 2001). When the different subunits of secretory IgA are synthesized in rice endosperm cells, the subcellular locale depends on the assembled status of the subunit proteins (Nicholson et al. 2005). More specifically, unassembled proteins (light chain, heavy chain and secretory component) accumulate predominantly within ER-derived protein bodies, while the assembled antibody, with antigen-binding capability, accumulates specifically in protein storage vacuoles.

5.3.4.3 Use of Proteins and Protein Motifs Sufficient for Stable Accumulation of Fusion Proteins in ER-Derived Protein Bodies

Recently, strategies have been developed to sequester recombinant proteins in ER-protein bodies as a means of improving their accumulation levels (production yields), and at the same time assist in their subsequent purification (reviewed in Floss et al. 2010; Conley et al. 2011). The unique intrinsic physico-chemical properties of the fusion partners appear to exploit general ER mechanisms that insulate the recombinant proteins, segregating them from both the secretory and the degradative (vacuolar or ERAD) pathways (Vitale and Boston 2008). Elastin-like polypeptides are thermally responsive biopolymers composed of a repeating pentapeptide 'VPGXG' sequence; to date, most studies in plant hosts have used between 25 and

125 multimers of this pentapeptide appended to the recombinant protein of interest. Some of their useful characteristics include an ability to undergo a thermally responsive reversible phase transition, which is retained when the elastin or elastin-like protein is fused to a recombinant protein. The temperature-dependent, reversible self-aggregation (referred to as an inverse phase transition), occurs within a 2–3°C temperature range, and can be monitored with a spectrophotometer, due to the resultant increase in the turbidity of the protein solution. Thus, below the transition temperature, elastin proteins are monomeric and soluble; at temperatures above the transition temperature, the proteins aggregate and become insoluble. Because of this property, some of the potential applications of these proteins include drug delivery (e.g. precise targeting of cytotoxic agents to tumour cells), formation of new molecules with unique biomechanical and chemical properties (e.g. silk-elastin like copolymer), tissue engineering or repair, and most relevant to this review, increasing the yields of recombinant proteins as well as providing a convenient mechanism for recombinant protein purification (reviewed in Floss et al. 2010; Conley et al. 2011). As noted in these recent reviews, some of the therapeutic targets of this strategy have included interleukin-10, murine interleukin-4, erythropoietin, single chain antibody fragments, and a full-size anti-human immunodeficiency virus-neutralizing antibody, primarily using tobacco leaves and transient systems based on *Nicotiana benthamiana* as hosts (see references therein). A systematic analysis characterizing the accumulation of GFP-elastin-like-protein fusions in relation to subcellular targeting (e.g. to the cytosol, chloroplasts, apoplast and ER), reveals that only protein targeting to, and retention in, the ER (via a C-terminal KDEL motif) leads to increased accumulation of the recombinant protein as a fusion (Conley et al. 2009). In this case, the fusion proteins, transiently expressed in *N. benthamiana* plants, induce the formation of novel protein bodies, which are similar in size and characteristics to the ER-derived protein bodies that accumulate prolamin storage proteins in the endosperm of seeds of rice and maize. The stable deposition of the fusion proteins in these organelles is very likely the reason for the positive effect of the elastin-like protein on recombinant protein accumulation – i.e. by sequestering the heterologous protein away from normal physiological turnover and perhaps the various regulated protein degradation pathways of the cell (e.g. ERAD and other pathways).

Although primarily used for expression of recombinant proteins in whole transgenic leaves or plants, there are reports of this strategy enhancing the yields of recombinant proteins in seeds. Addition of an elastin-like protein (ELP) to a single-chain Fv antibody fragment (scFv) leads to a 40-fold increase in the accumulation of the fusion protein, in which the levels approach 25% of the total soluble seed protein in transgenic tobacco; further, the antigen-binding properties of the antibody fragments are not compromised (Scheller et al. 2006). A recombinant 'ELPylated' [(Val-Pro-Gly-Xaa-Gly)$_{100}$] antibody with neutralizing activity against HIV-1 accumulates to a greater extent that the non-ELPylated control when expressed in transgenic tobacco seeds (Floss et al. 2009). The N-glycan profiles are congruent with the fusion protein residing primarily in putative ER-derived protein bodies, while the control ("free") recombinant antibody accumulates primarily in protein storage vacuoles. Importantly the protein is stable in the mature dry and stored seeds.

Equally important, regardless of the purification method, the addition of the elastin-like protein does not affect the binding affinity of the antibody, but does have a slightly negative effect on the HIV neutralizing capacity of the light chain (Floss et al. 2009, 2010). The size, orientation and peptide sequence or composition of the elastin-like protein all influence the efficiency of recombinant protein recovery from plant extracts (Conley et al. 2009, 2011). Manipulating the size of the elastin-like protein, fine-tuning the process for purification of the recombinant protein, and the use of self-cleaving elastin-like protein-intein tags (Banki et al. 2005; Fong et al. 2009) are among some of the improvements that can be made to this strategy, and may increase the biological activities of the target recombinant protein, and improve the efficiency of purification and recovery (Floss et al. 2009, 2010).

In a similar vein, a domain of the maize seed storage protein γ-zein, can induce the formation of protein storage bodies, thus facilitating the recovery of fused proteins using density-based separation methods (reviewed in Conley et al. 2011). γ-Zein is a prolamin-type storage protein that accumulates (along with α- and β- zeins) in the endosperm storage tissues of the maize seed during grain maturation. The γ-zein storage proteins have unique biophysical properties that in part lead to their retention in the ER and assembly into ER-derived protein bodies: a highly repetitive proline-rich sequence $(PPPVHL)_8$ and a Pro-X motif; the former is able to adopt an amphipathic helical conformation which can self-assemble, and may participate in a biophysically-mediated ER-retention process (Kogan et al. 2001; Geli et al. 1994). When fused to a recombinant protein of interest, the 112 amino acid-N-terminal proline-rich domain of γ-zein (the so-called 'Zera' sequence) leads to the stable accumulation of the fusion protein in ER-derived protein bodies of transgenic plant leaves and seeds (Torrent et al. 2009a, b). γ-Zein fusions (i.e. involving the 'Zera' sequence) have been used to facilitate the accumulation of various recombinant therapeutic proteins in leaves of transgenic tobacco; human growth hormone is increased by ~13-fold (to a max. of 3.2 g/kg fresh weight), and epidermal growth factor is increased by ~100-fold (to a max. of 0.5 g/kg fresh weight) (Torrent et al. 2009a, b). Perhaps surprisingly, the γ-zein domain also functions to stably sequester recombinant proteins in ER-derived organelles in several eukaryotic systems, not just plant cells, including insect cells, fungal cells, CHO cells, and other mammalian-cell systems (Torrent et al. 2009a). Capitalizing on some of the positive attributes conferred by ER-PB sequestration in established eukaryotic host systems for recombinant protein production may create unique opportunities for improving manufacturing efficiencies (Torrent et al. 2009a). Using *N. benthamiana* as a host, and a cyan fluorescent marker protein, a detailed characterization of the proline-rich domains of γ-zein reveals that two Cys residues (Cys7 and Cys9) of the proline-rich domain are critical for oligomerization, which is the first step towards protein body formation; the hydrophobic central region appears to facilitate lateral protein-protein interactions to allow for alignment of the proline-rich domain ('hydrophobic packing'), an arrangement also stabilized by intermolecular disulphide binding (Llop-Tous et al. 2010). The 8 units of PPPVHL in *Zera* appear to provide the optimal length for self-assembly.

A recombinant "fusion" storage protein ('zeolin') comprised of a portion of maize γ-zein and the bean storage protein phaseolin accumulates to high amounts in leaves of transgenic tobacco and forms ER-protein bodies ('zeolin protein bodies'). An association of the 'zeolin' protein bodies with BiP via an ATP-sensitive mechanism may partly account for the improved and stable deposition of the chimeric protein, which accumulates to 3.5% total soluble protein (Mainieri et al. 2004). A comparative analysis of γ-zein- versus 'zeolin'- fusions has been undertaken in relation to improving the yields of human immunodeficiency virus antigen Nef, a recombinant protein that is prone to instability (de Virgilio et al. 2008). The 'zeolin'-Nef fusion protein accumulates in small protein bodies to a maximum of 1.5% TSP in tobacco plants; in contrast, the γ-zein-Nef fusion is degraded by ER quality control, perhaps because it is recognized as structurally defective in contrast to the 'zeolin'-Nef fusion. Similar to the transgenic tissues of dicots, in transgenic rice, a prolamin-GFP fusion does not require seed-specific factors for aggregation; its interaction with BiP results in the formation of ER-protein bodies regardless of the tissue – i.e., in seeds, roots and leaves (Saito et al. 2009).

Yet another class of proteins with similar advantages is the hydrophobins, small fungal proteins capable of altering the hydrophobicity of their respective fusion partner. Part of the hydrophobin contains an exposed hydrophobic patch, and 8 conserved Cys residues that form 4 intramolecular disulphide bridges (Hakanpaa et al. 2004); the various properties of the hydrophobins make them prone to self-assembly into an amphipathic protein membrane at hydrophilic-hydrophobic interfaces (Wang et al. 2005). These small proteins are able to facilitate efficient purification by surfactant-based aqueous two-phase systems; this includes the recovery of hydrophobin-fusions from plant cell extracts (Joensuu et al. 2010; reviewed in Conley et al. 2011). Increased yields of recombinant proteins as hydrophobin fusions have been reported in the leaves of transgenic plants (e.g. in the transient expression system using *N. benthamina*) (Joensuu et al. 2010); as with γ-zein- and elastin-fusion proteins, the hydrophobin fusion proteins accumulate in ER-protein bodies. This stable accumulation appears to be beneficial on several levels. For example, under some circumstances non-chimeric ER-targeted proteins (e.g. those engineered to contain a C-terminal H/KDEL) accumulate to high levels; the result is a triggering of toxic responses in the ER (Joensuu et al. 2010).

Although some testing of the biological activities of target recombinant proteins (as fusions with elastin-like proteins, γ-zein, or hydrophobins) has begun, more extensive testing will likely be necessary to establish the feasibility of these fusion technologies for clinical or industrial applications (reviewed in Conley et al. 2011).

5.3.4.4 Targeting of Recombinant Proteins to Protein Storage Vacuoles

In some instances, targeting of human recombinant proteins to the protein storage vacuole of seeds appears to a suitable locale for their accumulation. The human insulin-like growth factor binding protein-3 is able to negatively regulate cell proliferation and induce apoptosis. The targeting of this recombinant protein to protein storage vacuoles of transgenic tobacco seeds, achieved by the use of a phaseolin signal peptide and C-terminal tetrapeptide AFVY, yields a high level of protein accumulation (800 lg/g dry weight) (Cheung et al. 2009). Targeting of human pro-insulin to the protein storage vacuoles of transgenic soybean seeds yields stable accumulation, in which the mature dry seeds stored at room temperature still retain high levels of the protein even after 7 years (Cunha et al. 2010).

As with ER lumen localization, the subcellular locale of a targeted protein is not always predictable (reviewed in Stoger et al. 2005). This appears to be true for expression in both cereal and dicot hosts. For example, when recombinant proteins are synthesized with a signal peptide and no additional targeting information, secretion to the apoplast is not consistently the outcome. This may be particularly true for expression of recombinant proteins in cereal endosperm cells, but may also occur in dicot hosts. For example human lysozyme synthesized in rice endosperm cells with an N-terminal signal peptide is accumulated in protein storage vacuoles (Yang et al. 2003); the same appears to be true for recombinant proteins synthesized in wheat endosperm cells (Arcalis et al. 2004). The human lysosomal enzyme glucocerebrosidase of transgenic tobacco seeds is targeted to the protein storage vacuole by the soybean basic 7S globulin signal peptide (Reggi et al. 2005).

There is a species- or tissue-dependent functionality of some protein sorting signals (Vitale and Hinz 2005); further, a reporter glycoprotein expressed in the endosperm of maize seeds reveals that there may be a shift in its intracellular trafficking route during seed development (Arcalis et al. 2010). A fungal phytase expressed in rice, is readily detected in the apoplastic environment of leaf tissues, but is retained in ER protein bodies and protein storage vacuoles in the seed endosperm (Drakakaki et al. 2006). As noted earlier, although the vacuole may be a suitable site for stable accumulation of recombinant proteins in seeds, often in plant vegetative tissues, proteins are quite unstable when targeted to this locale. For example, accumulation levels of the spider dragline silk protein (DP1B) expressed in Arabidopsis (Yang et al. 2005) show the impact of vacuolar targeting on the stability and yield of a recombinant protein. Although this silk protein is found at levels reaching 8% of TSP in seed storage vacuoles, it is undetectable in leaf cell vacuoles. In a similar manner, targeting of the same protein to the apoplast provides good yields in leaves, but poor yields in seeds (Yang et al. 2005). Thus, there is the need to conduct an empirical case-by-case assessment for each recombinant protein of subcellular targeting within the context of the target tissue for expression (Benchabane et al. 2008).

Other protein storage vacuole sorting pathways are being exploited in seeds. For example a target recombinant protein can be modified to contain the BP-80 transmembrane domain, and the cytosolic tail of α-TIP (tonoplast intrinsic protein); these act as a membrane anchor and a vacuolar sorting signal respectively. Their appendage onto a target recombinant protein results in its localization to a sub-compartment of the protein storage vacuole for stable accumulation (Jiang et al. 2000; Jiang and Sun 2002). This has been accomplished for the lysosomal enzyme glucocerebrosidase, which is targeted to the protein storage vacuoles of Arabidopsis seeds. However the levels of the recombinant enzyme are very low (0.02% of total soluble seed protein) (He et al. unpublished).

5.3.4.5 Targeting of Recombinant Proteins to Oil Bodies by Oleosin Tethering

When targeted to the oil bodies of transgenic *Arabidopsis* seeds as an oleosin-insulin fusion protein, recombinant human insulin precursor accumulates to significant levels (0.13% total seed protein) and can be enzymatically treated *in vitro* to generate a product with a mass identical to that of the predicted product (Nykiforuk et al. 2006). Human growth hormone has also been produced in oilseeds using an oil-body-targeting approach in which the range of protein accumulation is ~0.44–1.58% of total seed protein; the yields of this recombinant protein are significantly improved by oil-body-targeting, in contrast to its apoplastic accumulation, in which the protein accumulates to only 0.28% of total seed protein (reviewed in Boothe et al. 2010). A liquid-liquid separation leads to a marked enrichment of the recombinant protein due to the removal of more than 90% of host cell proteins; at this point in the recovery process all of the oleosin fusion proteins are still part of the oil body particle; treatment of the purified oil bodies with trypsin to cleave a trypsin-sensitive site between the oleosin and the recombinant protein fusion partner, enables a convenient recovery of the recombinant protein. Typically the partially purified extracts require some further downstream approach to purify the protein to homogeneity (Boothe et al. 2010).

A modification of the oleosin-tethering strategy involves adding an N-terminal fusion partner to the recombinant protein of interest to provide for the *in vivo* targeting or post-extraction capture of the fusion protein on seed oil bodies, thereby simplifying and reducing the cost of downstream purification (Van Roojen and Moloney 1995). This has been attempted with insulin in which the fusion partner is an anti-oleosin single chain antibody (ScFv) (reviewed in Boothe et al. 2010).

5.4 Controlling N-Glycosylation of Recombinant Proteins in Seeds

Manipulation of the post-translational processing machinery of plant cells is a challenging area. Human proteins that are of pharmaceutical importance undergo several complex post-translational processes in order to be functional (e.g. disulphide bond formation, proteolytic processing, N- or O-linked glycosylation and other modifications). Although glycosylation is only one of these processes, it is of prime importance to ensure that the protein's glycans are similar to those on the authentic protein. New strategies have emerged to control N-linked glycosylation in plants and seeds, particularly in relation to controlling N-glycan processing. However, significant challenges remain in this area. Elucidating the functions of N-glycan heterogeneity and microheterogeneity, especially with respect to protein function, stability and transport, are poorly understood and this represents an important area of cell biology. In addition, there are lines of evidence trickling in that N-glycosylation patterns are dependent upon the developmental stage of the plant or seed, and are further influenced by tissue-specific, host-specific and environmental factors.

It is perhaps important to note that these challenges are not unique to plant-based production systems. For example even in the system most commonly used to generate human protein therapeutics – Chinese hamster ovary (CHO) cells – the therapeutic quality can be compromised under certain conditions, for example, when the cultured cells over-express a recombinant protein. Further, the glycoforms of a recombinant protein vary depending on the culture conditions and from one cell line to another, despite great efforts toward glycoengineering different cell lines (reviewed in Hossler et al. 2009).

The central importance of glycosylation is underscored by the fact that at least one-third of approved biopharmaceuticals are glycoproteins (Walsh and Jefferis 2006); a large proportion of these are subjected to N-glycosylation (Gomord et al. 2010). While transgenic plants and seeds can effect many of the necessary post-translational modifications to proteins, the different glycosylation patterns mediated by Golgi-localized enzymes have hampered the use of plants as biofactories for production of therapeutic proteins. The differences in N-glycan processing in plant and mammalian cells occur during endomembrane transport as proteins transit through the Golgi apparatus. Within this compartment, enzymes convert the original high mannose N-glycans of proteins to complex N-glycans by a series of sequential reactions that rely on the accessibility of the glycan chain(s) to the Golgi-processing-machinery. Bisecting β1,2 xylose and core α1,3 fucose residues are assembled onto the trimmed N-glycans of plant-synthesized proteins (Fig. 5.3a and b); core α1,6 fucose residues and terminal sialic acid residues are added in mammalian cells. In addition, β1,3 galactose and fucose, α1,4 linked to the terminal GlcNAc (N-acetylglucosamine) of plant N-glycans, form the so-called Lewis^{a} (Le^{a}) oligosaccharide structure; in constrast, a β1,4 galactose residue is often combined with sialic acids in mammals (Lerouge et al. 1998). The typical kinds of N-glycans found on plant glycoproteins are shown in Fig. 5.3a. In addition, the typical N-glycans of human native antibodies, in comparison to those produced in different heterologous hosts, including plants are shown in Fig. 5.3b (Gomord et al. 2010; Beck et al. 2008).

In terms of therapeutic production is plants and seeds, the differences in N-glycan processing are reflected on at least two levels: (1) In the absence of control of N-glycan maturation, the plant or seed-made recombinant glycoprotein will contain undesired sugars (e.g. β1,2 xylose-, α1,3 fucose- and Le^{a} glycoepitopes) that render the glycoprotein immunogenic in mammals. This feature is of paramount importance, particularly for cases in which a therapeutic is to be parenterally administered (Gomord et al. 2010); (2) Recombinant therapeutic glycoproteins with high-mannose N-glycans may be susceptible to rapid clearance from the blood stream, thus compromising their efficacy. Note however, that there are cases in which the shorter half-life of a recombinant protein therapeutic may be advantageous (reviewed in Gomord et al. 2010). Further, the efficacy of some therapeutics relies upon mannose-terminated N-glycans (e.g. glucocerebrosidase). A shorter *in vivo* half-life may avoid cell toxicity or immune responses. This is important for antibodies used for passive immunotherapy, in which antibody persistence can interfere with active immunization in the patient's circulatory system. Under these circumstances

a Plant N-glycans

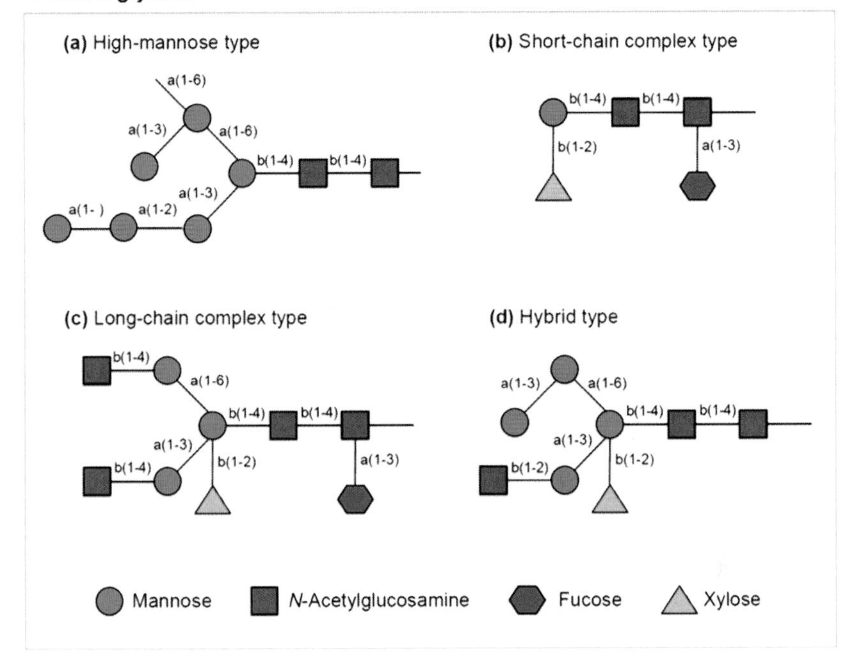

b N-glycans of human native & recombinant antibodies produced in different hosts

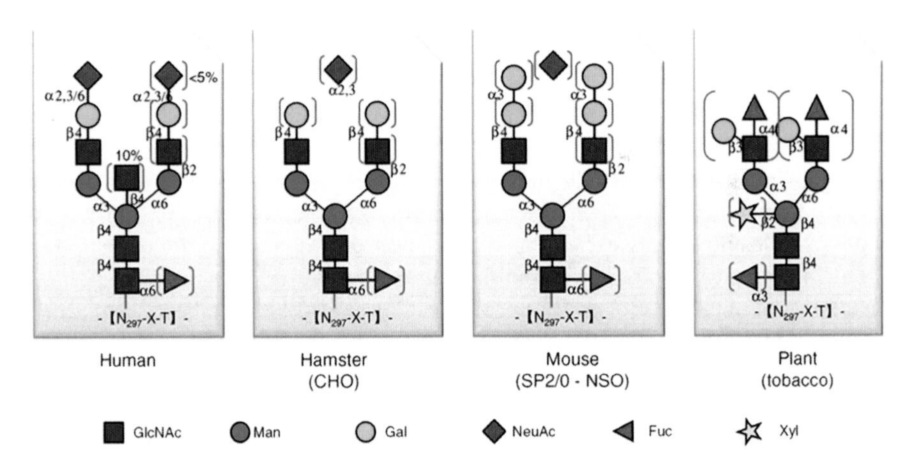

Fig. 5.3 (**a**) Examples of N-linked glycans found on plant glycoproteins. (*a*) High-mannose type. (*b*) Short-chain complex type. (*c*) Long-chain complex type. (*d*) Hybrid-type. In (*a*)-(*d*): a = α; b = β. From Twyman et al. 2003, with permission from Elsevier. (**b**) N-glycans of human native antibodies, and recombinant antibodies produced in different heterologous hosts, including CHO cells, murine cells (SP2/0 or NSO cell lines), or tobacco plants, as indicated (Adapted from Beck et al. 2008. From Gomord et al. 2010. With permission from John Wiley and Sons)

a mannose-terminated antibody may reduce this interference (Ko et al. 2003). For antibody conjugates used in cancer immunotherapy, rapid clearance from the bloodstream would reduce their nonspecific toxicity (Kogelberg et al. 2007). In the case of enzyme replacement therapies (ERTs) for lysosomal storage diseases, there are reports of ERT infusions causing immune responses to the administrated protein in patients (de Vries et al. 2010). Circulation of any cross-reacting IgG antibodies in the bloodstream may reduce the therapeutic effect by capturing newly administered enzyme, and a shorter *in vivo* half-life for some recombinant lysosomal enzymes could turn into an advantage as long as the correction of tissue pathology is not jeopardized.

The various strategies toward manipulating the N-glycan status of plant-produced recombinant glycoproteins have been directed toward two different goals or end-products – one is the production of human recombinant proteins containing exclusively high mannose- or oligomannosidic-N-glycans (–GlcNac$_2$-Man$_{5–9}$), which relies on the target glycoprotein being stable and biologically active in this form, or being amenable to downstream processing as appropriate. The second goal, is the production of human recombinant proteins containing complex N-glycans that lack xylose and fucose, but contain terminal N-glycan sugars that are characteristic of human glycoproteins, such as sialic acid and galactose.

5.4.1 Production of High-Mannose-Terminated Human Glycoproteins

5.4.1.1 Generation of Gene Knock-Outs or Knock-Downs to Avoid the Addition of Undesirable Sugar Residues

Strategies in which gene inactivation or silencing has been used to reduce or eliminate the activity of plant-specific glycosyltransferases (i.e. α1,3 fucosyltransferase and β1,2 xylosyltransferase) have been most successful in the moss *Physcomitrella patens*, in which homologous recombination can be performed (Koprivova et al. 2004), and in the aquatic plant *Lemna minor* (Cox et al. 2006). In these species the N-glycan chains of glycoproteins are devoid of α1,3 fucosyl and/or β1,2 xylosyl residues, e.g., on a human monoclonal antibody (Cox et al. 2006), and on a recombinant human vascular endothelial growth factor (Koprivova et al. 2004). This is in contrast to the incomplete efficacy of down-regulated expression of the glucosyltransferases in RNAi lines of *Medicago sativa* (alfalfa) and *Nicotiana benthamiana* (Sourrouille et al. 2008; Strasser et al. 2008). Notably the extent to which the undesired (α1,3 fucosyl and/or 1,2 xylosyl) residues are present on recombinant proteins produced in these hosts is protein-specific. This may reflect the accessibility of the N-glycan chains of the target protein to the processing enzymes, or may occur for some other reason (Gomord et al. 2010). For example, even the incomplete down-regulation of α1,3-fucosyltransferase- and β1,2 xylosyltransferase activities is sufficient to allow for the generation of IgG lacking α1,3-fucose and β1,2 xylose residues; thus the IgG-Fc and IgG antibodies may not represent the best reporter proteins to assess the overall efficiency of glycoengineering strategies (reviewed in Gomord et al. 2010).

Mutant transgenic seeds have been used to generate monoclonal antibodies with a controlled N-glycosylation pattern (Loos et al. 2011a). A comparative study of two antiviral monoclonal antibodies, one against the hepatitis A virus and another against HIV, was conducted in transgenic seeds of Arabidopsis wild-type plants and a triple glycosylation mutant lacking plant-specific N-glycan residues; both proteins were efficiently secreted. The specific glycosylation mutant lacks both plant-specific α1,3-fucosyltransferases and the β1,2 xylosyltransferase; it is referred to as 'triple knock out' ('TKO'), and exhibits a 'human-like' GnGn N-glycosylation pattern (Schähs et al. 2007). Interestingly the N-glycosylation patterns are dependent on the target protein. Both the wild-type, and the glycosylation mutant generate the anti-HIV monoclonal antibody with complex N-glycans, which are comprised of a single dominant N-glycan species, GnGnXF and GnGn, respectively. In contrast the anti-hepatitis A monoclonal antibody contains both complex N-glycans and oligo-mannosidic N-glycans. A KDEL-tagged version of the anti-HIV monoclonal antibody exhibits an ER-typical N-glycosylation pattern (i.e. primarily oligo-mannosidic N-glycans) but the recombinant protein transits to the protein storage vacuoles. A characterization of recombinant scFv antibodies in the same triple mutant reveals aberrant subcellular targeting of the proteins. One of the ScFv antibodies (anti-HIV scFv-Fc) is deposited in newly formed ER-vesicles irrespective of whether or not it contains a C-terminal KDEL sequence; it contains exclusively oligomannosidic N-glycans but is inactive with respect to its virus neutralizing capabilities (Loos et al. 2011b).

In a different approach, down-regulation of a GDP-D-mannose 4,6-dehydratase gene, encoding an enzyme associated with GDP-L-fucose biosynthesis in N. *benthamiana* plants successfully reduces the levels of core α-1,3-fucose and α-1,4-fucose residues on the N-glycans of a recombinant mouse granulocyte-macrophage colony-stimulating factor (Matsuo and Matsumura 2011).

5.4.1.2 Expression of Recombinant Proteins in Mutant Seeds Defective in Early Golgi Modifying Enzymes

There are other types of complex-glycan-deficient mutant plants or seeds (e.g. those of Arabidopsis) (Downing et al. 2006; Strasser et al. 2004; Frank et al. 2008), in which the first enzyme involved in complex glycan formation, N-acetylglucosamine transferase I, is either absent or deficient (Fig. 5.4). Some of the Golgi-modifying enzymes in plants include α-mannosidase I, N-acetylglucosaminyl transferase I, α-mannosidase II, N-acetylglucosaminyl transferase II and fucosyl-, galactosyl- and xylosyl-transferases (Lerouge et al. 1998). Fucosyl- and xylosyl-transferases require at least one terminal N-acetylglucosamine residue on the N-linked glycan (boxed GlcNAc, Fig. 5.4). Thus, a deficiency in N-acetylglucosaminyl transferase I (GnT I), as occurs in the *Arabidopsis cgl* mutant, is predicted to generate

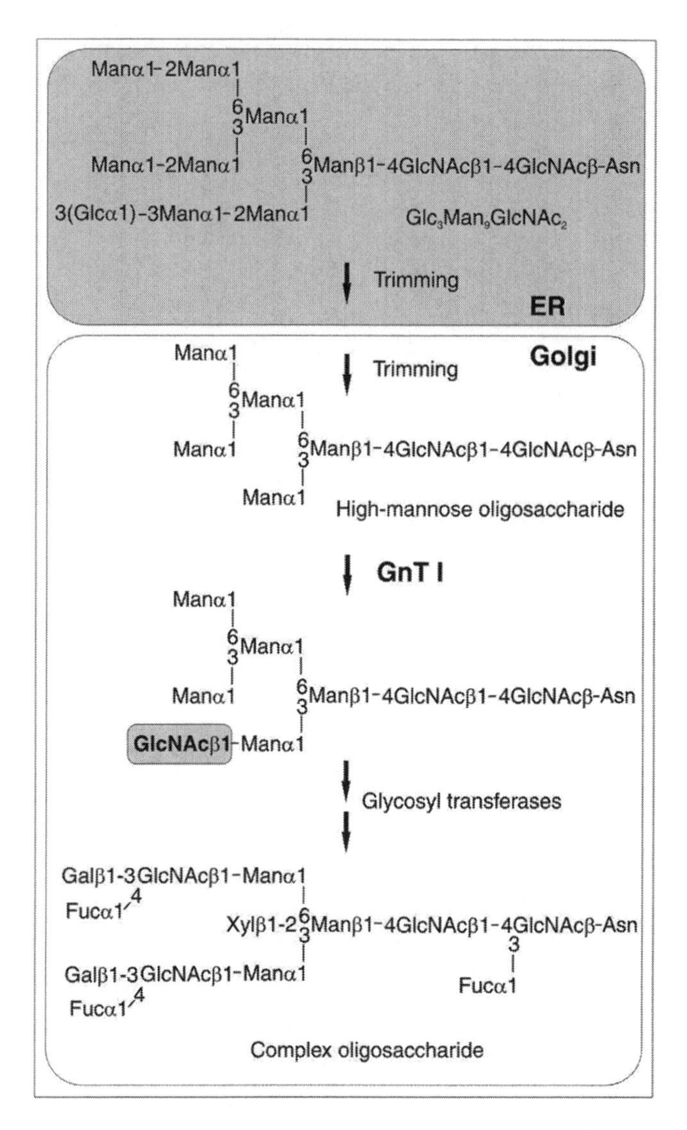

Fig. 5.4 Role of GnT I in the first step of complex N-glycan formation. In a deficient mutant, such as *cgl*, GlcNAc (N-acetylglucosamine) is not added to the trimmed N-linked glycan (*box*); thus, the various glycosyltransferases (e.g. xylosyl- and fucosyl-transferases) do not add their respective sugars (e.g. xylose and fucose, respectively). Note that for simplicity, the N-linked complex glycan shown is a plant Lewis[a]-type complex glycan. Complex glycan structures of plant-made biopharmaceuticals can show heterogeneity from one system to another; however, all plant species have the capacity to add bisecting β(1,2)-xylose and core α(1,3)-fucose residues onto the trimmed glycan (From Downing et al. 2006. With permission from John Wiley and Sons)

recombinant proteins that contain high mannose N-glycans (predominantly Man_5-$GlcNAc_2$ with small amounts of Man_6, Man_7 and Man_8) (Fig. 5.4) (von Schaewen et al. 1993); the point mutation underlying the *cgl* phenotype (referred to as *cgl* C5) has been characterized (Strasser et al. 2004). This approach has been used to synthesize active human α-iduronidase (Downing et al. 2006, 2007), in which approximately 93% of the N-glycans on the recombinant protein are in high-mannose or oligomannosidic form (He et al. unpublished). This particular Arabidopsis mutant arises because of a point mutation generating an additional site for N-glycosylation in the GnT I enzyme. This in turn creates a deficiency of the enzyme because the protein undergoes incorrect folding and is thus unstable (Frank et al. 2008). The GnT I activity can be restored, however, if transfer of the destabilizing *N*-glycan to GnT I is inhibited by tunicamycin treatment or by mutation of the oligosaccharyltransferase complex (Frank et al. 2008). The *cgl* C5 line is therefore a conditional mutant. The *cgl* C6 line, another Arabidopsis GnT I mutant (Frank et al. 2008), contains an intron splice site mutation that results in a frameshift in the protein product; this line is therefore an unconditional mutant. The conditional nature of the *cgl* C5 mutation likely explains why there are residual N-glycan xylose and fucose residues on the seed-produced recombinant human α-iduronidase and glucocerebrosidase (He et al. unpublished); for example, there may be some GnT I activity in this mutant during stages of seed development when storage protein synthesis is very active.

The *ALG3* gene of Arabidopsis is an α1,3-mannosyl transferase, which is involved in the build-up of dolichol-linked high-mannose type glycans in the ER (Henquet et al. 2008; Kajiura et al. 2010). A homozygous T-DNA insertion mutant, *alg3-2*, exhibits very low levels of transferase activity, and mostly truncated aberrant $Man_5GlcNAc_2$ N-glycans, rather than $Man_9GlcNAc_2$ N-glycans, are transferred from the dolichol intermediate to the Asn residue(s) of nascent glycoproteins. Thus, ER resident glycoproteins of the mutant plants are almost uniformly modified by an abnormal $Man_5GlcNAc_2$ glycan (Henquet et al. 2008). The N-glycan heterogeneity of ER resident glycoproteins is also reduced in seeds. Recently a KDEL-tagged scFv-Fc was expressed in seeds of wild-type and *alg3-2* plants (Henquet et al. 2010). The N-glycans on these antibodies from mutant seeds were predominantly comprised of $Man_5GlcNAc_2$, as compared to $Man_8GlcNAc_2$ and $Man_7GlcNAc_2$ isoforms on the recombinant protein produced in wild-type seeds; antigen-binding was not adversely affected by the aberrant N-glycan composition. Interestingly the proportion of both recombinant and endogenous proteins that were underglycosylated was more prevalent in the mutant than in seeds of wild-type plants (Henquet et al. 2010).

5.4.1.3 Avoiding Golgi Transport via Manipulation of Subcellular Targeting

The manipulation of subcellular targeting within the endomembrane system, in which transit of recombinant proteins through the Golgi complex is prevented, represents another strategy that has been used to manipulate N-glycan structures on recombinant proteins. One example is the use of HDEL/KDEL sequences (or extended versions of these sequences) for ER retention (Tekoah et al. 2004; Ko et al. 2003; Sriraman et al. 2004; Petruccelli et al. 2006; reviewed in Gomord et al. 2010). Plant-derived monoclonal antibodies synthesized with a C-terminal KDEL, do not contain any of the known antigenic glycan epitopes found on mammalian-derived monoclonal antibodies or on plant-derived recombinant proteins when no mechanism for control of glycosylation is used. However, in some cases, a significant proportion of complex (i.e. matured) N-glycans are still detected on KDEL-tagged antibodies, either as a result of escape from the normal retrieval mechanism, or possibly due to some degree of proteolytic cleavage of the KDEL tag (Tekoah et al. 2004; Gomord et al. 2010). A KDEL-tagged anti-HIV antibody expressed in the endosperm of transgenic maize seeds contains some xylose and fucose on its N-glycans; unexpectedly, most of the N-glycans contained a single GlcNac residue. This may be indicative of extensive trimming by unknown glucosidases (Rademacher et al. 2008), or may be due to a mass-spectrometry-mediated artifact (Karnoup et al. 2007; Karg and Kallio 2009). An ER retention SEKDEL sequence is effective in significantly diminishing the amounts of N-glycan xylose and fucose residues of the recombinant lysosomal enzyme α-L-iduronidase produced in seeds of transgenic tobacco and *Brassica napus* (Galpin et al. 2010). Thus, the proportion of synthesized recombinant proteins in which their constituent N-linked glycans undergo maturation (and hence, the addition of potentially immunogenic sugars) is reduced, but not completely eliminated by the presence of the ER retention motif. Notably, HDEL/KDEL sequences on the carboxy-termini of recombinant proteins do not necessarily guarantee their ER retention in plant cells via the retrieval mechanism (see earlier), and recombinant proteins may accumulate in post-ER compartments in which N-glycan maturation and further processing can take place. Aberrant subcellular trafficking of KDEL-tagged recombinant proteins in seeds, e.g. in which the recombinant protein is accumulated in the protein storage vacuole has been noted earlier; this is often reflected in the protein's constituent N-glycan sugars.

Unique trans-membrane domain and cytoplasmic tail sequences can be used as anchors for delivering recombinant proteins via distinct vesicular transport pathways to specific vacuolar compartments in dicot seeds, allowing for their stable accumulation (Jiang and Sun 2002) (see earlier). The mechanism bypasses protein transit through the Golgi complex and thus Golgi-specific N-glycan maturation can be avoided. Characterization of the N-glycan structures of recombinant glucocerebrosidase targeted to the protein storage vacuole in this manner is underway.

5.4.2 Production of Recombinant Proteins with 'Humanized' (Complex) N-Glycans

Considerable efforts have gone toward the production of recombinant proteins with galactosylated and sialylated N-glycans as most circulating human glycoproteins contain N-glycans that are capped by neuraminic acid on penultimate β1,4 galactose residues.

5.4.2.1 In Vitro Modification of Plant-Derived Recombinant Proteins

When both terminal GlcNac residues are present on a plant-produced recombinant protein, the N-glycans can be further modified by *in vitro* galactosylation (e.g. using purified human β1,4-galactosyltransferase) to produce N-glycans with galactose residues (reviewed in Karg and Kallio 2009). *In vitro* sialylation of plant-derived galactosylated recombinant proteins has also been demonstrated (Misaki et al. 2003).

5.4.2.2 Heterologous Expression of Human β1,4-Galactosyltransferase

This strategy has been used in transgenic plants to produce recombinant proteins (e.g. antibodies) with galactose-extended glycans. β1,4-galactosyltransferase is the first glycosyltransferase of mammalian cells that initiates the further branching of complex N-linked glycans after the action of N-acetylglucosaminyltransferases I and II (Palacpac et al. 1999). This strategy successfully generates galactosylated glycoproteins in tobacco suspension-culture BY2 cells and tobacco plants (Palacpac et al. 1999; Bakker et al. 2001), but the process is somewhat inefficient. Tobacco BY suspension cells engineered in this manner generate glycoproteins that possess glycans that react with *Ricinus communis* agglutinin 120 (specific for β1,4-linked galactose), but do not react with an antibody specific for complex glycans containing β1,2-xylose residues and have no detectable α1,3-fucose residues (determined by HPLC and IS-MS/MS). The efficiency of the human β1,4-galactosyltransferase is greatly improved by fusing its catalytic domain to the Golgi-localization domain of a plant glucosyltransferase (Bakker et al. 2006; Vezina et al. 2009; Gomord et al. 2010).

5.4.2.3 Heterologous Expression of Human Genes to Produce Sialylated N-Glycans on Human Glycoproteins

Recently, six genes of the mammalian sialic acid biosynthetic pathway (including a CMP-N-acetylneuraminic acid synthetase, and a CMP-sialic acid transporter) were expressed alongside a recombinant monoclonal antibody in *N. benthamiana* to yield the sialylated product (Castilho et al. 2010). This demonstrates the utility of *N. benthamiana* as an ideal expression system for rapid proof-of-principle studies involving engineering of the plant N-glycan biosynthetic pathway.

5.4.2.4 Heterologous Expression of GnT III

In most mammalian cells, the enzyme β-1,4-N-acetylglucosaminyltransferase III (GnT III) catalyzes the addition of a bisecting GlcNAc to the β-mannose of N-linked glycans, creating a bisected complex or hybrid oligosaccharide. A strategy that has been used to shift the pool of N-glycans toward hybrid structures, and further to reduce the addition of plant-specific core xylose and fucose residues is the expression of a heterologous *GnT III* gene in plant cells. Toward this end the catalytically active portion of rat GnT III was expressed in transgenic tobacco as a chimeric protein containing the Golgi-localization domains of an *A. thaliana* mannosidase II (Frey et al. 2009). MALDI-TOF MS analysis (Karg et al. 2009) revealed a strong reduction in the level of xylosylation and fucosylation of plant N-glycans. The fraction of sugars devoid of xylose and fucose drastically rose to ~60% in the transgenic plants as compared to the ~13% characteristic of wild-type plants; the pool of N-glycan structures was shifted toward hybrid N-glycan structures from complex N-glycans.

5.5 Conclusions

These are exciting times in the field of plant molecular pharming. The first plant-based therapeutic for a lysosomal storage disease intended for parenteral administration is being pursued for commercialization through the joint efforts of Protalix and Pfizer. This is a carrot-suspension-culture-derived recombinant glucocerebrosidase enzyme for treatment of the lysosomal storage disease, Gaucher disease (Shaaltiel et al. 2007). Presently, additional testing of this plant-based therapeutic is being sought by the FDA before it can be approved; it was it given fast-track status by the FDA due to Genzyme's CHO-cell-culture production problems in 2009–2010. As well as improving the yields of recombinant proteins in plant- and seed-based systems, manipulation of posttranslational processing in plant cells to produce recombinant proteins that are similar or identical to the native protein and are bioactive, stable and functional still represents a critical challenge. The functions of N-glycan heterogeneity and microheterogeneity, especially with respect to protein function, stability and transport, are poorly understood in all eukaryotic systems. The problems associated with plant-specific N-glycan maturation, and the manipulation of O-linked glycosylation, and other forms of protein processing are not trivial, despite recent progress in these exciting areas. The development of *in vitro* methods for specific downstream processing of recombinant proteins is awaited with interest, as are improvements in the efficiency of purification of recombinant proteins, and in methods to generate native mature recombinant proteins with no foreign amino acid motifs. These further developments will help to bring to fruition the full emergence of plant- and seed-based systems as suitable, viable alternatives to the current systems for large-scale production of biopharmaceuticals.

Acknowledgement This work was supported in part by a Natural Sciences and Engineering Research Council (NSERC) Strategic grant and by a Michael Smith Foundation for Health Research Senior Scholar Award. Dedicated to the loving memory of Virginia.

References

Ali S, Taylor WC (2001) The 3′ non-coding region of a C_4 photosynthesis gene increases transgene expression when combined with heterologous promoters. Plant Mol Biol 46:325–333

Allison M (2010) As Genzyme flounders, competitors and activist investors swoop in. Nat Biotechnol 28:3–4

Arcalis E, Marcel S, Altmann F, Kolarich D, Drakakaki G, Fischer R, Christou P, Stoger E (2004) Unexpected deposition patterns of recombinant proteins in post-endoplasmic reticulum compartments of wheat endosperm. Plant Physiol 136:3457–3466

Arcalis E, Stadlmann J, Marcel S, Drakakaki G, Winter V, Rodriguez J, Fischer R, Altmann F, Stoger E (2010) The changing fate of a secretory glycoprotein in developing maize endosperm. Plant Physiol 153:693–702

Bakker H, Bardor M, Molthoff JW, Gomord V, Elbers I, Stevens LH, Jordi W, Lommen A, Faye L, Lerouge P, Bosch D (2001) Galactose-extended glycans of antibodies produced by transgenic plants. Proc Natl Acad Sci USA 98:2899–2904

Bakker H, Rouwendal GA, Karnoup AS, Florack DEA, Stoopen GM, Helsper JPF, Van Ree R, Van Die I, Bosch D (2006) An antibody produced in tobacco expressing a hybrid beta-1,4-galactosyltransferase is essentially devoid of plant carbohydrate epitopes. Proc Natl Acad Sci USA 103:7577–7582

Banki MR, Feng L, Wood DW (2005) Simple bioseparations using self-cleaving elastin-like polypeptide tags. Nat Methods 2:659–661

Barbante A, Irons S, Hawes C, Frigerio L, Vitale A, Pedrazzini E (2008) Anchorage to the cytosolic face of the endoplasmic reticulum membrane: a new strategy to stabilize a cytosolic recombinant antigen in plants. Plant Biotechnol J 6:560–575

Beck A, Wagner-Rousset E, Bussat MC, Lokteff M, Klinguer-Hamour C, Haeuw JF, Goetsch L, Van Dorsselaer A, Corvaïa N (2008) Trends in glycosylation, glycoanalysis and glycoengineering of therapeutic antibodies and Fc-fusion proteins. Curr Pharm Biotechnol 9:482–501

Benchabane M, Goulet C, Rivard D, Faye L, Gomord V, Michaud D (2008) Preventing unintended proteolysis in plant protein biofactories. Plant Biotechnol J 6:633–648

Boothe J, Nykiforuk C, Shen Y, Zaplachinski S, Szarka S, Kuhlman P, Murray E, Morck D, Moloney MM (2010) Seed-based expression systems for plant molecular farming. Plant Biotechnol J 8:588–606

Castilho A, Strasser R, Stadlmann J, Grass J, Jez J, Gattinger P, Kunert R, Quendler H, Pabst M, Leonard R, Altmann F, Steinkellner H (2010) *In planta* protein sialylation through overexpression of the respective mammalian pathway. J Biol Chem 285:15923–15930

Cheung SC, Sun SS, Chan JC, Tong PC (2009) Expression and subcellular targeting of human insulin-like growth factor binding protein-3 in transgenic tobacco. Transgenic Res 18: 943–951

Chikwamba R, McMurray J, Shou H, Frame B, Pegg SE, Scott P, Mason H, Wang K (2002) Expression of a synthetic E. coli heat-labile enterotoxin B sub-unit (LT-B) in maize. Mol Breed 10:253–265

Clarke LA (2008) The mucopolysaccharidoses: a success of molecular medicine. Expert Rev Mol Med 10:e 1

Conley AJ, Joensuu JJ, Menassa R, Brandle JE (2009) Induction of protein body formation in plant leaves by elastin-like polypeptide fusions. BMC Biol 7:48

Conley AJ, Joensuu JJ, Richman A, Menassa R (2011) Protein body-inducing fusions for high-level production and purification of recombinant proteins in plants. Plant Biotechnol J 9: 419–433

Cox KM, Sterling JD, Regan JT, Gasdaska JR, Frantz KK, Peele CG, Black A, Passmore D, Moldovan-Loomis C, Srinivasan M, Cuison S, Cardarelli PM, Dickey LF (2006) Glycan optimization of a human monoclonal antibody in the aquatic plant *Lemna minor*. Nat Biotechnol 24:1591–1597

Cunha NB, Araújo ACG, Leite A, Murad AM, Vianna GR, Rech EL (2010) Correct targeting of proinsulin in protein storage vacuoles of transgenic soybean seeds. Genet Mol Res 9: 1163–1170

De Jaeger G, Scheffer S, Jacobs A, Zambre M, Zobell O, Goossens A, Depicker A, Angenon G (2002) Boosting heterologous protein production in transgenic dicotyledonous seeds using *Phaseolus vulgaris* regulatory sequences. Nat Biotechnol 20:1265–1268

De Muynck B, Navarre C, Boutry M (2010) Production of antibodies in plants: status after twenty years. Plant Biotechnol J 8:529–563

de Virgilio M, De Marchis F, Bellucci M, Mainieri D, Rossi M, Benvenuto E, Arcioni S, Vitale A (2008) The human immunodeficiency virus antigen Nef forms protein bodies in leaves of transgenic tobacco when fused to zeolin. J Exp Bot 59:2815–2829

de Vries JM, van der Beek NAME, Kroos MA, Özkan L, van Doorn PA, Richards SM, Sung CCC, Brugma J-D C, Zandbergen AAM, van der Ploeg AT, Reuser AJJ (2010) High antibody titer in an adult with Pompe disease affects treatment with alglucosidase alfa. Mol Genet Metab 101:338–345

Desnick RJ, Schuchman EH (2002) Enzyme replacement and enhancement therapies: lessons from lysosomal disorders. Nat Rev Genet 3:954–966

Downing WL, Galpin JD, Clemens S, Lauzon SM, Samuels AL, Pidkowich MS, Clarke A, Kermode AR (2006) Synthesis of enzymatically active human alpha-L-iduronidase in *Arabidopsis* cgl (complex glycan-deficient) seeds. Plant Biotechnol J 4:169–181

Downing WL, Hu X, Kermode AR (2007) Post-transcriptional factors are important for high-level expression of the human α-L-iduronidase gene in Arabidopsis *cgl* (*complex-glycan-deficient*) seeds. Plant Sci 172:327–334

Drakakaki G, Marcel S, Arcalis E, Altmann F, Gonzalez-Melendi P, Fischer R, Christou P, Stoger E (2006) The intracellular fate of a recombinant protein is tissue dependent. Plant Physiol 141: 578–586

Duret L, Mouchiroud D (1999) Expression pattern and, surprisingly, gene length shape codon usage in *Caenorhabditis*, *Drosophila* and *Arabidopsis*. Proc Natl Acad Sci USA 96: 4482–4487

Faye L, Gomord V (2010) Success stories in molecular farming – a brief overview. Plant Biotechnol J 8:525–528

Fischer R, Stoger E, Schillberg S, Christou P, Twyman RM (2004) Plant-based production of biopharmaceuticals. Curr Opin Plant Biol 7:152–158

Floss DM, Sack M, Arcalis E, Stadlmann J, Quendler H, Rademacher T, Stoger E, Scheller J, Fischer R, Conrad U (2009) Influence of elastin-like peptide fusions on the quantity and quality of a tobacco-derived human immunodeficiency virus-neutralizing antibody. Plant Biotechnol J 7:899–913

Floss DM, Schallau K, Rose-John S, Conrad U, Scheller J (2010) Elastin-like polypeptides revolutionize recombinant protein expression and their biomedical application. Trends Biotechnol 28:37–45

Fong BA, Wu W-Y, Wood DW (2009) Optimization of ELP-intein mediated protein purification by salt substitution. Protein Expr Purif 66:198–202

Fox JL (2006) Turning plants into protein factories. Nat Biotechnol 24:1191–1193

Frank J, Kaulfürst-Soboll H, Rips S, Koiwa H, von Schaewen A (2008) Comparative analyses of Arabidopsis complex glycan1 mutants and genetic interaction with staurosporin and temperature sensitive3a. Plant Physiol 148:1354–1367

Frey AD, Karg SR, Kallio PT (2009) Expression of rat beta(1,4)-N-acetylglucosaminyltransferase III in *Nicotiana tabacum* remodels the plant-specific *N*-glycosylation. Plant Biotechnol J 6:33–48

Fu LH, Miao Y, Lo SW, Seto TC, Sun SSM, Xu Z-F, Clemens S, Clarke LA, Kermode AR, Jiang L (2009) Production and characterization of soluble human lysosomal enzyme α-iduronidase with high activity from culture media of transgenic tobacco BY-2 cells. Plant Sci 177: 668–675

Gallie DR (2002) The 5′-leader of tobacco mosaic virus promotes translation through enhanced recruitment of eIF4F. Nucleic Acids Res 30:3401–3411

Gallie DR, Sleat DE, Watts JW, Turner PC, Wilson TM (1987) A comparison of eukaryotic viral 5′-leader sequences as enhancers of mRNA expression in vivo. Nucleic Acids Res 15: 8693–8711

Galpin JD, Clemens S, Kermode AR (2010) The carboxy-terminal ER-retention motif, SEKDEL, influences the N-linked glycosylation of recombinant human α-L-iduronidase but has little effect on enzyme activity in seeds of *Brassica napus* and *Nicotiana tabacum*. Plant Sci 178: 440–447

Geli MI, Torrent M, Ludevid D (1994) Two structural domains mediate two sequential events in [Gamma]-zein targeting: protein endoplasmic reticulum retention and protein body formation. Plant Cell 6:1911–1922

Giddings G, Allison G, Brooks D, Carter A (2000) Transgenic plants as factories for biopharmaceuticals. Nat Biotechnol 18:1151–1155

Gomord V, Faye L (2004) Post-translational modification of therapeutic proteins in plants. Curr Opin Plant Biol 7:171–181

Gomord V, Fischette A-C, Menu-Bouaouiche L, Saint-Jore-Dupas C, Plasson C, Michaud D, Faye L (2010) Plant-specific glycosylation patterns in the context of therapeutic protein production. Plant Biotechnol J 8:564–587

Goossens A, Dillen W, De Clercq J, Van Montagu M, Angenon G (1999a) The *arcelin-5* gene of *Phaseolus vulgaris* directs high seed-specific expression in transgenic *Phaseolus acutifolius* and Arabidopsis plants. Plant Physiol 20:1095–1104

Goossens A, Van Montagu M, Angenon G (1999b) Co-introduction of an antisense gene for an endogenous seed storage protein can increase expression of a transgene in *Arabidopsis thaliana* seeds. FEBS Lett 456:160–164

Gustafsson C, Govindarajan S, Minshull J (2004) Codon bias and heterologous protein expression. Trends Biotechnol 22:346–353

Hakanpää J, Paananen A, Askolin S, Nakari-Setälä T, Parkkinen T, Penttilä M, Linder MB, Rouvinen J (2004) Atomic resolution structure of the hfbii hydrophobin, a self-assembling amphiphile. J Biol Chem 279:534–539

Henquet M, Lehle L, Schreuder M, Rouwendal G, Molthoff J, Helsper J, Van Der Krol S, Bosch D (2008) Identification of the gene encoding the alpha1,3-mannosyltransferase (ALG3) in *Arabidopsis* and characterization of downstream *N*-glycan processing. Plant Cell 20:1652–1664

Henquet M, Eigenhuijsen J, Hesselink T, Spiegel H, Schreuder M, van Duijn E, Cordewener J, Depicker A, van der Krol A, Bosch D (2010) Characterization of the single-chain Fv-Fc antibody MBP10 produced in Arabidopsis *alg3* mutant seeds. Transgenic Res. doi:10.1007/s11248-010-9475-5

Hossler P, Khattak SF, Li ZJ (2009) Optimal and consistent protein glycosylation in mammalian cell culture. Glycobiology 19:936–949

Huang J, Wu L, Yalda D, Adkins Y, Kelleher SL, Crane M, Lonnerdal B, Rodriguez RL, Huang N (2002) Expression of functional recombinant human lysozyme in transgenic rice culture. Transgenic Res 11:229–239

Jiang L, Sun SS (2002) Membrane anchors for vacuolar targeting: application in plant bioreactors. Trends Biotechnol 20:99–102

Jiang L, Phillips TE, Rogers SW, Rogers JC (2000) Biogenesis of the protein storage vacuole crystalloid. J Cell Biol 150:755–770

Joensuu JJ, Conley AJ, Lienemann M, Brandle JE, Linder MB, Menassa R (2010) Hydrophobin fusions for high-level transient protein expression and purification in *Nicotiana Benthamiana*. Plant Physiol 152:622–633

Joshi CP, Zhou H, Huang X, Chiang VL (1997) Context sequences of translation initiation codon in plants. Plant Mol Biol 35:993–1001

Kajiura H, Seki T, Fujiyama K (2010) *Arabidopsis thaliana ALG3* mutant synthesizes immature oligosaccharides in the ER and accumulates unique N-glycans. Glycobiology 20:736–751

Karg SR, Kallio PT (2009) The production of biopharmaceuticals in plant systems. Biotechnol Adv 27:879–894

Karg SR, Frey AD, Ferrara C, Streich DK, Umaña P, Kallio PT (2009) A small-scale method for the preparation of plant N-linked glycans from soluble proteins for analysis by MALDI-TOF mass spectrometry. Plant Physiol Biochem 47:160–166

Karnoup AS, Kuppannan K, Young SA (2007) A novel HPLC-UV-MS method for quantitative analysis of protein glycosylation. J Chromatogr B Anal Technol Biomed Life Sci 859:178–191

Kawaguchi R, Bailey-Serres J (2002) Regulation of translation initiation in plants. Curr Opin Plant Biol 5:460–465

Kawaguchi R, Bailey-Serres J (2005) mRNA sequence features that contribute to translational regulation in Arabidopsis. Nucleic Acids Res 33:955–965

Kermode AR (1996) Mechanisms of intracellular protein transport and targeting. Crit Rev Plant Sci 15:285–423

Kermode AR (2006) Plants as factories for production of biopharmaceutical and bioindustrial proteins: lessons from cell biology. Can J Bot 84:679–694

Kermode AR, Zeng Y, Hu X, Lauson S, Abrams SR, He X (2007) Ectopic expression of a conifer *Abscisic Acid Insensitive3* transcription factor induces high-level synthesis of recombinant human α-L-iduronidase in transgenic tobacco leaves. Plant Mol Biol 63:763–776

Ko K, Tekoah Y, Rudd PM, Harvey DJ, Dwek RA, Spitsin S, Hanlon CA, Rupprecht C, Dietzschold B, Golovkin M, Koprowski H (2003) Function and glycosylation of plant-derived antiviral monoclonal antibody. Proc Natl Acad Sci USA 100:8013–8018

Kogan MJ, Dalcol I, Gorostiza P, Lopez-Iglesias C, Pons M, Sanz F, Ludevid D, Giralt E (2001) Self-assembly of the amphipathic helix $(VHLPPP)_8$. A mechanism for zein protein body formation. J Mol Biol 312:907–913

Kogelberg H, Tolner B, Sharma SK, Lowdell MW, Qureshi U, Robson M, Hillyer T, Pedley RB, Vervecken W, Contreras R, Begent RH, Chester KA (2007) Clearance mechanism of a mannosylated antibody-enzyme fusion protein used in experimental cancer therapy. Glycobiology 17:36–45, Erratum in: Glycobiol 17: 1030

Koprivova A, Stemmer C, Altmann F, Hoffmann A, Kopriva S, Gorr G, Reski R, Decker EL (2004) Targeted knockouts of *Physcomitrella* lacking plant-specific immunogenic N-glycans. Plant Biotechnol J 2:517–523

Kozak M (1991) A short leader sequence impairs the fidelity of initiation by eukaryotic ribosomes. Gene Expr 1:111–115

Lau OS, Sun SSM (2009) Plant seeds as bioreactors for recombinant protein production. Biotechnol Adv 27:1015–1022

Lerouge P, Cabanes-Macheteau M, Rayon C, Fischette-Lainé A-C, Gomord V, Faye L (1998) *N*-glycoprotein biosynthesis in plants: recent developments and future trends. Plant Mol Biol 38:31–48

Liu WX, Liu HL, Chai ZJ, Xu XP, Song YR, Qu LQ (2010) Evaluation of seed storage-protein gene 5′ untranslated regions in enhancing gene expression in transgenic rice seed. Theor Appl Genet 121:1267–1274

Llop-Tous I, Madurga S, Giralt E, Marzabal P, Torrent M, Ludevid MD (2010) Relevant elements of a maize gamma-zein domain involved in protein body biogenesis. J Biol Chem 285: 35633–35644

Loos A, Van Droogenbroeck B, Hillmer S, Grass J, Kunert R, Cao J, Robinson DG, Depicker A, Steinkellner H (2011a) Production of monoclonal antibodies with a controlled *N*-glycosylation pattern in seeds of *Arabidopsis thaliana*. Plant Biotechnol J 9:179–192

Loos A, Van Droogenbroeck B, Hillmer S, Grass J, Pabst M, Castilho A, Kunert R, Liang M, Arcalis E, Robinson DG, Depicker A, Steinkellner H (2011b) Expression of antibody fragments with a controlled *N*-glycosylation pattern and induction of endoplasmic reticulum-derived vesicles in seeds of *Arabidopsis thaliana*. Plant Physiol 155:2036–2048

Ma JK-C, Drake PMW, Christou P (2003) The production of recombinant pharmaceutical proteins in plants. Nat Rev Genet 4:794–805

Ma S, Huang Y, Davis A, Yin Z, Mi Q, Menassa R, Brandle JE, Jevnikar AM (2005) Production of biologically active human interleukin-4 in transgenic tobacco and potato. Plant Biotechnol J 3:309–318

Mainieri D, Rossi M, Archinti M, Bellucci M, De Marchis F, Vavassori S, Pompa A, Arcioni S, Vitale A (2004) Zeolin. A new recombinant storage protein constructed using maize γ-zein and bean phaseolin. Plant Physiol 136:3447–3456

Matsuo K, Matsumura T (2011) Deletion of fucose residues in plant N-glycans by repression of the GDP-mannose 4,6-dehydratase gene using virus-induced gene silencing and RNA interference. Plant Biotechnol J 9:264–281

Misaki R, Kimura Y, Palacpac NQ, Yoshida S, Fujiyama K, Seki T (2003) Plant cultured cells expressing human beta-(1,4)-galactosyltransferase secrete glycoproteins with galactose-extended N-linked glycans. Glycobiology 13:199–205

Morandini F, Avesani L, Bortesi L, Van Droogenbroeck B, De Wilde K, Arcalis E, Bazzoni F, Santi L, Brozzetti A, Falorni A, Stoger E, Depicker A, Pezzotti M (2011) Non-food/feed seeds as biofactories for the high-yield production of recombinant pharmaceuticals. Plant Biotechnol J 1–11. doi:10.1111/j.1467-7652.2011.00605.x

Moravec T, Schmidt MA, Herman EM, Woodford-Thomas T (2007) Production of *Escherichia coli* heat labile toxin (LT) B subunit in soybean seed and analysis of its immunogenicity as an oral vaccine. Vaccine 25:1647–1657

Nicholson L, Gonzales-Melendi P, van Dolleweerd C, Tuck H, Perrin Y, Ma JK-C, Fischer R, Christou P, Stoger E (2005) A recombinant multimeric immunoglobin expressed in rice shows assembly-dependent subcellular localization in endosperm cells. Plant Biotechnol J 3:115–127

Niimura Y, Terabe M, Gojobori T, Miura K (2003) Comparative analysis of the base biases at the gene terminal portions in seven eukaryote genomes. Nucleic Acids Res 31:5195–5201

Nykiforuk CL, Boothe JG, Murray EW, Keon RG, Goren HJ, Markley NA, Moloney MM (2006) Transgenic expression and recovery of biologically active recombinant human insulin from *Arabidopsis thaliana* seeds. Plant Biotechnol J 4:77–85

Palacpac NQ, Yoshida S, Sakai H, Kimura Y, Fujiyama K, Yoshida T, Seki T (1999) Stable expression of human beta1,4-galactosyltransferase in plant cells modifies N-linked glycosylation patterns. Proc Natl Acad Sci USA 96:4692–4697

Petruccelli S, Otegui MS, Lareu F, Tran Dinh O, Fitchette A-C, Circosta A, Rumbo M, Bardor M, Carcamo R, Gomord V, Beachy RN (2006) A KDEL-tagged monoclonal antibody is efficiently retained in the endoplasmic reticulum in leaves, but is both partially secreted and sorted to protein storage vacuoles in seeds. Plant Biotechnol J 4:511–527

Pooggin MM, Skryabin KG (1992) The 5′-untranslated leader sequence of potato virus X RNA enhances the expression of a heterologous gene in vivo. Mol Gen Genet 234:329–331

Rademacher T, Sack M, Arcalis E, Stadlmann J, Balzer S, Altmann F, Quendler H, Stiegler G, Kunert R, Fischer R, Stoger E (2008) Recombinant antibody 2G12 produced in maize endosperm efficiently neutralizes HIV-1 and contains predominantly single-GlcNAc N-gycans. Plant Biotechnol J 6:189–201

Reggi S, Marchetti S, Patti T, De Amicis F, Cariati R, Bembi B, Fogher C (2005) Recombinant human acid β-glucosidase stored in tobacco seed is stable, active and taken up by fibroblasts. Plant Mol Biol 57:101–113

Richter LJ, Thanavala Y, Arntzen CJ, Mason HS (2000) Production of hepatitis B surface antigen in transgenic plants for oral immunization. Nat Biotechnol 18:1167–1171

Saito Y, Kishida K, Takata K, Takahashi H, Shimada T, Tanaka K, Morita S, Satoh S, Masumura T (2009) A green fluorescent protein fused to rice prolamin forms protein body-like structures in transgenic rice. J Exp Bot 60:615–627

Sawant SV, Kiran K, Singh PK, Tuli R (2001) Sequence architecture downstream of the initiator codon enhances gene expression and protein stability in plants. Plant Physiol 126:1630–1636

Schähs M, Strasser R, Stadlmann J, Kunert R, Rademacher T, Steinkellner H (2007) Production of a monoclonal antibody in plants with a humanized N-glycosylation pattern. Plant Biotechnol J 5:657–663

Scheller J, Leps M, Conrad U (2006) Forcing single-chain variable fragment production in tobacco seeds by fusion to elastin-like polypeptides. Plant Biotechnol J 4:243–249

Schinkel H, Schiermeyer A, Soeur R, Fischer R, Schillberg S (2005) Production of an active recombinant thrombomodulin derivative in transgenic tobacco plants and suspension cells. Transgenic Res 14:251–259

Schmidt MA, Herman EM (2008) Proteome rebalancing in soybean seeds can be exploited to enhance foreign protein accumulation. Plant Biotechnol J 6:832–842

Semenyuk EG, Schmidt MA, Beachy RN, Moravec T, Woodford-Thomas T (2010) Adaptation of an ecdysone-based genetic switch for transgene expression in soybean seeds. Transgenic Res 19:987–999

Shaaltiel Y, Bartfeld D, Hashmueli S, Baum G, Brill-Almon E, Galili G, Dym O, Boldin-Adamsky SA, Silman I, Sussman JL, Futerman AH, Aviezer D (2007) Production of glucocerebrosidase with terminal mannose glycans for enzyme replacement therapy of Gaucher's disease using a plant cell system. Plant Biotechnol J 5:579–590

Sijmons PC, Dekker BM, Schrammeijer B, Verwoerd TC, van den Elzen PJ, Hoekema A (1990) Production of correctly processed human serum albumin in transgenic plants. Biotechnology 8:217–221

Sourrouille C, Marquet-Blouin E, D'Aoust MA, Kiefer-Meyer M-C, Séveno M, Pagny-Salehabadi S, Bardor M, Durambur G, Lerouge P, Vezina L, Gomord V (2008) Down-regulated expression of plant-specific glycoepitopes in alfalfa. Plant Biotechnol J 6:702–721

Sriraman R, Bardor M, Sack M, Vaquero C, Faye L, Fischer R, Finnern R, Lerouge P (2004) Recombinant anti-hCG antibodies retained in the endoplasmic reticulum of transformed plants lack core-xylose and core-alpha(1,3)-fucose residues. Plant Biotechnol J 2:279–287

Stoger E, Ma JK-C, Fischer R, Christou P (2005) Sowing the seeds of success: pharmaceutical proteins from plants. Curr Opin Biotechnol 16:167–173

Strasser R, Altmann F, Mach L, Glössl J, Stadlmann J, Steinkellner H (2004) Generation of *Arabidopsis thaliana* plants with complex N-glycans lacking beta1,2-linked xylose and core alpha1,3-linked fucose. FEBS Lett 561:132–136

Strasser R, Stadlmann J, Schähs M, Stiegler G, Quendler H, Mach L, Glössl J, Weterings K, Pabst M, Steinkellner H (2008) Generation of glyco-engineered *Nicotiana benthamiana* for the production of monoclonal antibodies with a homogeneous human-like *N*-glycan structure. Plant Biotechnol J 6:392–402

Tekoah Y, Ko K, Koprowski H, Harvey DJ, Wormald MR, Dwek RA, Rudd PM (2004) Controlled glycosylation of therapeutic antibodies in plants. Arch Biochem Biophys 426:266–278

Torrent M, Llompart B, Lasserre-Ramassamy S, Llop-Tous I, Bastida M, Marzabal P, Westerholm-Parvinen A, Saloheimo M, Heifetz PB, Ludevid MD (2009a) Eukaryotic protein production in designed storage organelles. BMC Biol 7:5

Torrent M, Llop-Tous I, Ludevid MD (2009b) Protein body induction: a new tool to produce and recover recombinant proteins in plants. Methods Mol Biol 483:193–208

Torres E, Gonzales-Melendi P, Stoger E, Shaw P, Twyman RM, Nicholson L, Vaquero C, Fischer R, Christou P, Perrin Y (2001) Native and artificial reticuloplasmins co-accumulate in distinct domains of the endoplasmic reticulum and in post-endoplasmic reticulum compartments. Plant Physiol 127:1212–1223

Twyman RM, Stoger E, Schillberg S, Christou P, Fischer R (2003) Molecular farming in plants: host systems and expression technology. Trends Biotechnol 21:570–578

Vain P, Finer KR, Engler DE, Pratt RC, Finer JJ (1996) Intron-mediated enhancement of gene expression in maize (*Zea Mays* L.) and bluegrass (*Poa pratensis* L.). Plant Cell Rep 15: 489–494

Van Hoof A, Green PJ (1997) Rare codons are not sufficient to destabilize a reporter gene transcript in tobacco. Plant Mol Biol 35:383–387

van Rooijen GJ, Moloney MM (1995) Plant seed oil-bodies as carriers for foreign proteins. Biotechnology (NY) 13:72–77

Vézina LP, Faye L, Lerouge P, D'Aoust MA, Marquet-Blouin E, Burel C, Lavoie PO, Bardor M, Gomord V (2009) Transient co-expression for fast and high-yield production of antibodies with human-like *N*-glycans in plants. Plant Biotechnol J 7:442–455

Vitale A, Boston RS (2008) Endoplasmic reticulum quality control and the unfolded protein response: insights from plants. Traffic 9:1581–1588

Vitale A, Hinz G (2005) Sorting of proteins to storage vacuoles: how many mechanisms? Trends Plant Sci 10:316–323

von Schaewen A, Sturm A, O'Neill J, Chrispeels MJ (1993) Isolation of a mutant *Arabidopsis* that lacks *N*-acetyl glucosaminyl transferase I and is unable to synthesize Golgi-mediated complex N-linked glycans. Plant Physiol 102:1109–1118

Walsh G, Jefferis R (2006) Post-translational modifications in the context of therapeutic proteins. Nat Biotechnol 24:1241–1252

Wang X, Shi F, Wösten HA, Hektor H, Poolman B, Robillard GT (2005) The SC3 hydrophobin self-assembles into a membrane with distinct mass transfer properties. Biophys J 88: 3434–3443

Werber Y (2004) Lysosomal storage diseases market. Nat Rev 3:9–10

Yang D, Wu L, Hwang Y-S, Chen L, Huang N (2001) Expression of the REB transcriptional activator in rice grains improves the yield of recombinant proteins whose genes are controlled by a *Reb*-responsive promoter. Proc Natl Acad Sci USA 98:11438–11443

Yang DC, Guo FL, Liu B, Huang N, Watkins SC (2003) Expression and localization of human lysozyme in the endosperm of transgenic rice. Planta 216:597–603

Yang J, Barr LA, Fahnestock SR, Liu ZB (2005) High yield recombinant silk-like protein production in transgenic plants through protein targeting. Transgenic Res 14:313–324

Chapter 6
Algae: An Alternative to the Higher Plant System in Gene Farming

Christoph Griesbeck and Anna Kirchmayr

Abstract Microalgae based systems have the potential to combine the advantages of plants with features of microorganisms, thus becoming an alternative for gene farming. Advantages such as a short time from gene to protein, inexpensive cultivation, fast growth and improved biosafety aspects make microalgae interesting candidates for novel molecular farming systems. As a model organism, the unicellular green alga *Chlamydomonas reinhardtii* is one of the best studied organisms in this field and provides established methods for transformation, markers and reporters. Its ability for expression of proteins with biopharmaceutical or biotechnological relevance, such as antibodies, enzymes or antigenic peptides, has been demonstrated in a number of cases. Although no commercialized product has been reported so far, the application of algal systems for certain fields, such as edible vaccines, are increasingly gaining interest. Aside from biopharmaceuticals, an additional field for the use of algae for pharmaceutical products could involve novel metabolites, improved by metabolic engineering.

6.1 Algae and Their Taxonomy

The term 'algae' does not refer to a specific taxon, but comprises a large number of phylogenetically very diverse groups of photosynthetically active organisms belonging to prokaryotes (cyanobacteria) or eukaryotes. Accounting for about half of the biologically fixed global carbon dioxide (Field et al. 1998), algae are of enormous ecological importance. In the context of molecular farming systems we will

---

C. Griesbeck (✉) • A. Kirchmayr
Biotechnology, MCI – Management Center Innsbruck, University of Applied Sciences, Maximilianstraße 2, 6020 Innsbruck, Austria
e-mail: christoph.griesbeck@mci.edu

A. Wang and S. Ma (eds.), *Molecular Farming in Plants: Recent Advances and Future Prospects*, DOI 10.1007/978-94-007-2217-0_6,

focus in this review on eukaryotic algae, as they are more comparable to higher plant systems in terms of cell architecture and genetics than prokaryotic cyanobacteria. Whereas macroscopic algae like the red alga *Porphyra* are used as a food source in some parts of the world, microalgae have received ever increasing attention for their biotechnological applications as indicated by a rising number of reviews (Franklin and Mayfield 2004, 2005; Leon-Banares et al. 2004; Mayfield and Franklin 2005; Walker et al. 2005b; Griesbeck et al. 2006; Potvin and Zhang 2010; Specht et al. 2010). Among the phylogenetically heterogeneous groups of eukaryotic algae, methods for genetic manipulation, the basis for gene farming, have been reported for green algae (Chlorophyta), diatoms (Bacillariophyta) and dinoflagellates (Dinophyceae) (Leon-Banares et al. 2004).

6.2 Methods for Biotechnology with Microalgae

In order to exploit microalgal systems for biotechnological use, methods for genetic engineering and cultivation have to be accessible. In this chapter we therefore provide an overview of biotechnological methods reported for microalgae including transformation, markers, reporters, gene structure and RNA interference. In addition to this, methods of microalgae cultivation are of particular importance due to the systems' requirements related to photoautotrophic growth.

6.2.1 Transformation

In general, algae possess three genetic systems: the nuclear, the mitochondrial and the plastid genome, each of which may be genetically manipulated. A number of microalgae have already been subject to studies on genetic modification as reviewed in Table 6.1. In this section the available transformation options are viewed in some detail.

6.2.1.1 Particle Bombardment

Biolistic transformation via microparticle bombardment (Boynton et al. 1988) is the most commonly chosen technique for the introduction of DNA into the chloroplast genome. For this method DNA-coated metal particles (usually gold or tungsten) are accelerated with a helium-driven pistol and impelled into the target recipient cells. Microparticle bombardment may also be used in the transformation of nuclear genomes. Unfortunately, this method results in the integration of multiple DNA copies at random sites throughout the genome which is often undesirable (Sodeinde and Kindle 1993).

Table 6.1 Transformation of algae

Organism	Transformation method	Reference
Chlamydomonas reinhardtii	Particle bombardment	Boynton et al. (1988)
Chlamydomonas reinhardtii	Glass beads	Kindle (1990)
Chlamydomonas reinhardtii	Electroporation	Brown et al. (1991)
Chlamydomonas reinhardtii	Silicon carbide whiskers	Dunahay (1993)
Volvox carteri	Particle bombardment	Schiedlmeier et al. (1994)
Phaeodactylum tricornutum	Particle bombardment	Apt et al. (1996)
Chlorella sorokiana	Particle bombardment	Dawson et al. (1997)
Chlorella ellipsoidea	Particle bombardment	Chen et al. (1998)
Chlorella vulgaris	Electroporation	Chow and Tung (1999)
Chlorella kessleri	Particle bombardment	El-Sheekh (1999)
Chlamydomonas reinhardtii	*Agrobacterium tumefaciens*	Kumar et al. (2004)
Dunaliella salina	Particle bombardment	Tan et al. (2005)
Dunaliella tertiolecta	Electroporation	Walker et al. (2005a)
Haematococcus pulvialis	Particle bombardment	Steinbrenner and Sandmann (2006)
Dunaliella viridis	Electroporation	Sun et al. (2006)
Dunaliella salina	Electroporation	Wang et al. (2007)
Closterium peracerosum-strigosum-littorale	Particle bombardment	Abe et al. (2008)
Lotharella amoebiformis	Particle bombardment	Hirakawa et al. (2008)
Cyanidioschyzon merolae	Glass beads	Ohnuma et al. (2008)
Nannochloropsis oculata	Electroporation	Chen et al. (2008)
Dunaliella salina	Glass beads	Feng et al. (2009)
Ulva pertusa	Particle bombardment	Kakinuma et al. (2009)
Gonium pectorale	Particle bombardment	Lerche and Hallmann (2009)
Haematococcus pulvialis	*Agrobacterium tumefaciens*	Kathiresan et al. (2009)

6.2.1.2 Glass Beads and Silicon Whiskers

The most popular method for introducing foreign DNA into the nuclear genome of algae was first described by Kindle in 1990. Glass beads are used to impact the cell wall and to allow DNA to enter the cell. To enhance this effect, cells are vortexed together with polyethylene glycol. This method is often applied because of its ease-of-use and the low cost of necessary materials and instruments. One disadvantage of this transformation procedure is the necessity of cell wall-deficient mutants or enzymatic cell treatment prior to the transformation step (Kindle 1990). Transformation with glass beads also results in lower copies of integrated DNA in comparison with particle bombardment (Kindle 1998).

Using silicon carbide whiskers instead of glass beads is yet another possible transformation method. This modified approach, however, is not commonly used due to the expensive and hazardous materials which are required. One notable advantage of the silicon carbide variation is the possible use of wild type strains instead of cell wall-deficient mutants. Lower cell lethality during vortexing could also be seen as an important advantage of this method (Dunahay 1993).

6.2.1.3 Electroporation

A further method for introducing DNA into algae cells is electroporation. Initial approaches to nuclear transformation via electroporation resulted only in low frequencies of stable transformation (Brown et al. 1991). Electroporation has been established for both wild type and cell wall-deficient cells of *C. reinhardtii* (Tang et al. 1995; Shimogawara et al. 1998).

6.2.1.4 *Agrobacterium tumefaciens* Mediated Transformation

Transformation of algae cells using T-DNA of *Agrobacterium tumefaciens*, as it is common for higher plants, was attempted with the unicellular green algae *Chlamydomonas reinhardtii* and *Haematococcus pluvialis* (Kumar et al. 2004; Kathiresan et al. 2009; Kathiresan and Sarada 2009). Co-cultivation of each alga with *Agrobacterium tumefaciens* resulted in the genomic integration of the genes for green fluorescent protein, hygromycin phosphotransferase and β-glucuronidase. For *C. reinhardtii*, transformation efficiency was stated to be 50-fold higher than that of the glass beads method (Kumar et al. 2004).

6.2.2 *Selection Markers, Reporter Genes and Promoters*

6.2.2.1 Selection Markers

Being a model organism, *Chlamydomonas reinhardtii* comes with the largest collection of selection markers, other algae species which are currently accessible for transformation have only made use of a small number of marker genes. Table 6.2 shows a detailed list of markers reported for microalgae. Most of the listed markers may be utilized for nuclear transformation, whereas some such as *aadA* (Goldschmidt-Clermont 1991; Cerutti et al. 1997) and *aphA-6* (Bateman and Purton 2000) confer resistance when integrated into the plastidic genome. The application of prototrophic markers such as *ARG7* (Debuchy et al. 1989), *NIA1* (Kindle et al. 1989), *NIC7* (Ferris 1995) and *THI10* (Ferris 1995) may be helpful in avoiding the integration of antibiotic resistance genes into algal genomes, but either restricts its use to the corresponding auxotrophic strains or requires the introduction of these mutations prior to transformation.

6.2.2.2 Reporters

Similarly to the situation of marker genes, *C. reinhardtii* offers the most comprehensive list of available reporters among algae. As pointed out below for transgenes in general, reporter genes have usually to be adapted to the codon usage of this species to allow for reproducible and measurable expression levels. In Table 6.3 an

Table 6.2 Selection markers for microalgae (Fuhrmann 2002; Griesbeck et al. 2006; Potvin and Zhang 2010)

Marker	Description	Organism	Reference
ARG7	Arginine prototrophy	*Chlamydomonas reinhardtii*	Debuchy et al. (1989)
NIT1 (NIA1)	Nitrate prototrophy	*Chlamydomonas reinhardtii*	Kindle et al. (1989)
nptII	Neomycin phosphotransferase	*Chlamydomonas reinhardtii, Symbiodinium sp., Phaeodactylum tricornutum, Amphidinium sp., Cyclotella crytica, Navicula saprophila*	Hall et al. (1993), Dunahay et al. (1995), Ten Lohuis and Miller (1998) and Zaslavskaia et al. (2000)
CRY1-1	Resistance to cryptop-leurine/emetine	*Chlamydomonas reinhardtii*	Nelson et al. (1994)
NIC7	Nicotinamide prototrophy	*Chlamydomonas reinhardtii*	Ferris (1995)
THI-10	Thiamine prototrophy	*Chlamydomonas reinhardtii*	Ferris (1995)
ble	Resistance to zeocin	*Chlamydomonas reinhardtii, Phaeodactylum tricornutum*	Apt et al. (1996) and Stevens et al. (1996) and Lumbreras et al. (1998)
aadA	Resistance to spectino-mycin/streptomycin	*Chlamydomonas reinhardtii*	Goldschmidt-Clermont (1991) and Cerutti et al. (1997)
PPX1	Resistance to porphyric herbicides	*Chlamydomonas reinhardtii*	Randolph-Anderson et al. (1998)
aphA-6	Resistance to kanamy-cin/amikacin	*Chlamydomonas reinhardtii*	Bateman and Purton (2000)
act-2	Resistance to cycloheximide	*Chlamydomonas reinhardtii*	Stevens et al. (2001)
aphVIII	Resistance to paromo-mycin/kanamycin	*Chlamydomonas reinhardtii*	Sizova et al. (2001)
ALS	Resistance to sulfometuronmethyl	*Chlamydomonas reinhardtii*	Kovar et al. (2002)
aph7′′	Resistance to hygromy-cin B	*Chlamydomonas reinhardtii*	Berthold et al. (2002)
oee-1	Oxygen-evolving enhancer protein	*Chlamydomonas reinhardtii*	Mayfield and Kindle (1990)
cat	Resistance to chloramphenicol	*Chlamydomonas reinhardtii, Phaeodactylum tricornuntum*	Tang et al. (1995) and Apt et al. (1996)
hpt	HygromycinB phosphotransferase	*Amphidinium, Symbiodinium*	Ten Lohuis and Miller (1998)
nat	Nourseothricin resistance	*Phaeodactylum tricornuntum*	Zaslavskaia et al. (2000)

(continued)

Table 6.2 (continued)

Marker	Description	Organism	Reference
sat-1	Nourseothricin resistance	*Phaeodactylum tricornuntum*	Zaslavskaia et al. (2000)
ARG9	plastid N-acetyl ornithine aminotransferase	*Chlamydomonas reinhardtii*	Remacle et al. (2009)
PDS	Phytoene desaturase	*Chlorella zofingiensis, H. pluvialis*	Steinbrenner and Sandmann (2006) and Huang et al. (2008)

Table 6.3 Reporter genes for microalgae (Fuhrmann 2002; Griesbeck et al. 2006; Potvin and Zhang 2010)

Reporter	Description	Organism	Reference
ARS	Arylsulfatase-colorimetric assay-not for sulfur starvation	*Chlamydomonas reinhardtii*	Davies et al. (1992)
crgfp	Nuclear codon-optimized GFP	*Chlamydomonas reinhardtii*	Fuhrmann et al. (1999)
rluc	Chloroplast-Luciferase from *Renilla reniformis*	*Chlamydomonas reinhardtii*	Minko et al. (1999)
gfpCt	Chloroplast codon-optimized GFP	*Chlamydomonas reinhardtii*	Franklin et al. (2002)
crluc	Nuclear codon-optimized Luciferase from *Renilla reniformis*	*Chlamydomonas reinhardtii*	Fuhrmann et al. (2004)
luxCt	Chloroplast codon-optimized Luciferase from *Vibrio harveyi*	*Chlamydomonas reinhardtii*	Mayfield and Schultz (2004)
lucCP	Chloroplast codon-optimized firefly Luciferase	*Chlamydomonas reinhardtii*	Matsuo et al. (2006)
cgluc	Nuclear codon-optimized *Gaussia princeps* luciferase	*Chlamydomonas reinhardtii*	Ruecker et al. (2008)
luc	Luciferase from *Horatia parvula*	*Phaeodactylum tricornutum*	Falciatore et al. (1999)
eGfp	GFP adapted to human codon usage	*Phaeodactylum tricornutum*	Zaslavskaia et al. (2001)
gus	β-Glucuronidase	*Phaeodactylum tricornutum, Amphidinium, Symbiodinium*	Ten Lohuis and Miller (1998) and Zaslavskaia et al. (2000)
ϵ-frustulin	Calcium-binding glycoprotein	*Cylindrotheca fusiformis*	Fischer et al. (1999)

overview of the frequently used reporter genes in different microalgae is given. GFP and luciferase as the most commonly used reporter genes have also been utilized for *C. reinhardtii*. The gene encoding the *green fluorescent protein* from *Aequorea victoria* was adapted to the nuclear and chloroplast codon usage and successfully expressed (Fuhrmann et al. 1999; Franklin et al. 2002). Two different luciferase genes from *Gaussia princeps* and *Renilla reniformis* were also codon-optimized and are commonly used for gene expression studies (Minko et al. 1999; Fuhrmann et al. 2004; Mayfield and Schultz 2004; Matsuo et al. 2006; Ruecker et al. 2008). In addition, the endogenous gene of arylsulfatase has also shown its effectiveness for the measurement of transgene expression (Davies et al. 1992). However, its activity can only be monitored when the endogenous enzyme is not expressed.

6.2.2.3 Promoters

Whereas the use of strong viral promoters in plant or mammalian systems is fairly common, such promoters are not available for algae to date due to the lack of suitable algal viruses. For some algae, typical plant virus promoters such as the cauliflower mosaic virus 35S promoter have proven their effectiveness (Ten Lohuis and Miller 1998). In the case of *Chlamydomonas reinhardtii*, strong endogenous promoters are usually chosen. Among the most frequently used nuclear promoters for this organism are the constitutive promoter of a small subunit of the ribulose bisphosphat carboxylase *RBCS2* (Stevens et al. 1996) and the β-2-tubulin promoter (Davies et al. 1992). When inducible expression is desired, *NIT1* and *CYC6* can be controlled by the nitrate and copper concentration in the medium, respectively (Loppes et al. 1999; Quinn et al. 2003). The upstream fusion of the heat shock promoter *HSP70A*, which can be further enhanced by heat shock, was found to be successful in improving neighboring promoter elements (Schroda et al. 2000). In Table 6.4 promoters are listed, which have been reported to function in *C. reinhardtii*.

6.2.3 Codon Usage and Gene Structure

Foreign gene expression in microalgae is often accompanied by low or barely detectable expression rates even when expressed under strong promoters. Several explanations could be given for this effect.

The influence of the codon usage was recognized as an important factor for transgene expression in studies with the microalgae *Phaeodactylum tricornuntum* (Zaslavskaia et al. 2000, 2001) and *Chlamydomonas reinhardtii*. For *C. reinhardtii* having a high GC content of 61%, expression of transgenes could only be achieved for genes with a comparable high GC content as demonstrated for *gfp* (Fuhrmann et al. 1999; Leon-Banares et al. 2004; Heitzer et al. 2007). Therefore, codon adaptation of transgenes is required for this alga as a general rule in order to permit significant expression.

Other strategies for improving transgene expression in *C. reinhardtii* include the integration of intron sequences into transgenes. The *Renilla*-luciferase gene was used as a reporter to study different intronic structures. Under the control of a constitutive chimeric *HSP70A/RBCS2* promoter, integration of *RBCS2* introns into the transgene influenced the expression levels significantly with the physiological number and order of the three introns having the best stimulating effect (Eichler-Stahlberg et al. 2009).

Table 6.4 Promoters used in *Chlamydomonas reinhardtii* (Griesbeck et al. 2006)

Promoter	Description	Genome	Reference
β-2-TUB	β-2-tubulin	Nuclear	Davies et al. (1992)
nos	Nopaline synthase from *Agrobacterium tumefaciens*	Nuclear	Hall et al. (1993)
CaMV	Cauliflower mosaic virus 35S	Nuclear	Tang et al. (1995)
RBCS2	Small subunit of ribulose bisphosphat carboxylase	Nuclear	Stevens et al. (1996)
NIA1	Nitrate reductase	Nuclear	Loppes et al. (1999)
COP	Chlamyopsin	Nuclear	Fuhrmann et al. (1999)
HSP70A	Heat shock protein 70A	Nuclear	Schroda et al. (2000)
psaD	Photosystem I complex protein	Nuclear	Fischer and Rochaix (2001)
cyc6	Cytochrome c_6	Nuclear	Quinn et al. (2003)
atpA	ATPase alpha subunit	Chloroplast	Mayfield et al. (2003), Sun et al. (2003) and Mayfield and Schultz (2004)
psbA	Photosystem II protein D1	Chloroplast	Mayfield and Schultz (2004)
rbcL	Ribulose bisphosphate carboxylase large subunit	Chloroplast	Franklin et al. (2002), Mayfield et al. (2003) and Mayfield and Schultz (2004)
cabII-1	Chlorophyll-ab binding protein	Chloroplast	Blankenship and Kindle (1992)

As using linearized DNA for transformation was also demonstrated to increase expression levels of transgenes in *C. reinhardtii* (Kindle 1998), a modular system for the fast and customizable creation of linear transgene constructs has been established, which allows the combination of different marker genes with the gene of interest. Fusion of two plasmids in vitro using Cre/lox site-specific recombination facilitates the rapid construction of large tandem vectors for transgene expression under the control of different promoter sequences. Bacterial vector sequences, which are not required at this point any more, can be removed from the linearized vector. The resulting constructs showed significantly higher coexpression rates of marker and gene of interest (Heitzer and Zschoernig 2007).

6.2.4 Posttranscriptional Gene Regulation

Posttranscriptional gene silencing is an important mechanism for gene regulation in many eukaryotes. MicroRNAs (miRNAs) and short interfering RNAs (siRNAs) have been demonstrated to be involved in posttranscriptional gene silencing mechanisms also in the unicellular alga *Chlamydomonas reinhardtii* (Molnar et al. 2007).

On the one hand, silencing can be responsible for unstable expression of transgenes and low yields of recombinant proteins when expressed from the nuclear genome, as effective recognition and silencing of transgenes has been shown for *C. reinhardtii* (Schroda 2005). On the other hand, these mechanisms can be utilized for targeted downregulation of endogenous genes. Methods for gene silencing have also been established for *C. reinhardtii*. The integration of an antisense construct induced silencing of a corresponding retinal protein (Fuhrmann et al. 2001). Inducible RNAi with the NIT1 promoter was reported in 2005 by Koblenz and Lechtreck (2005). Artificial miRNAs have been employed for highly specific gene silencing and could be useful for high-throughput studies (Molnar et al. 2009). Such approaches could be helpful not only for investigations of gene function, but also for metabolic engineering, for example if glycosylation patterns have to be modified by downregulation of genes of the glycosylation machinery.

6.2.5 Cultivation

Photoautotrophic growth of microalgae makes cultivation a challenging issue. Especially in the context of upscaling processes, illumination and carbon dioxide supply create most difficulties. Generally, microalgae can be cultivated both in open and closed systems, either associated with a set of advantages and drawbacks.

6.2.5.1 Open Systems

Open cultivation systems such as tanks, circular ponds, raceway ponds or inclined-surface photobioreactors represent a simple and inexpensive way to cultivate microalgae (Borowitzka 1999; Pulz 2001). Major disadvantages are the insufficient mass transfer due to the usual lack of stirring and the limited illumination at higher layer thickness. Moreover, cells can be easily damaged by high light intensities during the day. Thus, the presently achievable cell densities of ca. 1 g/l (Pulz 2001) are lower compared to stirred and artificially illuminated closed bioreactors. Because of the low cell densities and layer thickness, this cultivation method requires large areas of land. As a matter of common sense, open systems cannot be run under sterile conditions, which allows only the cultivation of algae growing under highly selective conditions e.g. in terms of salinity. Production of biopharmaceuticals under such high risk of contamination is in effect impracticable.

6.2.5.2 Closed Systems

Due to the drawbacks of open systems mentioned above, cultivation of microalgae is more and more carried out in closed photobioreactors. These forms of reactors cut contamination risk to a minimum. Due to the closed construction, cultivation can be performed under elevated carbon dioxide concentrations. In general, growth conditions such as temperature can be adjusted more precisely and are more reproducible compared to open systems (Pulz 2001). Whereas experience from conventional bioreactor systems can be used for the optimization of issues such as gassing, mixing or sterilization, the most challenging task in designing photobioreactors is still the provision of an optimum illumination, which can be achieved by artificial light, sun light or a combination of both. Different types of closed photobioreactors have been described: flat-plate, horizontal/serpentine tubular airlift, tubular photobioreactors, bubble column, airlift column, stirred tank, helical tubular, conical, torus and seaweed-type reactors (Eriksen 2008; Ugwu et al. 2008). Rising complexity of such photobioreactor systems leads to increasing costs of investment and operations, but on the other hand to higher achievable cell densities.

6.3 Recombinant Proteins Expressed in Microalgae

6.3.1 Proteins Expressed in Chlamydomonas reinhardtii

In order to demonstrate the potential of novel expression systems, different proteins for pharmaceutical or biotechnological use have to be expressed and tested for their functions and properties. In the case of *Chlamydomonas reinhardtii* as the best studied alga, a number of diverse proteins have been expressed successfully, but to date none has reached economical or clinical relevance. Table 6.5 shows a detailed overview of the most important recombinant proteins which have successfully been expressed in *C. reinhardtii*. According to their function, these proteins act as antigenic proteins, antibodies, enzymes or hormones.

Within the past few years, the number and diversity of expressed and tested proteins have increased significantly indicating a rising interest in this system as biotechnological tool. A recent study demonstrated successful expression of several human therapeutic proteins in the chloroplast. These include a cytokine, antibody mimicking proteins and a vascular endothelial growth factor (Rasala et al. 2010).

Besides the accumulation of recombinant proteins, the functionality and bioactivity of different protein classes expressed in *Chlamydomonas* could be shown. An example of this can be observed in the binding affinity of an algal expressed antibody against anthrax protective antigen 83. It was shown to be similar to that of an antibody expressed in mammalian cells (Tran et al. 2009). Another interesting example for biofarming with *C. reinhardtii* is the stable expression of the D2 fibronectin-binding domain of *Staphylococcus aureus* fused with the cholera toxin B subunit. Algae-based vaccinations lead to the induction of specific mucosal and systemic immune responses and protected mice against lethal doses of this pathogenic bacterium. Moreover, the algal-based vaccine was stable for more than a year at ambient temperature (Dreesen et al. 2010).

Table 6.5 Recombinant proteins expressed in *Chlamydomonas reinhardtii* (Griesbeck et al. 2006; Specht et al. 2010)

Recombinant protein	Location	Protein class	Yield/expression level	Reference
Avian metallothionein type II	Nucleus/periplasm	Metal binding	Not specified	Cai et al. (1999)
Mothbean Δ (1)-pyroline-5-carboxylate synthetase	Nucleus	Enzyme	Not specified	Siripornadulsil et al. (2002)
Antigenic peptide P57 of the pathogenic *Rennibacterium salmoninarium*	Chloroplast, nucleus/periplasm	Antigenic protein	Not specified	Sayre et al. (2003)
Antigenic proteins VP19, 24, 26, 28 of the white spot syndrome virus	Chloroplast/nucleus/periplasm/cytoplasm	Antigenic protein	Not specified	Sayre et al. (2003), Surzycki et al. (2009)
Foot-and-mouth disease virus VP1 protein fused with cholera toxin B subunit	Chloroplast	Antigenic protein	Not specified	Sun et al. (2003)
Anti-glycoprotein D of herpes simplex virus large single-chain antibody (human IgA)	Chloroplast	Antibody	Not specified	Mayfield et al. (2003)
Human metallothionine-2	Nucleus	Metal binding	Not specified	Zhang et al. (2006)
Anti-rabbit-IgG single-chain antibody fused with luciferase	Nucleus/medium	Antibody/enzyme	Not specified	Griesbeck et al. (2006)
Human tumor necrosis factor-related apoptosis-inducing ligand	Chloroplast	Ligand	0.43–0.67%	Yang et al. (2006)
Bovine mammary-associated serum amyloid	Chloroplast	Serum protein	~5% TSP	Manuell et al. (2007)
Classical swine fever virus E2 viral protein	Chloroplast	Antigenic protein	~2% TSP	He et al. (2007)
Human glutamic acid decarboxylase 65	Chloroplast	Antigenic protein	~0.3% TSP	Wang et al. (2008)
Human erythropoietin	Nucleus	Hormone	100 μg/l culture	Eichler-Stahlberg et al. (2009)
Anti-anthrax protective antigen 83 antibody	Chloroplast	Antibody	0.01% dry algal biomass	Tran et al. (2009)
D2 fibronectin-binding domain of Staphylococcus aureus fused with cholera toxin B subunit	Chloroplast	Antigenic protein	0.7% TSP	Dreesen et al. (2010)
Humane fibronectin (domains 10 and 14)	Chloroplast	Antibody mimic	14FN3: 3% TSP 10FN3: detectable	Rasala et al. (2010)
Proinsuline	Chloroplast	Hormone	Detectable	Rasala et al. (2010)
Human vascular endothelial growth factor isoform 121 (VEGF)	Chloroplast	Hormone	2% TSP	Rasala et al. (2010)
High mobility group protein B1 (HMGB1)	Chloroplast	Cytokine	2.5% TSP	Rasala et al. (2010)

TSP total soluble protein

Altogether, these results can be seen as proof of principle for the *Chlamydomonas* expression system. Whereas for expression within the chloroplast the observed yields are in a range sufficient for commercial production as documented by several publications (He et al. 2007; Manuell et al. 2007; Rasala et al. 2010), for nuclear expression the productivity is still quite low and requires significant enhancement. Another question to be solved remains the glycosylation patterns of nuclear expressed and secreted proteins from *C. reinhardtii*. Studies on the posttranslational machinery of this alga would be helpful in order to exploit this as an option of algal protein production.

6.3.2 Recombinant Proteins in Other Microalgae

Whereas the vast majority of recombinant proteins have been expressed in the model alga *C. reinhardtii*, publications concerning transgene expression in other algae are scarce, as methods and tools for genetic engineering of other algae have not been developed in the same degree. Because of the phylogenetic and structural diversity of algae, methods established for *C. reinhardtii* cannot be easily transferred to other species, but usually require major adaptations. In addition, helpful information such as genome sequences is available only for a limited number of algae (Grossman 2007). One example of a recombinant protein is the human growth hormone (hGH) expressed in the nucleus of *Chlorella vulgaris*. The product was secreted into the medium at a yield of 200–600 ng/ml (Hawkins and Nakamura 1999). Another example is the expression of fish growth hormone (GH) in *Nannochloropsis oculata* with an amount of 0.42–0.27 μg/ml. When these algae were fed to red-tilapia larvae, their body length and weight increased (Chen et al. 2008).

6.3.3 Advantages of Algae as Heterologous Expression Systems

Production of biopharmaceuticals in algae has diverse advantages. On the one hand, algae are comparable to plant systems including their ability to produce proteins with post-translational modifications. Moreover, many algae are generally regarded as safe (GRAS), which designates them as organisms generally accepted not to contain toxins or human pathogens (Mayfield and Franklin 2005; Walker et al. 2005b). This could allow for the reduction of necessary purification steps during downstream processes.

As microorganisms on the other hand, microalgae display high growth rates and need only a short time from transformation to product formation. This process including scale up could be implemented within a few weeks. Another advantage of algae compared to higher plants is vegetative reproduction, leading to uniform clones with comparable production rates. Cultivation can be inexpensive due to the low costs of typical mineral media for algae. One striking advantage compared to plant systems is the use of closed photobioreactors. This method reduces the risk of contamination and prevents transgene dispersing into the environment.

Therapeutic proteins can be produced either in the chloroplast or in the nuclear genome of microalgae. These two systems have distinct advantages. The expression of proteins in the chloroplast currently makes accumulation of far higher product quantities possible, the reason being that transgenes are usually integrated in multiple copies in the chloroplast and gene silencing has not been observed in this organelle (Specht et al. 2010). Posttranslational modifications and secretion, however, are not possible when producing proteins in the chloroplast. These options, in turn, can be obtained as a great benefit when expressing transgenes from the nuclear genome, where the current expression rates are still quite low and silencing of transgenes can occur.

6.4 Future Prospects of Algae for Molecular Farming

6.4.1 Recombinant Proteins

Although a significant number of proteins have been successfully expressed in transgenic microalgae, essentially *Chlamydomonas reinhardtii*, algae will not replace the established and commercialized bacterial and mammalian expression systems. Especially in the context of regulatory aspects in the pharmaceutical sector, novel expression systems have to offer enormous advantages over conventional systems to be chosen for new products. There are two options to make transgenic microalgae systems competitive in the field of pharmaceutical proteins.

On the one hand, advantages can be related to product quality, e.g. demonstrated as certain beneficial posttranslational modifications, product stability or biosafety. On the other hand, there is the matter of possible cost savings during production processes, which could play a role in special fields, where large quantities of products are required at low costs. This could hold true for example for recombinant antibodies or veterinary products.

Edible vaccines are a possible field of application for algal expression systems, combining biosafety issues with inexpensive production and storage. Expression of antigenic peptides and proteins in algae has already been demonstrated as outlined above. Furthermore many algae are GRAS certified (Mayfield and Franklin 2005; Walker et al. 2005b). Vaccination concepts for a large number of diseases prevalent in developing nations based on recombinant antigen expression in microalgae could result in inexpensive production and distribution as well as long term storage at room temperature (Dreesen et al. 2010; Specht et al. 2010).

6.4.2 Other Products from Microalgae

Whereas production of therapeutic proteins in transgenic microalgae in general is still to come of age, algae have already been established as biotechnological production systems for a number of secondary metabolites useful as food additives or cosmetics (Plaza et al. 2009). Being photosynthetically active organisms, algae have to deal with light and oxygen and have therefore developed diverse pigments and antioxidants. One commercially relevant example is the production of carotene using *Dunaliella salina* (Hosseini Tafreshi and Shariati 2009). As screening of diverse algal species for bioactive substances has revealed a number of novel compounds with medical or nutritional relevance, algae have gained increasing attention within the past few years as potential producers, which is illustrated by a few examples. Lutein as an antioxidant and food colorant has been found in several microalgae species in substantially larger amounts compared to conventional sources (Fernandez-Sevilla et al. 2010). Antiviral activities have been identified in a considerable number of marine species. Many substances are already in preclinical and clinical stages as potential antiviral drugs (Rechter et al. 2006; Yasuhara-Bell and Lu 2010). The significance of algal systems for pharmaceutical products will presumably increase most in the field of novel bioactive substances, when substances are unique and cannot feasibly be manufactured in other systems. Exploiting the full potential of algae for pharmaceutical use will include metabolic engineering of interesting production strains of bioactive substances in order to tailor these compounds and to optimize the yield.

Acknowledgement This work was supported by the MCI Doctoral Grant Program. We would like to thank Ian Wallace for critical reviewing the manuscript.

References

Abe J, Hiwatashi Y, Ito M, Hasebe M, Sekimoto H (2008) Expression of exogenous genes under the control of endogenous HSP70 and CAB promoters in the *Closterium peracerosum-strigosum-littorale* complex. Plant Cell Physiol 49:625–632

Apt KE, Kroth-Pancic PG, Grossman AR (1996) Stable nuclear transformation of the diatom *Phaeodactylum tricornutum*. Mol Gen Genet 252:572–579

Bateman JM, Purton S (2000) Tools for chloroplast transformation in *Chlamydomonas*: expression vectors and a new dominant selectable marker. Mol Gen Genet 263:404–410

Berthold P, Schmitt R, Mages W (2002) An engineered *Streptomyces hygroscopicus aph 7´´* gene mediates dominant resistance against hygromycin B in *Chlamydomonas reinhardtii*. Protist 153:401–412

Blankenship JE, Kindle KL (1992) Expression of chimeric genes by the light-regulated cabII-1 promoter in *Chlamydomonas reinhardtii*: a cabII-1/nit1 gene functions as a dominant selectable marker in a nit1- nit2- strain. Mol Cell Biol 12:5268–5279

Borowitzka MA (1999) Commercial production of microalgae: ponds, tanks, tubes and fermenters. J Biotechnol 70:313–321

Boynton JE, Gillham NW, Harris EH, Hosler JP, Johnson AM, Jones AR, Randolph-Anderson BL, Robertson D, Klein TM, Shark KB, Sanford JC (1988) Chloroplast transformation in *Chlamydomonas* with high velocity microprojectiles. Science 240:1534–1538

Brown LE, Sprecher SL, Keller LR (1991) Introduction of exogenous DNA into *Chlamydomonas reinhardtii* by electroporation. Mol Cell Biol 11:2328–2332

Cai XH, Brown C, Adhiya J, Traina SJ, Sayre R (1999) Growth and heavy metal binding properties of transgenic *Chlamydomonas* expressing a foreign metallothionein gene. Int J Phytoremediation 1:53–65

Cerutti H, Johnson AM, Gillham NW, Boynton JE (1997) A eubacterial gene conferring spectinomycin resistance on *Chlamydomonas reinhardtii*: integration into the nuclear genome and gene expression. Genetics 145:97–110

Chen Y, Li WB, Bai QH, Sun YR (1998) Study on transient expression of GUS gene in *Chlorella ellipsoidea* (Chlorophyta), by using biolistic particle delivery system. Chin J Oceanol Limnol 47:9–16

Chen HL, Li SS, Huang R, Tsai HJ (2008) Conditional production of a functional fish growth hormone in the transgenic line of *Nannochloropsis oculata* (Eustigmatophyceae). J Phycol 44: 768–776

Chow KC, Tung WL (1999) Electrotransformation of *Chlorella vulgaris*. Plant Cell Rep 18: 778–780

Davies JP, Weeks DP, Grossman AR (1992) Expression of the arylsulfatase gene from the beta 2-tubulin promoter in *Chlamydomonas reinhardtii*. Nucleic Acids Res 20:2959–2965

Dawson HN, Burlingame R, Cannons AC (1997) Stable transformation of *Chlorella*: rescue of nitrate reductase-deficient mutants with the nitrate reductase gene. Curr Microbiol 35:356–362

Debuchy R, Purton S, Rochaix JD (1989) The argininosuccinate lyase gene of *Chlamydomonas reinhardtii*: an important tool for nuclear transformation and for correlating the genetic and molecular maps of the ARG7 locus. EMBO J 8:2803–2809

Dreesen IA, Charpin-El Hamri G, Fussenegger M (2010) Heat-stable oral alga-based vaccine protects mice from Staphylococcus aureus infection. J Biotechnol 145:273–280

Dunahay TG (1993) Transformation of *Chlamydomonas reinhardtii* with silicon carbide whiskers. Biotechniques 15:452–455

Dunahay TG, Eric E, Jarvis EE, Roessler PG (1995) Genetic transformation of the diatons *Cyclotella cryptica* and *Navicula saprophila*. J Phycol 31:1004–1012

Eichler-Stahlberg A, Weisheit W, Ruecker O, Heitzer M (2009) Strategies to facilitate transgene expression in *Chlamydomonas reinhardtii*. Planta 229:873–883

El-Sheekh M-M (1999) Stable transformation of the intact cells of *Chlorella kessleri* with high velocity microprojectiles. Biol Plant Prague 42:209–216

Eriksen NT (2008) The technology of microalgal culturing. Biotechnol Lett 30:1525–1536

Falciatore A, Casotti R, Leblanc C, Abrescia C, Bowler C (1999) Transformation of nonselectable reporter genes in marine diatoms. Mar Biotechnol (NY) 1:239–251

Feng S, Xue L, Liu H, Lu P (2009) Improvement of efficiency of genetic transformation for *Dunaliella salina* by glass beads method. Mol Biol Rep 36:1433–1439

Fernandez-Sevilla JM, Acien Fernandez FG, Molina Grima E (2010) Biotechnological production of lutein and its applications. Appl Microbiol Biotechnol 86:27–40

Ferris PJ (1995) Localization of the *nic-7*, *ac-29* and *thi-10* genes within the mating-type locus of *Chlamydomonas reinhardtii*. Genetics 141:543–549

Field CB, Behrenfeld MJ, Randerson JT, Falkowski P (1998) Primary production of the biosphere: integrating terrestrial and oceanic components. Science 281:237–240

Fischer N, Rochaix JD (2001) The flanking regions of PsaD drive efficient gene expression in the nucleus of the green alga *Chlamydomonas reinhardtii*. Mol Genet Genomics 265:888–894

Fischer H, Robl I, Sumper M, Kröger N (1999) Targeting and covalent modification of cell wall and membrane proteins heterologously expressed in the diatom *Cylindrotheca fusiformis* (Bacillariophyceae). J Phycol 35:113–120

Franklin SE, Mayfield SP (2004) Prospects for molecular farming in the green alga *Chlamydomonas reinhardtii*. Curr Opin Plant Biol 7:159–165

Franklin SE, Mayfield SP (2005) Recent developments in the production of human therapeutic proteins in eukaryotic algae. Expert Opin Biol Ther 5:225–235

Franklin S, Ngo B, Efuet E, Mayfield SP (2002) Development of a GFP reporter gene for *Chlamydomonas reinhardtii* chloroplast. Plant J 30:733–744

Fuhrmann M (2002) Expanding the molecular toolkit for *Chlamydomonas reinhardtii* – from history to new frontiers. Protist 153:357–364

Fuhrmann M, Oertel W, Hegemann P (1999) A synthetic gene coding for the green fluorescent protein (GFP) is a versatile reporter in *Chlamydomonas reinhardtii*. Plant J 19:353–361

Fuhrmann M, Stahlberg A, Govorunova E, Rank S, Hegemann P (2001) The abundant retinal protein of the *Chlamydomonas* eye is not the photoreceptor for phototaxis and photophobic responses. J Cell Sci 114:3857–3863

Fuhrmann M, Hausherr A, Ferbitz L, Schödl T, Heitzer M, Hegemann P (2004) Monitoring dynamic expression of nuclear genes in *Chlamydomonas reinhardtii* by using a synthetic luciferase reporter gene. Plant Mol Biol 55:869–881

Goldschmidt-Clermont M (1991) Transgenic expression of aminoglycoside adenine transferase in the chloroplast: a selectable marker of site-directed transformation of *Chlamydomonas*. Nucleic Acids Res 19:4083–4089

Griesbeck C, Kobl I, Heitzer M (2006) *Chlamydomonas reinhardtii*: a protein expression system for pharmaceutical and biotechnological proteins. Mol Biotechnol 34:213–223

Grossman AR (2007) In the grip of algal genomics. Adv Exp Med Biol 616:54–76

Hall LM, Taylor KB, Jones DD (1993) Expression of a foreign gene in *Chlamydomonas reinhardtii*. Gene 124:75–81

Hawkins RL, Nakamura M (1999) Expression of human growth hormone by the eukaryotic alga, *Chlorella*. Curr Microbiol 38:335–341

He DM, Qian KX, Shen GF, Zhang ZF, Li YN, Su ZL, Shao HB (2007) Recombination and expression of classical swine fever virus (CSFV) structural protein E2 gene in *Chlamydomonas reinhardtii* chroloplasts. Colloids Surf B Biointerfaces 55:26–30

Heitzer M, Zschoernig B (2007) Construction of modular tandem expression vectors for the green alga *Chlamydomonas reinhardtii* using the Cre/lox-system. Biotechniques 43:324–332

Heitzer M, Eckert A, Fuhrmann M, Griesbeck C (2007) Influence of codon bias on the expression of foreign genes in microalgae. Adv Exp Med Biol 616:46–53

Hirakawa Y, Kofuji R, K-i I (2008) Transient transformation of a chlorarachinophyta alga, Lotharella amoebiformis (Chlorarachiophyceae), with uidA and egfp reporter genes. J Phycol 44:814–820

Hosseini Tafreshi A, Shariati M (2009) *Dunaliella* biotechnology: methods and applications. J Appl Microbiol 107:14–35

Huang J, Liu J, Li Y, Chen F (2008) Isolation and characterization of the phytoene desaturase gene as a potential selective marker for genetic engineering of the astaxanthin-producing green alga Chlorella zofingiensis (Chlorophyta). J Phycol 44:684–690

Kakinuma M, Ikeda M, Coury D, Tominaga H, Kobayashi T, Amano H (2009) Isolation and characterization of the rbcS genes from a sterile mutant of *Ulva pertusa* (Ulvales, Chlorophyta) and transient gene expression using the rbcS gene promoter. Fish Sci 75:1015–1028

Kathiresan S, Sarada R (2009) Towards genetic improvement of commerically important microalga *Haematococcus pluvialis* for biotech applications. J Appl Phycol 21:553–558

Kathiresan S, Chandrashekar A, Ravishankar A, Sarada R (2009) *Agrobacterium*-mediated transformation in the green alga *Haematococcus pluvialis* (Chlorophyceae volvocales). J Phycol 45: 642–649

Kindle KL (1990) High-frequency nuclear transformation of *Chlamydomonas reinhardtii*. Proc Natl Acad Sci USA 87:1228–1232

Kindle KL (1998) Nuclear transformation: technology and applications. In: Rochaix JD, Goldschmidt-Clermont M, Merchant S (eds) The molecular biology of chloroplasts and mitochondria in *Chlamydomonas*. Kluwer Academic, Dordrecht pp 42–61

Kindle KL, Schnell RA, Fernandez E, Lefebvre PA (1989) Stable nuclear transformation of *Chlamydomonas* using the *Chlamydomonas* gene for nitrate reductase. J Cell Biol 109: 2589–2601

Koblenz B, Lechtreck K-F (2005) The *NIT1* promoter allows inducible and reversible silencing of centrin in *Chlamydomonas reinhardtii*. Eukaryot Cell 4:1959–1962

Kovar JL, Zhang J, Funke RP, Weeks DP (2002) Molecular analysis of the acetolactate synthase gene of *Chlamydomonas reinhardtii* and development of a genetically engineered gene as a dominant selectable marker for genetic transformation. Plant J 29:109–117

Kumar SV, Misquitta RW, Reddy VS, Rao BJ, Rajam MV (2004) Genetic transformation of the green alga-*Chlamydomonas reinhardtii* by *Agrobacterium tumefaciens*. Plant Sci 166:731–738

Leon-Banares R, Gonzalez-Ballester D, Galvan A, Fernandez E (2004) Transgenic microalgae as green cell-factories. Trends Biotechnol 22:45–52

Lerche K, Hallmann A (2009) Stable nuclear transformation of *Gonium pectorale*. BMC Biotechnol 9:64–81

Loppes R, Radoux M, Ohresser MC, Matagne RF (1999) Transcriptional regulation of the *Nia1* gene encoding nitrate reductase in *Chlamydomonas reinhardtii*: effects of various environmental factors on the expression of a reporter gene under the control of the *Nia1* promoter. Plant Mol Biol 41:701–711

Lumbreras V, Stevens DR, Purton S (1998) Efficient foreign gene expression in *Chlamydomonas reinhardtii* mediated by an endogenous intron. Plant J 14:441–447

Manuell A, Beligni MV, Elder JH et al (2007) Robust expression of a bioactive mammalian protein in *Chlamydomonas* chloroplast. Plant Biotechnol J 5:402–412

Matsuo T, Onai K, Okamoto K, Minagawa J, Ishiura M (2006) Real-time monitoring of chloroplast gene expression by a luciferase reporter: evidence for nuclear regulation of chloroplast circadian period. Mol Cell Biol 26:863–870

Mayfield SP, Franklin SE (2005) Expression of human antibodies in eukaryotic micro-algae. Vaccine 23:1828–1832

Mayfield SP, Kindle KL (1990) Stable nuclear transformation of Chlamydomonas reinhardtii by using a C. reinhardtii gene as the selectable marker. Proc Natl Acad Sci USA 87:2087–2091

Mayfield SP, Schultz J (2004) Development of a luciferase reporter gene, *luxCt*, for *Chlamydomonas reinhardtii* chloroplast. Plant J 37:449–458

Mayfield SP, Franklin SE, Lerner RA (2003) Expression and assembly of a fully active antibody in algae. Proc Natl Acad Sci USA 100:438–442

Minko I, Holloway SP, Nikaido S, Carter M, Odom OW, Johnson CH, Herrin DL (1999) *Renilla* luciferase as a vital reporter for chloroplast gene expression in *Chlamydomonas*. Mol Gen Genet 262:421–425

Molnar A, Schwach F, Studholme DJ, Thuenemann EC, Baulcombe DC (2007) miRNAs control gene expression in the single-cell alga Chlamydomonas reinhardtii. Nature 447:1126–1129

Molnar A, Bassett A, Thuenemann E, Schwach F, Karkare S, Ossowski S, Weigel D, Baulcombe D (2009) Highly specific gene silencing by artificial microRNAs in the unicellular alga *Chlamydomonas reinhardtii*. Plant J 58:165–174

Nelson JAE, Savereide PB, Lefebvre PA (1994) The CRY1 gene in *Chlamydomonas reinhardtii*: structure and use as a dominant selectable marker for nuclear transformation. Mol Cell Biol 14:4011–4019

Ohnuma M, Yokoyama T, Inouye T, Sekine Y, Tanaka K (2008) Polyethylene glycol (PEG)-mediated transient gene expression in a red alga, *Cyanidioschyzon merolae 10D*. Plant Cell Physiol 49:117–120

Plaza M, Herrero M, Cifuentes A, Ibanez E (2009) Innovative natural functional ingredients from microalgae. J Agric Food Chem 57:7159–7170

Potvin G, Zhang Z (2010) Strategies for high-level recombinant protein expression in transgenic microalgae: a review. Biotechnol Adv 28:910–918

Pulz O (2001) Photobioreactors: production systems for phototrophic microorganisms. Appl Microbiol Biotechnol 57:287–293

Quinn JM, Kropat J, Merchant S (2003) Copper response element and *Crr1*-dependent Ni(2+)-responsive promoter for induced, reversible gene expression in *Chlamydomonas reinhardtii*. Eukaryot Cell 2:995–1002

Randolph-Anderson BL, Sato R, Johnson AM, Harris EH, Hauser CR, Oeda K, Ishige F, Nishio S, Gillham NW, Boynton JE (1998) Isolation and characterization of a mutant protoporphyrinogen oxidase gene from *Chlamydomonas reinhardtii* conferring resistance to porphyric herbicides. Plant Mol Biol 38:839–859

Rasala BA, Muto M, Lee PA, Jager M, Cardoso RM, Behnke CA, Kirk P, Hokanson CA, Crea R, Mendez M, Mayfield SP (2010) Production of therapeutic proteins in algae, analysis of expression of seven human proteins in the chloroplast of *Chlamydomonas reinhardtii*. Plant Biotechnol J 8:719–733

Rechter S, Konig T, Auerochs S, Thulke S, Walter H, Dornenburg H, Walter C, Marschall M (2006) Antiviral activity of Arthrospira-derived spirulan-like substances. Antiviral Res 72:197–206
Remacle C, Cline S, Boutaffala L, Gabilly S, Larosa V, Barbieri MR, Coosemans N, Hamel PP (2009) The ARG9 gene encodes the plastid-resident N-acetyl ornithine aminotransferase in the green alga Chlamydomonas reinhardtii. Eukaryot Cell 8:1460–1463
Ruecker O, Zillner K, Groebner-Ferreira R, Heitzer M (2008) Gaussia-luciferase as a sensitive reporter gene for monitoring promoter activity in the nucleus of the green alga *Chlamydomonas reinhardtii*. Mol Genet Genomics 280:153–162
Sayre R, Wagner R, Siripornadulsil S, Farias C (2003). Transgenic algae for delivering antigens to an animal. US Patent 7,410,637
Schiedlmeier B, Schmitt R, Muller W, Kirk MM, Gruber H, Mages W, Kirk DL (1994) Nuclear transformation of *Volvox carteri*. Proc Natl Acad Sci USA 91:5080–5084
Schroda M (2005) RNA silencing in *Chlamydomonas*: mechanisms and tools. Curr Genet 49:69–84
Schroda M, Blocker D, Beck CF (2000) The HSP70A promoter as a tool for the improved expression of transgenes in *Chlamydomonas*. Plant J 21:121–131
Shimogawara K, Fujiwara S, Grossman A, Usuda H (1998) High-efficiency transformation of *Chlamydomonas reinhardtii* by electroporation. Genetics 148:1821–1828
Siripornadulsil S, Traina S, Verma DP, Sayre RT (2002) Molecular mechanisms of proline-mediated tolerance to toxic heavy metals in transgenic microalgae. Plant Cell 14:2837–2847
Sizova I, Fuhrmann M, Hegemann P (2001) A *Streptomyces rimosus aphVIII* gene coding for a new type phosphotransferase provides stable antibiotic resistance to *Chlamydomonas reinhardtii*. Gene 277:221–229
Sodeinde OA, Kindle KL (1993) Homologous recombination in the nuclear genome of *Chlamydomonas reinhardtii*. Proc Natl Acad Sci USA 90:9199–9203
Specht E, Miyake-Stoner S, Mayfield S (2010) Micro-algae come of age as a platform for recombinant protein production. Biotechnol Lett 32:1373–1383
Steinbrenner J, Sandmann G (2006) Transformation of the green alga *Haematococcus pluvialis* with a phytoene desaturase for accelerated astaxanthin biosynthesis. Appl Environ Microbiol 72:7477–7484
Stevens DR, Rochaix JD, Purton S (1996) The bacterial phleomycin resistance gene *ble* as a dominant selectable marker in *Chlamydomonas*. Mol Gen Genet 251:23–30
Stevens DR, Atteia A, Franzen LG, Purton S (2001) Cycloheximide resistance conferred by novel mutations in ribosomal protein L41 of *Chlamydomonas reinhardtii*. Mol Gen Genet 264: 790–795
Sun M, Qian K, Su N, Chang H, Liu J, Chen G (2003) Foot-and-mouth disease virus VP1 protein fused with cholera toxin B subunit expressed in *Chlamydomonas reinhardtii* chloroplast. Biotechnol Lett 25:1087–1092
Sun Y, Gao X, Li Q, Zhang Q, Xu Z (2006) Functional complementation of a nitrate reductase defective mutant of a green alga *Dunaliella viridis* by introducing the nitrate reductase gene. Gene 377:140–149
Surzycki R, Greenham K, Kitayama K, Dibal F, Wagner R, Rochaix JD, Ajam T, Surzycki S (2009) Factors effecting expression of vaccines in microalgae. Biologicals 37:133–138
Tan C, Qin S, Zhang Q, Jiang P, Zhao F (2005) Establishment of a micro-particle bombardment transformation system for Dunaliella salina. J Microbiol 43:361–365
Tang DK, Qiao SY, Wu M (1995) Insertion mutagenesis of *Chlamydomonas reinhardtii* by electroporation and heterologous DNA. Biochem Mol Biol 36:1025–1035
Ten Lohuis MR, Miller DJ (1998) Genetic transformation of dinoflagellates (Amphidinium and Symbiodinium): expression of GUS in microalgae using heterologous promoter constructs. Plant J 13:427–435
Tran M, Zhou B, Pettersson PL et al (2009) Synthesis and assembly of a full-length human monoclonal antibody in algal chloroplasts. Biotechnol Bioeng 104:663–673
Ugwu CU, Aoyagi H, Uchiyama H (2008) Photobioreactors for mass cultivation of algae. Bioresour Technol 99:4021–4028

Walker TL, Becker DK, Dale JL, Collet C (2005a) Towards the development of a nuclear transformation system for *Dunaliella tertiolecta*. J Appl Phycol 17:363–368

Walker TL, Purton S, Becker DK, Collet C (2005b) Microalgae as bioreactors. Plant Cell Rep 24: 629–641

Wang T, Xue L, Hou W, Yang B, Chai Y, Ji X, Wang Y (2007) Increased expression of transgene in stably transformed cells of *Dunaliella salina* by matrix attachment regions. Appl Microbiol Biotechnol 76:651–657

Wang X, Brandsma M, Tremblay R, Maxwell D, Jevnikar AM, Huner N, Ma S (2008) A novel expression platform for the production of diabetes-associated autoantigen human glutamic acid decarboxylase (hGAD65). BMC Biotechnol 8:87–99

Yang Z, Li Y et al (2006) Expression of human soluble TRAIL in *Chlamydomonas reinhardtii* chloroplast. Chin Sci Bull 51:1703–1709

Yasuhara-Bell J, Lu Y (2010) Marine compounds and their antiviral activities. Antiviral Res 86: 231–240

Zaslavskaia LA, Lippmeier JC, Kroth PG, Grossman AR, Apt KE (2000) Transformation of the diatom *Phaeodactylum tricornutum* (*Bacillariophyceae*) with a variety of selectable marker and reporter genes. J Appl Phycol 36:379–386

Zaslavskaia LA, Lippmeier JC, Shih C, Grossman AR, Apt KE (2001) Trophic conversion of an obligate photoautotrophic organism through metabolic engineering. Science 292:2073–2075

Zhang Y-K, Shen G-F, Ru B-G (2006) Survival of human metallothionein-2 transplastomic *Chlamydomonas reinhardtii* ultraviolet B exposure. Acta Biochim Biophys Sin 38:187–193

Chapter 7
The Production of Vaccines and Therapeutic Antibodies in Plants

Richard M. Twyman, Stefan Schillberg, and Rainer Fischer

Abstract Biopharmaceuticals such as antibodies and recombinant subunit vaccines are generally produced on a commercial basis by process-scale fermentation in bacteria, yeast or animal cells. Plants and plant cells have joined the exclusive club of commercial production platforms comparatively recently, but they offer certain advantages over the more established systems particularly in terms of economy, scalability, response times and formulation options. After a promising start and then a rocky transition from early R&D to preclinical and clinical development, plants are now becoming more firmly established as an alternative to microbes and mammalian cells for the production of pharmaceutical proteins. Several plant-derived pharmaceuticals have undergone clinical trials and the first products for human use are approaching market authorization, with antibodies and vaccines strongly represented among these front runners. Although scientific advances have played an important role in the maturation of plant-based production technology, perhaps the most critical development has been the definition of a workable regulatory framework which has recently yielded the first processes for biopharmaceutical production in plants that comply with good manufacturing practice. In this chapter, we consider

R.M. Twyman (✉)
Department of Biological Sciences, University of Warwick, Coventry, CV4 7AL, UK
e-mail: richard@writescience.com

S. Schillberg
Fraunhofer Institute for Molecular Biology and Applied Ecology,
Forckenbeckstrasse 6, 52074 Aachen, Germany

R. Fischer
Fraunhofer Institute for Molecular Biology and Applied Ecology,
Forckenbeckstrasse 6, 52074 Aachen, Germany

RWTH, Institute for Molecular Biotechnology, Worringerweg 1, 52074 Aachen, Germany

A. Wang and S. Ma (eds.), *Molecular Farming in Plants: Recent Advances and Future Prospects*, DOI 10.1007/978-94-007-2217-0_7,

the state of the art in plant-based production systems for antibodies and vaccines and discus the development issues which remain to be addressed before plants become an acceptable mainstream production technology.

7.1 Introduction

Humans have relied on plants as a source of medicines since antiquity and it is estimated that up to 25% of drugs on the market today contain active pharmaceutical ingredients derived from plants (Raskin et al. 2002). More recently, plants have also been considered as heterologous production platforms for recombinant pharmaceutical proteins, which are usually produced by process-scale fermentation in bacteria, yeast or animal cells (Twyman et al. 2005; Desai et al. 2010). The first pharmaceutical proteins produced in plants were human serum albumin expressed in tobacco and potato leaves and suspension cells (Sijmons et al. 1990), and a monoclonal antibody that was expressed in tobacco leaves (Hiatt et al. 1989). These pioneering studies demonstrated that plants could produce stable and functional human proteins, leading to the concept of molecular farming, i.e. the commercial production of valuable recombinant proteins in plants on an agricultural scale (Schillberg et al. 2003). Since then, hundreds of pharmaceutical proteins have been produced in a variety of plants and plant-based systems, amassing a vast literature of proof-of-principle studies demonstrating the virtues of different plant species, cells/tissues and expression strategies (Ma et al. 2003; Twyman et al. 2003, 2005; De Muynck et al. 2010; Rybicki 2010).

These studies to a greater or lesser degree all emphasized the advantages of plants in three main areas: economy, scalability and safety. Plants are inexpensive compared to fermenter systems, they are also much more scalable, and in terms of safety they do not produce endotoxins like bacteria, nor do they support the proliferation of human viruses and prions like mammalian cells. Many companies were set up to exploit these advantages using an array of diverse plant systems, and the world looked forward to a new era of inexpensive medicines. Unfortunately, these early pioneers had not reckoned on the difficulties that would be encountered when translating their research into practical, industrial processes. There were technical limitations in that the yields from plants were not in the same league as those achieved using industry standards such as *Escherichia coli* and Chinese hamster ovary (CHO) cells, there were previously unrecognized differences between recombinant proteins produced in plants and animals because of different glycan structures, but most importantly there was a lack of regulations governing the use of plant systems which resulted in a huge roadblock in clinical development (Spok et al. 2008). The re-emergence of molecular farming as a viable production technology in the last few years has resulted from a decade of work to overcome technical limitations and to define and develop a suitable regulatory pathway for the production of pharmaceuticals in plants according to good manufacturing practice (Fischer et al. 2011).

7.2 Lead Products – Why Antibodies and Vaccines?

There are still no approved plant-derived pharmaceutical products for human use, although at the time of writing the FDA is on the brink of approving a recombinant form of glucocerebrosidase produced in carrot cells. A tobacco-derived subunit vaccine against Newcastle disease was approved by the USDA in February 2006 for use in poultry. Furthermore, a recombinant antibody that binds the *Hepatitis B virus* surface antigen was approved in Cuba by the Center for State Control of Medication Quality (CECMED) in June 2006. This is not used directly as a pharmaceutical product, but as an affinity reagent to purify the viral surface antigen (which is produced using conventional fermentation technology). However, it is important to include this product because the approval process is as rigorous as that for an active pharmaceutical ingredient.

With one plant-derived pharmaceutical product perched on the brink of approval, it is important to consider the clinical pipeline and what products are likely to follow. This reveals that there are two major classes of product in development – plant-derived antibodies and plant-derived subunit vaccines. The popularity of these products as development targets reflects their regulatory status and the strategic advantages of using plants to produce them. Antibodies are popular targets because they currently dominate the biopharmaceutical pipeline, representing up to 30% of biopharmaceuticals in development (Sheridan 2010). There is a large body of literature confirming that antibodies in a variety of formats can be expressed successfully in plants and remain functional after isolation from plant tissue (De Muynck et al. 2010). Perhaps most importantly, antibodies tend to have a stoichiometric mechanism of action and are therefore required in large doses. This means that for some antibodies, particularly those envisaged as topical reagents (e.g. as microbicides) production on the 100–1,000 kg scale would be necessary to cope with demand (Gottschalk 2009). Currently, only plants have the scalability to meet this challenge. Antibodies as topical reagents also attract less regulatory scrutiny than injectables, so it is no coincidence that topical antibodies envisaged as microbicides were chosen as fast-track candidates during the development of regulatory guidelines for plant-derived pharmaceuticals (Ma et al. 2005a, b).

Much the same reasoning applies to the development of subunit vaccines expressed in plants because many of the pioneers were developed as oral vaccines, to be administered in partly processed plant food (e.g. mashed potato or tomato paste), therefore sidestepping much of the regulatory scrutiny that would apply to vaccines administered by injection. Where oral vaccination is possible, antigens presented as part of a food matrix tend to remain more stable and are therefore more efficacious, although dosing can be difficult to control (Nochi et al. 2007). The development of human vaccines produced in plants has also been boosted by the large number of studies showing the efficacy of plant-derived vaccines for the prevention of diseases in other animals, culminating with the USDA approval of the Newcastle disease vaccine discussed above (Rybicki 2010). More recently, the development of transient expression platforms based on *Agrobacterium tumefaciens*, plant viruses, or hybrids combining the advantages of both, has led to a resurgence of interest in plant-derived vaccines because transient expression in plants can be scaled up much more quickly than fermentation or egg-based platforms, therefore offering a rapid response to emerging pandemics or bioterrorism (D'Aoust et al. 2010; Rybicki 2010).

7.2.1 Antibodies Produced in Plants

Many antibodies have been expressed in a variety of different plant species and tissues, but it is important to distinguish between those produced as pharmaceutical proteins with the intention of at least partial extraction and purification, and those intended to function *in planta* as a strategy to tackle plant diseases (Schillberg et al. 2001). Pharmaceutical antibodies in plants often represent proof-of-principle studies with no subsequent development, but others are heading towards the clinic, and several plant-derived antibodies have already been evaluated in clinical trials or are poised to enter clinical trials. The ones we consider are listed below:

- Avicidin from maize (NeoRx/Moinsanto). Indicated for colorectal cancer, but withdrawn from phase II trials in 1998 because of adverse effects.
- CaroRx from tobacco (Planet Biotechnology). Indicated for dental caries, the antibody binds specifically to *Streptococcus mutans*, the bacterium that causes dental caries. Phase II trials completed and product is licensed in the EU as a medical device.
- BLX-301, an anti-CD20 optimized antibody for the treatment of non-Hodgkin B-cell lymphoma expressed in the aquatic plant *Lemna minor* by Biolex Inc.
- MAPP66, a combination of antibodies envisaged as a HSV/HIV microbicide and expressed in *N. benthamiana* using Bayer's MagnICON (magnifection) technology.
- 2G12, an HIV-neutralizing antibody expressed in tobacco (Fraunhofer IME/ Pharma-Planta).

The two plant-derived antibody products that have completed phase II trials are Avicidin and CaroRx. Avicidin is a full length IgG specific for EpCAM (a marker of colorectal cancer) produced in maize and developed jointly by NeoRx and Monsanto. Although Avicidin demonstrated therapeutic efficacy in patients with advanced colon and prostate cancers, it was withdrawn from phase II trials in 1998 because treatment resulted in a high incidence of diarrhea (Gavilondo and Larrick 2000). The same issue arose with the equivalent antibody produced in mammalian cells and in all other respects the two antibodies were comparable (physicochemical properties, serum clearance, urine clearance and dosimetry). Therefore, the adverse effects were not linked in any way to the use of maize as the production platform.

CaroRx is a chimeric secretory IgA/G produced in transgenic tobacco plants which has completed phase II trials sponsored by Planet Biotechnology Inc. (Ma et al. 1998). Secretory antibody production requires the expression of four separate polypeptides, which in this case were initially expressed in four different plant lines that were crossed over two generations to stack all the transgenes in one line. The antibody binds to the major adhesin SA I/II of *Streptococcus mutans*, the bacterium that causes tooth decay in humans. Topical application following elimination of bacteria from the mouth helped to prevent recolonization by *S. mutans* and led to the replacement of this pathogenic organism with harmless endogenous flora. To circumvent the regulatory vacuum surrounding plant-derived pharmaceuticals in the late 1990s and early 2000s, CaroRx was registered as a medical device.

BLX-301 has completed phase I trials. BLX-301 is an anti-CD20 antibody indicated for non-Hodgkin lymphoma which produced in the Lex system developed by the US biotechnology company Biolex Inc. and based on duckweed (*Lemna minor*). This plant has a number of significant advantages for the production of recombinant pharmaceutical proteins as it can be grown in sealed, aseptic tanks under constant conditions with a chemically defined medium, which ensures batch-to-batch consistency and allows clones to be scaled up rapidly in the company's GMP facility. The Lex system also allows the glycan structures of the antibody to be controlled and optimized (Cox et al. 2006), thus Biolex claims that BLX-301 is more potent and efficacious than equivalent antibodies produced in mammalian cells and has fewer adverse effects.

MAPP66 is an antibody cocktail envisaged as a microbicide to prevent the transmission of HIV and HSV, and it has also completed phase I trials. MAP66 is produced by transient expression in *Nicotiana benthamiana* using the magnifection technology. The principle of magnifection, developed by Icon Genetics (now part of Bayer) is that genes are delivered as part of a recombinant plant virus (in this case *Tobacco mosaic virus*) but the systemic spreading of the virus is made unnecessary through the use of *A. tumefaciens* as a delivery vehicle (Marillonnet et al. 2005; Gleba et al. 2005). The bacterium delivers the viral genome to so many cells that local spreading is sufficient for the entire plant to be infected and this overcomes an often critical limitation of plant viruses which is that the host range, efficiency of infection and ultimately the speed at which recombinant proteins accumulate all depend on its natural ability to spread systemically through the plant. Taking the systemic spreading function away from the virus and relying instead on the bacterium to deliver the viral genome to a large number of cells allows the same viral vector to be used in a wide range of plants (Gleba et al. 2004).

Four monoclonal antibodies with potent HIV-neutralizing activity have been identified (b12, 2G12, 2F5 and 4E10) and have been shown to prevent virus transmission in animal models (Cardoso et al. 2005). Combinations of 2–4 of these antibodies would therefore make effective microbicides, and the demand for such products particularly in endemic regions such as sub-Saharan Africa would mean multi-ton scale production, which would only be possible in plants. Tobacco and maize plants have been used to produce 2G12 (Ramessar et al. 2008; Rademacher et al. 2008) and 2F5 (Sack et al. 2007), and tobacco plants producing 2G12 have been developed within the EU Framework Program 6 project Pharma-Planta as a pioneer to help define the regulatory pathway for plant-derived pharmaceuticals (Ma et al. 2005a). The objective of this project was to take candidate pharmaceutical products from gene to clinic, defining the regulatory pathway in concert with the appropriate regulatory bodies along the way. This process has been largely successful and a phase I clinical study of the plant-derived 2G12 antibody commenced in July 2011.

7.2.2 *Vaccine Candidates Produced in Plants*

Plant-derived vaccines can be divided into two categories – those designed for veterinary use and those designed for medical use. The Newcastle disease vaccine for poultry was the first plant-derived pharmaceutical product to be approved, and there is a large body of both immunogenicity and challenge data to support the efficacy of such vaccines, including numerous clinical studies (Twyman et al. 2005).

Seven human clinical trials involving plant-derived subunit vaccines have also been reported. Tacket et al. (1998) performed the first such trial with transgenic potatoes expressing the enterotoxigenic *E. coli* (ETEC) labile toxin B-subunit (LTB), one of the most potent known oral immunogens. The LTB content of the tubers varied between 3.7 and 15.7 μg g^{-1} fresh weight. Fourteen volunteers were given either transgenic or non-transgenic potato on days 0, 7, and 21 of the trial. Almost all of those consuming the transgenic potatoes showed at least four-fold increases in serum IgG against LTB while no such increase was seen in those consuming the non-transformed potatoes. Five of these individuals also demonstrated at least a four-fold increase in anti-LTB IgA, detected in stool samples. There were few side effects, such as nausea and diarrhea. A more recent trial using LTB expressed in processed corn seed produced similar results to the potato study (Tacket et al. 2004). The same group also described the results of a clinical trial performed using transgenic potato tubers expressing the Norwalk virus capsid protein (NVCP) (Tacket et al. 2000). Twenty adult volunteers were given two or three 150-g doses each of raw transgenic potato tuber containing 215–750 μg NVCP. Although only 50% of the NVCP subunits assembled into virus-like particles in the potato cells, thus reducing the effective dose of the vaccine, nearly all of the volunteers showed significant increases in the numbers of IgA antibody-forming cells (AFCs), and six of these individuals also showed increases in IgG AFCs. There were also noticeable increases in serum antibodies against NVCP and stool IgA antibodies in a few of the participants.

A clinical trial has also been carried out using orally delivered HBV surface antigen produced in lettuce (Kapusta et al. 1999). Two of three volunteers who were given two 150-g doses of transgenic lettuce containing about 2 μg of the antigen per dose, produced protective serum antibodies after the second dose, although the titers declined in a few weeks. Even so, the study confirmed that naïve subjects could be seroconverted by the oral delivery of a plant-derived viral antigen. A similar trial in the United States involved the HBV surface antigen expressed in transgenic potatoes although participants in this trial had already been seroconverted with the standard, yeast-derived vaccine (Richter et al. 2000). Of 33 participants given either two or three 1-mg doses of the antigen, about half showed increased serum IgG titers against the virus.

Yusibov et al. (2002) have carried out a trial involving 14 volunteers given spinach infected with *Alfalfa mosaic virus* vectors expressing the rabies virus glycoprotein and nucleoprotein. Five of these individuals had previously received a conventional rabies vaccine. Three of those five and all nine of the initially naïve subjects produced antibodies against rabies virus while no such response was seen in those given normal spinach.

In all the above cases, the subunit vaccine was delivered orally as part of a matrix of plant material, and the advantage of using plants was convenience of the oral delivery route and the efficacy mediated by presentation of the antigen as a bioencapsulated formulation. In contrast, McCormick et al. (1999) produced a single chain antibody fragment designed to confer passive immunity against non-Hodgkin lymphoma using *Tobacco mosaic virus* as a vector in tobacco plants, but purified the vaccine for administration by injection. In preclinical development, a vaccine based on the well-characterized mouse lymphoma cell line 38C13 stimulated the production of anti-idiotype antibodies capable of recognizing 38C13 cells, providing immunity against lethal challenge with the lymphoma. Here the advantage of plants was not the potential for oral delivery, but the rapid production that can be achieved by transient expression, allowing the development of patient-specific vaccines for personalized B-cell lymphoma therapy. At least 12 such vaccines have been tested in phase II trials (McCormick et al. 2008).

The rapid production that is possible in plants is useful not only for personalized medicines but also for scaling up production as a rapid response to pandemic threats. Researchers at Medicago Inc. in Canada and at Fraunhofer CMB in Newark, Delaware, recently achieved the impressive feat of producing gram quantities of vaccine against an emerging influenzavirus strain within 1 month of isolating the hemagglutinin sequence (Rybicki et al. 2010). The Medicago vaccine was subsequently tested in a phase I clinical study to demonstrate safety (Landry et al. 2010).

7.3 Recent Advances in Production Technology

7.3.1 Optimizing Yields and Recovery

The yields of recombinant proteins produced in plants are generally much lower than those routinely achieved in fermenter systems and this has been a challenge both in terms of establishing the credibility of molecular farming and also in the development of economically viable processes and (in the case of oral vaccines) efficacious products. The recombinant protein itself plays an important role in determining the yield because some proteins are inherently less stable than others, and some proteins produce unanticipated negative effects on plant growth. In general terms, yields can be improved by optimizing expression constructs to maximize transcription, mRNA stability and protein synthesis, increasing the transgene copy number and introducing transgenes into germplasm that is best suited for high-level expression (reviewed by Desai et al. 2010). However, many of the problems with low expression levels are restricted to nuclear transgenic plants, and very high expression levels can be achieved using transient expression platforms based on Agrobacterium and/or plant viruses (Giritch et al. 2006; Sainsbury and Lomonossoff 2008; Vézina et al. 2009; Huang et al. 2010; Pogue et al. 2010). For the time being, transient expression is likely to be the first choice for products required rapidly in large quantities. Even so, transgenic systems have a number of advantages despite their longer development times, slower scaling-up and the increased regulatory burden that comes with the GM label – these include the permanent genetic resource represented by an integrated transgene (no requirement for the reintroduction of bacteria or viruses in every production generation), the absence of genetically modified bacteria and viruses, and the batch-to-batch consistency. Transgenic plants are likely to remain the system of choice for products required in large quantities but not as an urgent response.

Because antibodies are complex multimeric glycoproteins, the stability (and therefore the yield) depends greatly on their ability to fold and assemble correctly, which in turn depends on their subcellular localization (this is not such a critical factor with vaccine candidates, which tend to be smaller and simpler proteins). The yield of an antibody is enhanced by adding a signal peptide to allow secretion to the apoplast, or even better a signal peptide and a KDEL/HDEL tetrapeptide so that secreted proteins are retrieved to the endoplasmic reticulum (ER), reflecting the favorable environment for protein folding, the presence of chaperones and the absence of significant protease activity (Sharma and Sharma 2009). In some cases, antibody stability can be increased by coexpressing the corresponding antigen although subsequent removal of the antigen after the complex has been extracted increases the complexity of downstream processing. Alternatively, stability can be increased by fusion to a stabilizing protein partner. For example, Floss et al. (2008) showed that the fusion of elastin-like peptides (ELPs) to the C-terminus of antibody 2F5 enhanced its stability without affecting its binding activity. The ELP also provides a convenient extraction method known as reverse transition cycling, which is based on reversible temperature-dependent precipitation (Conley et al. 2009, 2011). A similar approach involves fusion with the seed storage protein γ-zein, which results in the assembly of new storage organelles and increases yields by up to 300% (Torrent et al. 2009).

7.3.2 *Dealing with Plant Glycans*

The other major technological challenge facing the pioneers of molecular farming was the impact of differences between plants and mammals in terms of protein glycosylation (Twyman et al. 2005). Plant-derived recombinant human glycoproteins tend to contain the carbohydrate groups β(1,2)xylose and α(1,3)fucose, which are not found in mammals, but they do not contain terminal galactose and sialic acid residues that are found in native human glycoproteins because the corresponding enzymes are not present in plants (Gomord et al. 2010). The impact of glycan differences is more important for antibodies than vaccine candidates because antibodies are expected to bind antigens and carry out their effector functions, whereas antigens are expected solely to generate an immune response. Glycans are necessary for the biological activity of some proteins, and incorrect glycans may influence the solubility, stability or activity of a protein and thus its pharmacokinetic properties. Plant glycans may be immunogenic in some mammals but there is no evidence for this in humans, and it should be noted that CHO cells, which are rodent in origin, also produce non-human glycans. Nevertheless, the potential impact of plant glycans on protein structure, activity and safety has resulted in regulatory pressure on researchers to either remove or 'humanize' plant glycans, e.g. by expressing aglycosylated derivates of proteins lacking the glycan attachment sites, by targeting proteins to the ER and thus avoiding Golgi-specific modifications to ensure all glycans are of the universal 'high-mannose' type (Sriraman et al. 2004; Triguero et al. 2005), and by glycoengineering to prevent the addition of plant-type glycans and/or to add human type glycans (Gomord et al. 2004). This has been achieved using gene knockout and RNA interference techniques in Arabidopsis, tobacco, duckweed and moss to abolish plant-specific glycosylases (Strasser et al. 2004, 2008; Decker and Reski 2004; Schahs et al. 2007) and through the expression of the entire mammalian pathway for sialic acid synthesis to allow protein galactosylation and sialylation (Strasser et al. 2009; Castilho et al. 2010).

It is important to state that non-authentic glycans are not necessarily always a disadvantage, and can affect the solubility, stability and biological activity of a protein positively as well as negatively, e.g. by extending its half life or increasing its affinity for a target. BLX-301 has already been cited as one example of beneficial glycan modification in plants, where the optimized glycan structures increase the potency of the antibody. In a different sense, Uplyso (taliglucerase alfa), the recombinant glucocerebrosidase produced in carrot cells which is poised for FDA approval, provides another example of beneficial plant glycans. Here the protein lacks sialic acid residues which are present in the equivalent protein produced in CHO cells (Cerezyme, imiglucerase). However, because sialic residues reduce the potency and longevity of the protein, they are removed from Cerezyme after purification by in vitro treatment with an exoglycosidase. It is ironic that plant glycans were once regarded as a serious bottleneck in the commercial development of molecular farming, but the extensive technology for glycan modification that has developed in response now makes plants the most versatile platform for the production of pharmaceutical proteins whose properties can be augmented or improved by glycan engineering.

7.3.3 Downstream Processing and Good Manufacturing Practice

The first decade of molecular farming research focused almost exclusively on upstream productivity, increasing yields and protein stability. Comparatively little attention was paid to downstream processing, and almost nothing was done to address the challenges of large-scale processing which would be an integral component of any commercial platform.

One company that did look into downstream processing in detail was Prodigene Inc., which developed maize as a commercial platform for the production of protein technical reagents used in research (Hood 2002). Much of the accepted wisdom about downstream processing in plant-based production systems came from their detailed studies of avidin and β-glucuronidase (GUS), which were commercially viable and marketed by Sigma-Aldrich Corp. alongside avidin from chicken eggs and GUS from *E. coli*. In both cases, the company demonstrated equivalence to the native product, stability during processing operations such as dry milling, fractionation and hexane extraction, and yields equivalent to 1–2% of soluble seed protein, which is the common benchmark cited for commercial viability in molecular farming (Hood et al. 1997; Witcher et al. 1998; Kusnadi et al. 1998). Before their demise, Prodigene Inc. was working on processing strategies for plant-derived antibodies and vaccines, and these processes provided foundations for the development of GMP processes that are now applied in other plants. Several aspects of downstream processing have to be customized specifically for plant systems, including the removal of fibers, oils and other by-products from certain crops, and process optimization for the treatment of different plant species and tissues (Menkhaus et al. 2004; Nikolov and Woodard 2004).

Two distinct strategies have arisen for creating a regulatory pathway for plant-derived pharmaceuticals, one based on the development of systems that are similar in concept to the industry standards, and the other based on the modification of existing regulations to deal with whole plants. In the first category, processes have been developed based on cultivated plant cells, analogous to processes for microbes and mammalian cells, e.g. the tobacco BY-2 system developed by Fraunhofer and Dow AgroSciences, and the carrot cell platform (ProCellEx) developed by Protalix BioTherapeutics (Aviezer et al. 2009). Further systems were developed based on algae, moss and aquatic plants, all linked by the ability to use closed vessels, defined synthetic media and controlled growth environments (Cox et al. 2006; Franklin and Mayfield 2005; Decker and Reski 2004). These systems provide containment, consistency and are similar to mammalian cells both conceptually and practically (Hellwig et al. 2004; Tiwari et al. 2009; Xu et al. 2011). In the second category, new concepts were developed such as the replacement of cell banking with seed banking, and new paradigms to embrace the biological differences between complex multicellular organisms and single cells (Ma et al. 2005a; Whaley et al. 2011; Fischer et al. 2011). After much preparatory work, this has resulted in the successful development and application of GMP processes involving contained systems such as carrot cells (Protalix Biotherapeutics), moss (greenovation, Frieberg) and Lemna (Biolex Therapeutics). Several organizations also now offer GMP manufacturing based on transient expression in tobacco or *Nicotiana benthamiana*, e.g. Kentucky BioProcessing (Owensboro, KY), Bayer/Icon Genetics (Halle, Germany), Fraunhofer CMB (Newark, DE), Medicago (Quebec, Canada) and Texas A&M University/G-Con LLC (College Station, TX). Finally, and uniquely, the Fraunhofer IME in Aachen, Germany, has a GMP facility for the production of recombinant antibodies in transgenic tobacco plants.

7.4 Conclusions and Outlook

A number of plant-derived pharmaceutical products are now very close to market authorization, thanks to the development of GMP-compliant production processes and the successful completion of clinical trials. The first product is likely to be Uplyso, the recombinant glucocerebrosidase produced in carrot cells by Protalix Biotherapeutics in concert with Pfizer, but many of the remaining products in the pipeline are either antibodies or vaccines, reflecting the key advantages of plants – economy, scalability and rapid response. The absence of a GMP system for whole plants until recently has resulted in some inventive approaches to innovate around the regulations, including registering a plant-derived antibody as a medical device and focusing on oral vaccination and topical prophylaxis as primary administration routes. Now GMP processes are available for whole plants as well as contained plant-based systems, it is likely that the pipeline will fill rapidly with candidates falling into three main categories: (a) antibodies and vaccines required in 100–1,000 kg amounts, which exceed the capacity of traditional platforms; (b) vaccine candidates required as a rapid response to bioterrorism threats or emerging pandemics; and (c) pharmaceutical proteins whose efficacy or pharmacokinetic properties can be improved by glycan modification in vivo, which is now more advanced in plants that any other production platform.

References

Aviezer D, Brill-Almon E, Shaaltiel Y, Hashmueli S, Bartfeld D, Mizrachi S, Liberman Y, Freeman A, Zimran A, Galun E (2009) A plant-derived recombinant human glucocerebrosidase enzyme – a preclinical and phase I investigation. PLoS One 4:e4792

Cardoso RMF, Zwick MB, Stanfield RL, Kunert R, Binley JM, Katinger H, Burton DR, Wilson IA (2005) Broadly neutralizing anti-HIV antibody, 4E10 recognizes a helical conformation of a highly conserved fusion-associated motif in gp41. Immunity 22:163–173

Castilho A, Strasser R, Stadlmann J, Grass J, Jez J, Gattinger P, Kunert R, Quendler H, Pabst M, Leonard R, Altmann F, Steinkellner H (2010) In planta protein sialylation through overexpression of the respective mammalian pathway. J Biol Chem 285:15923–15930

Conley AJ, Joensuu JJ, Jevnikar AM, Menassa R, Brandle JE (2009) Optimization of elastin-like polypeptide fusions for expression and purification of recombinant proteins in plants. Biotechnol Bioeng 103:562–573

Conley AJ, Joensuu JJ, Richman A, Menassa R (2011) Protein body-inducing fusions for high-level production and purification of recombinant proteins in plants. Plant Biotechnol J 9:419–433

Cox KM, Sterling JD, Regan JT, Gasdaska JR, Frantz KK, Peele CG, Black A, Passmore D, Moldovan-Loomis C, Srinivasan M, Cuison S, Cardarelli PM, Dickey LF (2006) Glycan optimization of a human monoclonal antibody in the aquatic plant *Lemna minor*. Nat Biotechnol 24:1591–1597

D'Aoust MA, Couture MM, Charland N, Trépanier S, Landry N, Ors F, Vézina LP (2010) The production of hemagglutinin-based virus-like particles in plants: a rapid, efficient and safe response to pandemic influenza. Plant Biotechnol J 8:1–13

De Muynck B, Navarre C, Boutry M (2010) Production of antibodies in plants: status after twenty years. Plant Biotechnol J 8:529–563
Decker EL, Reski R (2004) The moss bioreactor. Curr Opin Plant Biol 7:166–170
Desai P, Shrivastava N, Padh H (2010) Production of heterologous protein in plants: strategies for optimal expression. Biotechnol Adv 28:427–435
Fischer R, Schillberg S, Hellwig S, Twyman RM, Drossard J (2011) GMP issues for plant-derived recombinant proteins. Biotechnol Adv (in press)
Floss DM, Sack M, Stadlmann J, Rademacher T, Scheller J, Stoger E, Fischer R, Conrad U (2008) Biochemical and functional characterization of anti-HIV antibody-ELP fusion proteins from transgenic plants. Plant Biotechnol J 6:379–391
Franklin SE, Mayfield SP (2005) Recent developments in the production of human therapeutic proteins in eukaryotic algae. Expert Opin Biol Ther 5:225–235
Gavilondo JV, Larrick JW (2000) Antibody engineering at the millennium. Biotechniques 29:128–138
Giritch A, Marillonnet S, Engler C, van Eldik G, Botterman J, Klimyuk V, Gleba Y (2006) Rapid high-yield expression of full-size IgG antibodies in plants coinfected with noncompeting viral vectors. Proc Natl Acad Sci USA 103:14701–14706
Gleba Y, Marillonnet S, Klimyuk V (2004) Engineering viral expression vectors for plants: the 'full virus' and the 'deconstructed virus' strategies. Curr Opin Plant Biol 7:182–188
Gleba Y, Klimyuk V, Marillonnet S (2005) Magnifection – a new platform for expressing recombinant vaccines in plants. Vaccine 23:2042–2048
Gomord V, Sourrouille C, Fitchette AC, Bardor M, Pagny S, Lerouge P, Faye L (2004) Production and glycosylation of plant-made pharmaceuticals: the antibodies as a challenge. Plant Biotechnol J 2:83–100
Gomord V, Fitchette AC, Menu-Bouaouiche L, Saint-Jore-Dupas C, Plasson C, Michaud D, Faye L (2010) Plant-specific glycosylation patterns in the context of therapeutic protein production. Plant Biotechnol J 8:564–587
Gottschalk U (2009) Process scale purification of antibodies. Wiley, New York
Hellwig S, Drossard J, Twyman RM, Fischer R (2004) Plant cell cultures for the production of recombinant proteins. Nat Biotechnol 22:1415–1422
Hiatt A, Cafferkey R, Bowdish K (1989) Production of antibodies in transgenic plants. Nature 342:76–78
Hood EE (2002) From green plants to industrial enzymes. Enzyme Microb Technol 30:279–283
Hood EE, Witcher DR, Maddock S, Meyer T, Baszczynski C, Bailey M, Flynn P, Register J, Marshall L, Bond D, Kulisek E, Kusnadi A, Evangelista R, Nikolov Z, Wooge C, Mehigh RJ, Herman R, Kappel WK, Ritland D, Li CP, Howard J (1997) Commercial production of avidin from transgenic maize: characterization of transformant, production, processing, extraction and purification. Mol Breed 3:291–306
Huang Z, Phoolcharoen W, Lai H, Piensook K, Cardineau G, Zeitlin L, Whaley KJ, Arntzen CJ, Mason HS, Chen Q (2010) High-level rapid production of full-size monoclonal antibodies in plants by a single-vector DNA replicon system. Biotechnol Bioeng 106:9–17
Kapusta J, Modelska A, Figlerowicz M, Pniewski T, Letellier M, Lisowa O, Yusibov V, Koprowski H, Plucienniczak A, Legocki AB (1999) A plant-derived edible vaccine against hepatitis B virus. FASEB J 13:1796–1799
Kusnadi AR, Evangelista RL, Hood EE, Howard JA, Nikolov ZL (1998) Processing of transgenic corn seed and its effect on the recovery of recombinant β-glucuronidase. Biotechnol Bioeng 60:44–52
Landry N, Ward BJ, Trépanier S, Montomoli E, Dargis M, Lapini G, Vézina LP (2010) Preclinical and clinical development of plant-made virus-like particle vaccine against avian H5N1 influenza. PLoS One 5:e15559
Ma JK, Hikmat BY, Wycoff K, Vine ND, Chargelegue D, Yu L, Hein MB, Lehner T (1998) Characterization of a recombinant plant monoclonal secretory antibody and preventive immunotherapy in humans. Nat Med 4:601–606

Ma JKC, Drake P, Christou P (2003) The production of recombinant pharmaceutical proteins in plants. Nat Rev Genet 4:794–805

Ma JKC, Barros E, Bock R, Christou P, Dale PJ, Dix PJ, Fischer R, Irwin J, Mahoney R, Pezzotti M, Schillberg S, Sparrow P, Stoger E, Twyman RM (2005a) Molecular farming for new drugs and vaccines. Current perspectives on the production of pharmaceuticals in transgenic plants. EMBO Rep 6:593–599

Ma JKC, Chikwamba R, Dale PJ, Fischer R, Mahoney R, Twyman RM (2005b) Plant-derived pharmaceuticals – the road forward. Trends Plant Sci 10:580–585

Marillonnet S, Thoeringer C, Kandzia R, Klimyuk V, Gleba Y (2005) Systemic *Agrobacterium tumefaciens*-mediated transfection of viral replicons for efficient transient expression in plants. Nat Biotechnol 23:718–723

McCormick AA, Kumagai MH, Hanley K, Turpen TH, Hakim I, Grill LK, Tuse D, Levy S, Levy R (1999) Rapid production of specific vaccines for lymphoma by expression of the tumor-derived single-chain Fv epitopes in tobacco plants. Proc Natl Acad Sci USA 96:703–708

McCormick AA, Reddy S, Reinl SJ, Cameron TI, Czerwinkski DK, Vojdani F, Hanley KM, Garger SJ, White EL, Novak J, Barrett J, Holtz RB, Tuse D, Levy R (2008) Plant-produced idiotype vaccines for the treatment of non-Hodgkin's lymphoma: safety and immunogenicity in a phase I clinical study. Proc Natl Acad Sci USA 105:10131–10136

Menkhaus TJ, Bai Y, Zhang C, Nikolov ZL, Glatz CE (2004) Considerations for the recovery of recombinant proteins from plants. Biotechnol Prog 20:1001–1014

Nikolov ZL, Woodard SL (2004) Downstream processing of recombinant proteins from transgenic feedstock. Curr Opin Biotechnol 15:479–486

Nochi T, Takagi H, Yuki Y, Yang L, Masumura T, Mejima M, Nakanishi U, Matsumura A, Uozumi A, Hiroi T, Morita S, Tanaka K, Takaiwa F, Kiyono H (2007) Rice-based mucosal vaccine as a global strategy for cold-chain- and needle-free vaccination. Proc Natl Acad Sci USA 104:10986–10991

Pogue GP, Vojdani F, Palmer KE, Hiatt E, Hume S, Phelps J, Long L, Bohorova N, Kim D, Pauly M, Velasco J, Whaley K, Zeitlin L, Garger SJ, White E, Bai Y, Haydon H, Bratcher B (2010) Production of pharmaceutical-grade recombinant aprotinin and a monoclonal antibody product using plant-based transient expression systems. Plant Biotechnol J 8:638–654

Rademacher T, Sack M, Arcalis E, Stadlmann J, Balzer S, Altmann F, Quendler H, Stiegler G, Kunert R, Fischer R, Stoger E (2008) Recombinant antibody 2G12 produced in maize endosperm efficiently neutralize HIV-1 and contains predominantly single-GlcNAc N-glycans. Plant Biotechnol J 6:189–201

Ramessar K, Rademacher T, Sack M, Stadlmann J, Platis D, Stiegler G, Labrou N, Altmann F, Ma J, Stöger E, Capell T, Christou P (2008) Cost-effective production of a vaginal protein microbicide to prevent HIV transmission. Proc Natl Acad Sci USA 105:3727–3732

Raskin I, Ribnicky DM, Komarnytsky S, Ilic N, Poulev A, Borisjuk N, Brinker A, Moreno DA, Ripoll C, Yakoby N, O'Neal JM, Cornwell T, Pastor I, Fridlender B (2002) Plants and human health in the twenty-first century. Trends Biotechnol 20:522–531

Richter LJ, Thanavala Y, Arntzen CJ, Mason HS (2000) Production of hepatitis B surface antigen in transgenic plants for oral immunization. Nat Biotechnol 18:1167–1171

Rybicki EP (2010) Plant-made vaccines for humans and animals. Plant Biotechnol J 8:620–637

Sack M, Paetz A, Kunert R, Bomble M, Hesse F, Stiegler G, Fischer R, Katinger H, Stoger E, Rademacher T (2007) Functional analysis of the broadly neutralizing human anti-HIV-1 antibody 2F5 produced in transgenic BY-2 suspension cultures. FASEB J 21:1655–1664

Sainsbury F, Lomonossoff GP (2008) Extremely high-level and rapid protein production in plants without the use of viral replication. Plant Physiol 148:1212–1218

Schähs M, Strasser R, Stadlmann J, Kunert R, Rademacher T, Steinkellner H (2007) Production of a monoclonal antibody in plants with a humanized N-glycosylation pattern. Plant Biotechnol J 5:657–663

Schillberg S, Zimmermann S, Zhang MY, Fischer R (2001) Antibody-based resistance to plant pathogens. Transgenic Res 10:1–12

Schillberg S, Fischer R, Emans N (2003) Molecular farming of recombinant antibodies in plants. Cell Mol Life Sci 60:433–445

Sharma AK, Sharma MK (2009) Plants as bioreactors: recent developments and emerging opportunities. Biotechnol Adv 27:811–832

Sheridan C (2010) Fresh from the biologic pipeline – 2009. Nat Biotechnol 28:307–310

Sijmons PC, Dekker BMM, Schrammeijer B, Verwoerd TC, Van Den Elzen PJM, Hoekema A (1990) Production of correctly processed human serum albumin in transgenic plants. Bio/Technol 8:217–221

Spok A, Twyman RM, Fischer R, Ma JKC, Sparrow PAC (2008) Evolution of a regulatory framework for plant-made pharmaceuticals. Trends Biotechnol 26:506–517

Sriraman R, Bardor M, Sack M, Vaquero C, Faye L, Fischer R, Finnern R, Lerouge P (2004) Recombinant anti-hCG antibodies retained in the endoplasmic reticulum of transformed plants lack core-xylose and core-α(1,3)fucose residues. Plant Biotechnol J 2:279–287

Strasser R, Altmann F, Mach L, Glossl J, Steinkellner H (2004) Generation of *Arabidopsis thaliana* plants with complex N-glycans lacking β1,2-linked xylose and core α1,3-linked fucose. FEBS Lett 561:132–136

Strasser R, Stadlmann J, Schähs M, Stiegler G, Quendler H, Mach L, Glössl J, Weterings K, Pabst M, Steinkellner H (2008) Generation of glycoengineered *Nicotiana benthamiana* for the production of monoclonal antibodies with a homogeneous human-like N-glycan structure. Plant Biotechnol J 6:392–402

Strasser R, Castilho A, Stadlmann J, Kunert R, Quendler H, Gattinger P, Jez J, Rademacher T, Altmann F, Mach L, Steinkellner H (2009) Improved virus neutralization by plant-produced anti-HIV antibodies with a homogeneous β1,4-galactosylated N-glycan profile. J Biol Chem 284:20479–20485

Tacket CO, Mason HS, Losonsky G, Clements JD, Levine MM, Arntzen CJ (1998) Immunogenicity in humans of a recombinant bacterial-antigen delivered in transgenic potato. Nat Med 4:607–609

Tacket CO, Mason HS, Losonsky G, Estes MK, Levine MM, Arntzen CJ (2000) Human immune responses to a novel Norwalk virus vaccine delivered in transgenic potatoes. J Infect Dis 182:302–305

Tacket CO, Pasetti MF, Edelman R, Howard JA, Streatfield S (2004) Immunogenicity of recombinant LT-B delivered orally to humans in transgenic corn. Vaccine 22:4385–4389

Tiwari S, Verma PC, Singh PK, Tuli R (2009) Plants as bioreactors for the production of vaccine antigens. Biotechnol Adv 27:449–467

Torrent M, Llompart B, Lasserre-Ramassamy S, Llop-Tous I, Bastida M, Marzabal P, Westerholm-Parvinen A, Saloheimo M, Heifetz PB, Ludevid MD (2009) Eukaryotic protein production in designed storage organelles. BMC Biol 7:5

Triguero A, Cabrera G, Cremata JA, Yuen CT, Wheeler J, Ramírez NI (2005) Plant-derived mouse IgG monoclonal antibody fused to KDEL endoplasmic reticulum-retention signal is N-glycosylated homogeneously throughout the plant with mostly high-mannose-type N-glycans. Plant Biotechnol J 3:449–457

Twyman RM, Stoger E, Schillberg S, Christou P, Fischer R (2003) Molecular farming in plants: host systems and expression technology. Trends Biotechnol 21:570–578

Twyman RM, Schillberg S, Fischer R (2005) Transgenic plants in the biopharmaceutical market. Expert Opin Emerg Drugs 10:185–218

Vézina LP, Faye L, Lerouge P, D'Aoust MA, Marquet-Blouin E, Burel C, Lavoie PO, Bardor M, Gomord V (2009) Transient co-expression for fast and high-yield production of antibodies with human-like N-glycans in plants. Plant Biotechnol J 7:442–455

Whaley KJ, Hiatt A, Zeitlin L (2011) Emerging antibody products and *Nicotiana* manufacturing. Hum Vaccine 7:349–356

Witcher D, Hood EE, Peterson D, Bailey M, Bond D, Kusnadi A, Evangelista R, Nikolov Z, Wooge C, Mehigh R, Kappel W, Register JC, Howard JA (1998) Commercial production of β-glucuronidase (GUS): a model system for the production of proteins in plants. Mol Breed 4:301–312

Xu J, Ge X, Dolan MC (2011) Towards high-yield production of pharmaceutical proteins with plant cell suspension cultures. Biotechnol Adv 29:278–299

Yusibov V, Hooper DC, Spitsin SV, Fleysh N, Kean RB, Mikheeva T, Deka D, Karasev A, Cox S, Randall J, Koprowski H (2002) Expression in plants and immunogenicity of plant virus-based experimental rabies vaccine. Vaccine 20:3155–3164

Chapter 8
Production of Industrial Proteins in Plants

Elizabeth E. Hood and Deborah Vicuna Requesens

Abstract The plant production system is advantageous for industrial enzymes. Enzymes with large scale products that demand low cost manufacturing are the markets of choice for plants. The plant production system is also advantageous for products that are harmful to single cell systems, for example oxidation/reduction (redox) enzymes. Four classes of enzymes are discussed in this chapter—xylanases, redox enzymes, amylases and cellulases. Examples of each of these classes of enzyme have been produced in plants—some as demonstration projects, others with the intent to sell the product. The authors have chosen specific examples to describe the advantages of the plant system, issues that have arisen, and potential for addressing markets. These case studies illustrate the value of using plants for production with simple agricultural inputs of sunlight, nutrients and water. With the developing demand for biofuels and biobased products, large volume enzyme markets for processing agricultural materials are rapidly becoming a demand. The logical system for producing those enzymes is in co-products of the feedstock materials. Our examples below illustrate the system.

8.1 Introduction

"Imagine a world in which any protein, either naturally occurring or designed by man, could be produced safely, inexpensively and in almost unlimited quantities using only simple nutrients, water and sunlight. This could one day become reality

E.E. Hood (✉) • D.V. Requesens
Arkansas Biosciences Institute, Arkansas State University, P.O. Box 639, State University, AR 72467, USA
e-mail: ehood@astate.edu; dvicuna@astate.edu

A. Wang and S. Ma (eds.), *Molecular Farming in Plants: Recent Advances and Future Prospects*, DOI 10.1007/978-94-007-2217-0_8,

as we learn to harness the power of plants for the production of recombinant proteins on an agricultural scale" (Ma et al. 2003).

Industrial enzymes are widely used in both manufacturing and household products. They often have unique advantages over chemicals with regard to specificity and impact on the environment (Senior et al. 1999; Bhat 2002; Gupta et al. 2002; van der Maarel 2002). The use of enzymes is growing, but is often blocked by performance issues, the need for large volumes of such proteins and the low cost of chemicals in established processes (Bergquist et al. 2002; Silveira and Jonas 2002).

The topic of plant-produced industrial enzymes has been reviewed a number of times recently (Howard and Hood 2005, 2007). More specifically, the concept of producing cellulases *in planta* for biomass conversion has been reviewed (Austin et al. 1994; Howard et al. 2011; Sticklen 2008). These timely reviews support the developing industry that could be producing plant-derived products for industrial applications in the near future.

Industrial enzymes can be isolated from natural sources, but most often they are obtained from recombinant microbial organisms, which facilitate both re-engineering and production of the protein of interest. Although traditional expression systems continue to improve in yield, thus lowering the cost of the enzymes, they can still fall short of cost and/or capacity requirements.

Plant-produced enzymes are best in addressing markets that require low prices and large volumes. Many plant systems offer the lowest price for production, particularly when produced in commodity crops where scale-up is easy and production is inexpensive (Howard and Hood 2005). Industrial applications such as pulp and paper, food and beverage, and animal feed are the current largest markets (Hayes et al. 2007). However, the largest market by far will be the lignocellulosic feedstock-supported biofuels and biobased products market. Unprecedented amounts of cellulases, hemicellulases and ligninases will be required to deconstruct materials for the biobased economy.

The first industrial enzyme manufactured and sold from a plant system was trypsin, brand name, Trypzean (Woodard et al. 2003). This bovine enzyme is purified from corn flour and marketed by Sigma Chemical Company (St. Louis, MO). This first product manufactured from transgenic corn has many features that indicate feasibility for new, more ambitious products to come on line through field production rather than from tanks of microbes. These features include benign impurities derived from food-based raw materials rather than from animals or microbes, and freedom from animal pathogens, important if used in pharmaceutical applications.

A number of technologies have been applied to the process of producing enzymes in plants—seed-specific expression (Hood et al. 2007; Hood and Howard 2008; Clough et al. 2006; Woodard et al. 2003) and constitutive expression in biomass (Sticklen 2008; Austin et al. 1994). Additional technologies include expressing the enzymes in chloroplasts to prevent pollen transmission to facilitate containment (Leelavathi et al. 2003) and incorporating inteins to prevent activity prior to the desired process conditions (Raab 2010). Any or all of these technologies can be applied to successfully produce the enzymes. The final challenge becomes producing them cost-effectively for a highly cost-sensitive environment.

This chapter focuses on four classes of enzymes—xylanases, redox enzymes, amylases and cellulases. Examples of each of these classes of enzymes have been produced in plants—some as demonstrations, others with the intent to sell. The authors have chosen specific examples to describe the advantages of the plant system, issues that have arisen, and potential for addressing markets. These case studies illustrate the value of using plants for production with simple agricultural inputs of sunlight, nutrients and water.

8.2 Enzyme Case Studies

8.2.1 Amylase Enzymes

On February 12, 2011, the US Department of Agriculture announced its decision to approve the commercial production of Enogen, Syngenta's genetically engineered corn expressing recombinant amylase (http://www.nytimes.com/2011/02/12/business/12corn.html; http://online.wsj.com/article/SB10001424052748703843004576138911297227814.html). This unprecedented decision makes Enogen the first GE plant engineered solely for industrial purposes to meet the statutory requirements to be grown without permits in the US. This "amylase corn" is an ideal candidate for a case study on production of industrial enzymes in plants.

This case was chosen because it is the only case to date that has obtained non-regulated status from the USDA and commercialized. We aim here to describe a protein that is a proven candidate for production and has passed beyond the research stage. For an industrial protein expressed in plants to be commercially successful many criteria must be met. First of all, high accumulation of the recombinant protein is crucial. Different parameters must be taken into account, such as expression levels, targeting, correct folding, and activity. Many strategies and technical approaches are used to maximize expression and obtain higher production (Streatfield 2007; Lau and Sun 2009; Hood and Vicuna Requesens 2011). Once a high level of accumulation of the desired protein is achieved, developers face the production and commercialization challenges linked to growing genetically engineered plants. Amylase-producing maize plants provide an excellent example to illustrate these challenges.

8.2.1.1 Characteristics of the Enzymes

Amylases are a class of enzymes that catalyze the hydrolysis of starch. They are glycoside hydrolases acting on α-1,4-glycosidic bonds. They are very widespread in nature and are produced by animals and plants as well as several microorganisms (van der Maarel 2002; Smith 1999; Pandey et al. 2000). There are two main categories of amylases, denoted alpha and beta amylases and they differ in the way they cleave the glycosidic bonds of the starch molecules (Table 8.1).

Table 8.1 Enzyme characteristics

Enzyme	EC #	Substrate	Reaction	Product	Reference
α-Amylase	3.2.1.1	Linear starch	Hydrolysis	Shorter starch polymers	Mitsui and Iton (1997)
β-Amylase	3.2.1.2	Ends of starch polymers	Hydrolysis	Glucose	Pandey et al. (2000)
β-D-xylosidase	3.2.1.37	Ends of xylan chains	Hydrolysis	Xylose	Polizeli et al. (2005)
Endo-β-1, 4-xylanase	3.2.1.8	Xylan polymers	Hydrolysis	Shorter xylan chains	Kulkarni et al. (1999)
Laccase	1.10.3.2	Lignin, phenolics	Oxidation/ reduction	Oxidized and reduced compounds	Loera Corral et al. (2006)
Peroxidase	1.11.1.7	Lignin, phenolics	Oxidation/ reduction	Oxidized and reduced compounds	Banci et al. (2003)
Endo-β-D-1,4 glucanase	3.2.1.4	Cellulose polymers	Hydrolysis	Shorter cellulose chains	Schülein (2000)
Cellobiohydrolase	3.2.1.91	Ends of cellulose polymers	Hydrolysis	Cellobiose and cellotriose	Schülein (2000)

α-Amylases: These endoamylases (EC 3.2.1.1) are calcium metalloenzymes that cleave the bonds present in the inner part of the amylase chain. They are the major form of amylases in mammals but they are also found in plants, fungi (ascomycetes and basidiomycetes) and bacteria (*Bacillus*).

β-Amylases: Also synthesized by plants, fungi and bacteria, these amylases (EC 3.2.1.2) act on the external glucose residues of amylose, working from the non-reducing end and catalyzing the hydrolysis of the second α-1,4 glycosidic bond, cleaving off two glucose units at a time.

8.2.1.2 Market Applications

Amylases and in particular α-amylases, are used in a number of industrial applications, such as production of glucose and fructose from starch (Crabb and Mitchinson 1997); in baking industry (van der Maarel 2002); to clarify haze in beer and fruit juices (Bhat 2002); in pretreatment of animal feed to improve digestibility (Rodrigues et al. 2008; Moharrery et al. 2009); and in laundry detergent and cleaners (Mukherjee et al. 2009). Value of these markets is shown in Table 8.2.

In recent years, importance of the fuel ethanol industry from crops such as corn and wheat has grown, increasing the need for enzymes like amylases in the conversion of starchy materials to ethanol. Large amounts of fuel ethanol from biomass are needed to substitute petroleum-based fuels and even though large efforts are being deployed to develop technologies to convert lignocellulosic biomass into fuel, starch-rich grains are still the major polymer used for ethanol production (Cardona and Sánchez 2007).

Table 8.2 Market applications of industrial enzymes

		Estimated world market size (million USD)		
Enzyme	Industrial applications	2011	2016	Comments
Amylase	Baking, ethanol, textiles, detergents, distilling, brewing	$821	$1028	Mostly widely used industrial enzyme
Xylanase	Pulp bleaching, biofuels, animal feed	$488	$584	Total market of "other" carbohydrases
Redox enzymes	Cleaning, pulp bleaching, adhesives, gasoline oxygenates, sealants	Unknown	Unknown	Rapid market growth depends on inexpensive supply
Cellulase enzymes	Lignocellulosic deconstruction for renewable resources	$318	$708	Gaining rapidly with introduction of lignocellulosic biofuels

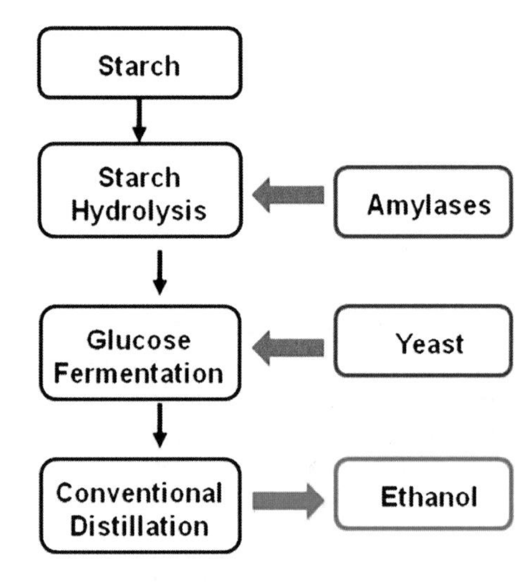

Fig. 8.1 Generic block diagram of ethanol production from starch

The conversion of starchy biomass into ethanol, not only for biofuels but also as the raw material for other chemicals, is a complicated process and still requires improvement in various steps. One of the main problems encountered in this industry is the need for large amounts of amylolytic enzymes able to hydrolyze the starch to obtain glucose syrup, which can be converted into ethanol by yeast. Figure 8.1 shows a generic block diagram of fuel ethanol production from starchy materials.

8.2.1.3 Expression Systems

Several approaches are being taken to improve the crucial step of enzyme production. The use of amylase-producing *Aspergillus awamori* has been tested to ferment part of the feedstock (wheat flour) and obtain degradation of the polymers (Koutinas et al. 2004). Other investigations have proposed the use of immobilized amylases as part of corn starch bioconversion (Krishnan 2000; Mandavilli 2000).

Another alternative to adding starch-decomposing enzymes is to use recombinant microbial strains that produce these enzymes. Yeast strains such as *Saccharomyces cerevisiae* and *Candida tropicalis* have been successfully genetically engineered to produce α-amylase and other glucoamylases and results have shown similar production of ethanol compared to the addition of starch decomposing enzymes to the medium (Jamai et al. 2007). The alternative approach to producing these industrial enzymes from micro-organisms is to express and recover the enzymes from genetically engineered plants. Several plants have been reported as suitable for the expression of amylases. Examples include the expression of an alpha-amylase from *Bacillus licheniformis* in tobacco and alfalfa (Pen et al. 1992; Austin et al. 1995) or the rice alpha-amylase over-expressed in tobacco plants at high levels (Kumagai 2000).

Since over 90% of the current fuel ethanol production is maize-based, efforts on developing highly-fermentable maize will improve the efficiency of conversion of grain to ethanol (Wolt and Karaman 2007). Transgenic amylase maize, developed by Syngenta, is the first genetically engineered plant expressing a product quality trait specifically intended for biofuels production.

8.2.1.4 A Specific Case: Corn Amylase "Enogen"

Syngenta's GE maize, named Enogen, expresses a thermostable α-amylase enzyme which accumulates in the grain. It has been engineered to suit the starch processing step of dry-grind ethanol production. The α-amylase expressed in this event is both stable at high temperatures and requires low levels of calcium, making it ideal for the starch ethanol industry. It is believed that the use of this GE corn amylase will improve the efficiency, and lower the cost as well as the environmental footprint of biofuels (Urbanchuk et al. 2009).

The α-amylase gene expressed in Enogen is derived from amylase sequences of three hyperthermophilic microorganisms of the archael order *Thermococcales,* and was codon-optimized for expression in maize. The gene was fused to a maize gamma-zein promoter and signal sequence, confining the expression uniquely to the endosperm tissue of the corn kernel, and a C-terminal SEKDEL sequence for targeting to and retention of the protein in the endoplasmic reticulum. Amylase levels of the recombinant protein in the kernel ranged from 838 to 1,627 μg/g dry weight (0.08–0.16% of dry weight).

Much time and effort are needed to produce a genetically engineered plant expressing a protein of interest at profitable levels that conforms to all the safety standards requested by the regulatory agencies. But the development of a product like Enogen does not stop there. Developers of genetically engineered crops must navigate the quite long and complicated regulatory framework prior to commercialization of their product, and the costs involved have been estimated to be anywhere from $20 to $100 million USD (McElroy 2003; Economist 2009).

Syngenta first introduced the petition for non-regulated status of its Corn Amylase to the US Department of Agriculture in September 2006. In January 2007, they submitted a response to the Animal and Plant Health Inspection Service/Biotechnology Regulatory Services' (APHIS/BRS) review of their original Petition for Determination of Non-Regulated Status for this corn event. By 2009, Syngenta had concluded its consultation on the food and feed safety of the amylase corn and had to wait until early 2011 for regulatory approval from the USDA.

In November 2008, APHIS announced the availability of Syngenta's petition to the public and solicited comments on the possibility of the amylase corn to pose a plant pest risk (73 FR 69602–69604, Docket No. APHIS–2007–0016). Within 60 days, APHIS received over 13,000 comments. There were 40 comments from organizations or individuals that supported the deregulation of this event and over 13,000 comments that were opposed to its deregulation! The majority of these comments against Syngenta's petition was essentially identical and was compiled by organizations generally opposed to any genetic engineering of plants. Many of these comments were based on general opposition to the development and use of GE plants, and did not cite or address any specific environmental issues. Other comments expressed nonspecific concerns over possible gene flow, disruption to organic farming practices, and concerns of food and environmental safety. Nonetheless, some of these opposing comments raised valid concerns regarding food safety and the potential for the Corn Amylase to be allergenic.

Most of the comments supporting nonregulated status for this amylase corn came from several national organizations representing corn farmers and ethanol production interests. They commented on the anticipated benefits for farmers and the ethanol industry as well as the biofuels production targets. Fortunately, many of the supportive statements were based on scientific facts, such as evidence of decreased water use in ethanol production, reduced greenhouse gas emissions and other reduced inputs in ethanol production, compared to current ethanol production practices. More importantly, many of these scientific studies supported the substantial equivalence of the amylase corn to currently grown corn lines in agronomic and nutritional qualities.

Some comments argued that the α-amylase could be a plant pest and interfere with corn starch processing and damage plants or plant products. APHIS agreed with Syngenta that enzymes are proteins that catalyze chemical reaction and cannot be regarded as "living" thus cannot be defined as plant pests.

Even though there were no clear and science-based comments against this event, APHIS decided to reopen the comment period to allow interested persons additional time to prepare and submit comments on the petition until July 2009.

It is believed that Syngenta spent several hundred million dollars to develop this corn event (Brock 2011). Part of this was due to the almost 6 years Enogen corn was caught up in USDA's regulatory review process fueled by criticism from the grain milling industry and food safety groups which fought its commercialization. After more than 5 years and millions of dollars, Syngenta has obtained non-regulated status for Enogen, and it was also reviewed and approved in other countries including Mexico, Russia, Australia, New Zealand, Canada, South Korea, Taiwan, Japan, Switzerland and the Philippines.

8.2.2 *Xylanase Enzymes*

8.2.2.1 Characteristics of the Enzymes

Xylanases are a class of enzyme which catalyzes the hydrolysis of xylans. Xylan is the most abundant type of hemicellulose, one of the major components of plant cell walls, and one of the most abundant biopolymers on earth, second only to cellulose. It is a linear polymer of β-D-xylopyranosyl units and xylanases have the ability to degrade the linear polysaccharide α-1,4-xylan into xylose. An enzyme complex is necessary for xylan degradation and the two main and most studied components of this complex are endo-β-1,4-xylanases (EC 3.2.1.8) and β-D-xylosidase (E.C. 3.2.1.37) (Kulkarni et al. 1999).

Xylanases are essentially produced by micro-organisms, where they play an important role in degradation of plant cell walls, but also in marine algae, protozoans, insects, snails, and crustaceans, although filamentous fungi are the most prominent commercial source of this enzyme (Polizeli et al. 2005).

8.2.2.2 Market Applications

The production of xylan has many commercial applications which include assistance in the bleaching process of cellulose pulp prior to papermaking, the production of bread and other foods and drinks, the improvement of silage digestibility for animal feed and the production of biofuels from lignocellulosic feedstocks.

Pulp and paper industry: One of the most important industrial applications of xylanases resides in the pulping and bleaching processes related to the paper industry. Chemical bleaching of Kraft pulp is traditionally based on the use of chlorine and sodium hydrosulfite. By-products of these chemicals are toxic and persistent in the environment, and a more environmental friendly alternative has been found in the use of enzymes such as xylanases (Bae et al. 2008; Zhao et al. 2006). Xylanases are used in the pre-bleaching of Kraft pulp and represent an economically attractive and clean option for this industry. They have the advantage of reducing the amount of chemicals used while still obtaining the desired brightness of the pulp. The main enzyme needed to enhance the delignification of the pulp is endo-β-xylanase. It works by changing the pulp structure, by partially removing the hemicellulose on the surface of the fibers (Beg et al. 2001). Xylanases have been used in this industry for more than two decades and are usually obtained from microbial fermentation processes, mainly from *Streptomyces*, *Bacillus* and *Aspergillus* (Polizeli et al. 2005). Xylanases from transgenic plants appear to be a low cost production system for this industry.

Animal feed: Even though xylans are a major component of several crops, it is an important anti-nutritional material in monogastric animals. The arabinoxylan found in grains has an anti-nutrient effect, particularly in poultry. Xylan can increase the viscosity of the feed, interfering with the digestion of other components and decreasing the absorption of nutrients. Adding xylanases to feed containing maize or sorghum improves the digestibility of the feed by creating a more digestible food mixture. Enzymes for food and animal feed represent the largest segment for industrial enzymes. More than 600 million tons of enzymes are produced worldwide for this industry and this robust market is likely to continue to grow (Polizeli et al. 2005; Yang et al. 2007).

Biofuels: Currently, biomass for fuel ethanol comes primarily from cornstarch but expansion of this industry requires the use of alternative lignocellulosic biomass. The production of ethanol from lignocellulosic biomass such as corn stover, switchgrass, and forestry by-products is a complicated process and requires the pretreatment of the feedstock prior to the fermentation of sugars by microorganisms. Since xylans are among the most abundant form of hemicellulose, xylanases represent a key enzyme in the conversion of lignocellulose into monomeric sugars (Li et al. 2007).

In order to obtain free fermentable sugars from cellulose and hemicellulose, pretreatments combining high temperature and an acid or base catalyst must be used. The hemicellulose fraction can then be processed with enzymes such xylanases for the efficient saccharification of the pretreated woody biomass. Nonetheless, the cost of producing these enzymes is limiting for the biofuels industry and strategies such as producing them in transgenic plants must be taken into account to lower their cost (Kim et al. 2010).

Food and beverage industries: White wheat flour used in bread-making contains hemicellulose. Several enzymes, among them xylanases, are added to the flour to break down the hemicellulose and allow redistribution of water, increasing the volume of the baked bread. This process leaves the dough softer, delays crumb formation, improves the resistance to fermentation and allows the dough to grow. The addition of purified endoxylanases from *Aspergillus* has been shown to cause a 30% increase in volume of the dough over a crude extract (Harbak and Thygesen 2002; Camacho and Aguilar 2002).

In the production of fruit and vegetable juices, xylanases, in combination with other enzymes such as cellulases and amylases, are used to improve yield of juice. They hydrolyze arabinoxylans and starch and lead to improvements in the liquefaction of fruit and vegetables, stabilization of the pulp, reduction of viscosity and general increased recovery of aromas of juices (Polizeli et al. 2005).

8.2.2.3 Plant Expression Systems

Many research groups have investigated the production of xylanases in GE plants and even though they have not reached a commercial stage yet, they provide an optimistic future for this technology. Recombinant xylanases have been produced in several plants including Arabidopsis, barley, rice, potato, canola and tobacco.

In Arabidopsis, xylanases have been expressed while being targeted to chloroplasts and peroxisomes (Hyunjong et al. 2006). These efforts were based on the fact that soluble proteins can be accumulated at high levels when targeted to organelles. They expressed a xylanase from *Trichoderma reesei* and showed higher expression when targeted to both of these organelles rather than either one alone. More recently, a recombinant xylanase also from *T. reesei* was expressed constitutively in Arabidopsis and tested for potential biotechnological applications of plant produced xylanases in the pulp bleaching industry. The authors demonstrated the efficiency of recombinant xylanases when compared to commercial enzymes and offered the transgenic system as a viable alternative approach to traditional xylanase production (Bae et al. 2008). An alkai- and thermostable xylanase gene from *Bacillus subtilis* was expressed in tobacco chloroplasts to avoid any undesirable interaction with the cell wall components (Leelavathi et al. 2003). Transgenic tobacco plants overexpressing active enzyme were obtained without any changes in the growth or development of the plants. When using GE plants as production systems, the process of transporting the plants from the farms to the site of the downstream processing unit must be taken into account. In the case of xylanase-expressing tobacco plants, the xylanases retained more than 85% activity after leaves were sun-dried or dried at high temperatures. Drying of leaves is a current practice on tobacco farms and the stability of the enzyme is an essential part of its production in plants.

To reach enzyme levels high enough to be useful in industry, plants other than tobacco and Arabidopsis have been transformed. Potato (*Solanum tuberosum* L.) has been used to express a xylanase gene from *Streptomyces olivaceoviridis* (Yang et al. 2007). As described above, xylanases play an important role in the poultry feeding industry and the xylanase-expressing potatoes could be an alternative feed for poultry and other monogastric animals or be used as a feed additive. Potatoes also have the advantage of having tubers that can be stored for a long time without special equipment and the possibility to be used for feeding without cooking, which can inactivate the enzyme, making this a cost efficient system.

Xylanases have been expressed in various other systems, such as canola seed oil-bodies (Liu et al. 1997) and barley endosperms (Patel et al. 2000). In canola, a fungal xylanase gene was fused to an oleosin gene to increase the accumulation of the enzyme in the oil-bodies of the plant. Recombinant xylanase was active and could be recovered and recycled by simple flotation of the oil bodies after being used in a reaction. This technique could decrease significantly the costs of xylanases in the pulp and bioconversion industries. Targeting the recombinant protein to the seed, in the case of barley, offers advantages regarding the stability, storage and transport of the enzyme. A recombinant fungal xylanase was targeted to the cytoplasm of the endosperm cells of developing barley grain using a rice glutelin-1 promoter or a barley B1 hordein promoter. Using the cereal grain as a bioreactor to produce enzymes for feed can become a by-product of the main feed component, potentially reducing costs as well as possible accidental microbial contamination of feed.

8.2.3 Oxidation/Reduction Enzymes

Oxidation-reduction (redox) reactions involve the transfer of electrons from an electron donor (the reducing agent) to an electron acceptor (the oxidizing agent) (Lehninger et al. 2005). In some cases, the electron is transferred as a hydrogen atom and is thus a dehydrogenation reaction. Numerous redox enzyme classes exist in nature and include but are not limited to: oxidases in the electron transport chains of mitochondria and chloroplasts, catalase to remove hydrogen peroxide from cells, ascorbate oxidase, and peroxidases and laccases, which are often involved in lignin formation and degradation. Tremendous interest in the industrial sector has developed for peroxidase and laccase as potential agents for the manufacture of useful products from biomass-derived lignin and its monomers.

8.2.3.1 Characteristics of the Enzymes

Laccase: Laccase is a blue copper oxidase that has four Cu++ ions per protein molecule (EC 1.10.3.2). It is capable of reacting with a large number of aromatic substrates. Widely distributed in white rot fungi (Elegir et al. 2007), it is believed to possess a key role in the degradation of lignin (Leonowicz et al. 2001). Laccase can perform the oxidation of free phenolic lignin moieties and, in the presence of aromatic redox mediators, also non-phenolic moieties (Bourbonnais and Paice 1990). Its activity will polymerize lignins or lignin monomers when mediators are not present, but degrades lignin in the presence of mediators (Mattinen et al. 2008).

Manganese Peroxidase: Manganese-dependent peroxidase (EC 1.11.1.7) is a heme-containing enzyme primarily from white rot fungi that catalyzes the degradation of lignin. Using peroxide as the electron acceptor, the first reaction is oxidation of divalent manganese to trivalent manganese, a diffusible oxidant capable of oxidizing a large variety of phenolic substrates (Kuan and Tien 1993). Organic acids such as oxalic acid can stimulate its activity in nature.

8.2.3.2 Market Applications

Laccase and peroxidase are valued for their potential in manufacturing and processing of phenolic monomers and polymers. Lignin-derived phenolic substrates can be used to make adhesives, gasoline oxygenates, precursors for low cost carbon fiber, and sealants, to name a few (Holladay et al. 2007). Most of the current uses of lignin are bulk density, low cost products without specificity because of the difficulty of manipulating the substrate. However, this situation could be rectified if enzymes that produce specific products could be produced in bulk quantities for manufacture of the higher value products. With the increase in use of lignocellulosic feedstocks for biofuels and biobased products, a market demand is developing for inexpensive enzymes that manipulate the lignin structure to control the product outcome.

Unfortunately, laccase and peroxidase are relatively difficult to produce in industrial quantities, largely due to their active oxygen-producing activities. Currently, laccases are used in small volumes for bio-bleaching applications in cosmetics and pulp and paper processing. Additionally, Novozymes (Bagsvaerd, Denmark) manufactures a laccase product (DENILITE) that is used in the textile industry for fabric processing. If and when cost-effective manufacture of redox enzymes can be accomplished, their use in industry will fulfill many unmet market opportunities. Laccases could be applied to many bleaching processes as well as lignin polymerization (Mattinen et al. 2008). Peroxidases are promising substitutes for formaldehyde in phenolic polymerization processes for such applications as binding plywood and particle board.

8.2.3.3 Plant Expression Systems

Redox enzymes are difficult to express because the active oxygen species they produce are quite detrimental to biological systems. This is particularly true of single cell systems where the active oxygen cannot be sequestered away and the organism's survival is compromised. Thus, multicellular organisms may likely be a more viable system for expression of these genes and thus accommodate high accumulation of the enzymes for industrial uses.

To test this concept, laccase was expressed in maize embryos (Hood et al. 2003). Plants were recovered that had protein accumulating to detectable levels, though not high levels, from protein that was targeted to specific subcellular compartments—the apoplast being the most effective. Any attempt to express the protein intracellularly failed to generate any transgenic plants, suggesting that the protein was toxic to cells. Extractable enzyme had to be reactivated by treating with copper sulfate, salt and heat (Bailey et al. 2004). Much of the enzyme remained associated with the plant tissue and did not appear to be extractable, but was active (Hood et al. 2003; Bailey et al. 2004). When these transgenic plants were bred to elite germplasm to improve agronomic performance and increase protein amounts, the plants began to show characteristics of damage, presumably from active oxygen species generated by the laccase. Even though the maize globulin-1 promoter is embryo-preferred (Belanger and Kriz 1991), low levels of expression can be detected in some vegetative tissues (Hood et al. 2003, 2011). Within the seed, expression is confined to the embryo and the pericarp (Fig. 8.2).

Expression of this enzyme may benefit from technologies that produce an inactive form of the enzyme using inteins (Raab 2010) or from expressing a thermotolerant form of the enzyme that is less active at ambient temperature (Uthandi et al. 2010). Active enzyme concentration in defatted germ from grain of these transgenic lines is approximately 0.2% of dry weight (Hood et al. 2003), an amount that would allow cost-effective production for applications (Howard and Hood 2005).

Manganese peroxidase (MnPOX) has also been expressed in maize seed from the maize globulin-1 promoter (Clough et al. 2006). The subcellular location had dramatic impact on the expression level of this enzyme as well, with the apoplast

Fig. 8.2 Embryo preferred expression of laccase in GE maize seed

being the best location for protein accumulation. When the protein is expressed from a constitutive promoter and secreted into the apoplast, significant leaf and stem lesions are seen (Clough et al. 2006). Indeed, over-expression of an anionic peroxidase constitutively in tobacco species caused chronic severe leaf wilting (Lagrimini et al. 1990). Thus, the seed production system has advantages in sequestering the protein and limiting plant damage. Accumulation level of this protein in fourth generation post-transformation seed is approximately 0.015% dry weight.

8.2.4 Cellulase Enzymes

To meet the 2022 goal of 36 billion gallons of liquid transportation fuels assuming that all comes from biomass, 1.1 million tonnes of cellulase will be required to meet US goals for bioenergy (Banerjee et al. 2010; Howard et al. 2011). This value is based on using 30 g cellulase to produce enough sugars from biomass for a gallon of ethanol or other liquid biofuel. If more enzyme is required to deconstruct the biomass, the amount of total enzyme needed goes up proportionately. Current enzyme production strategies focus on fungal-based production in fermentation tanks. Enzyme production at the scale and cost required to meet bioenergy goals is not viable using microbe-based systems due to the significant up-front capitalization costs of the necessary bioreactors (tanks/fermenters), the operational costs for running those bioreactors, as well as the time to build them. This being said, the lowest cost of cellulase enzymes is currently by extraction from fungal cultures and years of experience using these cultures have allowed optimization for cellulase production. In the near term while the market is developing, it is expected that these fungal systems will continue to provide cellullase enzymes, though not at the target price of $0.12–0.17 per gallon (Kabir Kazi et al. 2010) and although continual improvements are being made to fungal strains to accumulate higher levels of enzymes, improvements are unlikely to be much greater than the already very high levels because of the maturity of the system. Similarly, optimization of production practices is likely to be incremental and unlikely to drive cost down.

8.2.4.1 Characteristics of the Enzymes

Thousands of glycosyl hydrolase enyzmes have been discovered and characterized (Cantarel et al. 2009). Multiple enzymes have been screened for efficient activity on crystalline cellulose, including groups of enzymes that work well together (Baker et al. 1998; Nieves et al. 1998). As shown in Table 8.1, endo-cellulases hydrolyze the cellulose chain internally, creating free ends on which the exo-cellulases are active. Two exo-cellulase activities are required for the most efficient deconstruction of cellulose—one working on the reducing end (cellobiohydrolase I) and one working on the non-reducing end (cellobiohydrolase II). The products of those two enzymes are small oligomers of glucose that can be hydrolyzed to glucose through the activity of β-glucosidase. The activities of the biomass conversion enzymes should be in the range of pH 5 and approximately 45–60°C to be compatible with process conditions in the industry. Increases in specific activity and thermostability have been sought by researchers to increase their process stability (Heinzelman et al. 2009).

8.2.4.2 Market Applications

Developing markets will require cellulases for the conversion of biomass to sugars for applications in biofuels and biobased products. Plant biomass is a complex matrix of polymers comprising the polysaccharides cellulose and hemicellulose, as well as the phenolic polymer, lignin. A strategy designed to use lignocellulose for fuel or bio-products must include the ability to efficiently convert the polysaccharide and lignin components of plant cell walls to simple sugars and phenolic monomers, respectively. Deconstruction of lignocellulose can be accomplished by heat and chemical means (Taherzadeh and Karimi 2007) but the preferred environmentally friendly method is to use enzymes.

With the current state of technology for biomass conversion, the overwhelming enzyme requirement is for endo-cellulase, exo-cellulase and β-glucosidase (Jørgensen et al. 2007; Merino and Cherry 2007). The specific activity of most cellulases is quite low (Jørgensen et al. 2007; Sticklen 2008) and much effort has focused on increasing their activity levels. However, even with improved enzymes and improved methods of fungal fermentation production, the amount of cellulase required to deconstruct the volumes of biomass necessary for 30% replacement of gasoline (the US goal of 36 billion gallons) is over a million tons.

This is an unprecedented challenge in terms of the amount of enzymes and the extremely low cost that is required for competitively priced liquid transportation fuels. Currently, microbial fermentation in bioreactors serves as the dominant source for cellulases, and likely will continue to do so in the short term during industry development. However, the amount of upfront capital required for tankage/fermenter capacity is problematic for the large scale deployment of the industry. If one assumes that the enzyme load onto biomass for deconstruction into sugars is 100 g cellulases per gallon of biofuel (i.e., ethanol or butanol) the investment numbers are staggering for tankage capacity to produce cellulases in fungal culture. A total capital investment of over $57 billion dollars would be required for more than 2,000 tanks with associated controls. At a price of $0.17/gallon of biofuel (NREL 2011 target), this capacity would return annual sales revenues of ~$6 billion (Banerjee et al. 2010; Howard et al. 2011). Clearly, significant innovation and alternative approaches are required to address enzyme costs at this scale to achieve the RFS2 goal (http://www.epa.gov/otaq/fuels/renewablefuels/index.htm).

8.2.4.3 Plant Expression Systems

Cellulases have been expressed in a number of plants and tissues including tobacco leaves, rice stems, corn stover and corn seed (Gray et al. 2009; Hood et al. 2007; Jin et al. 2003; Mei et al. 2009; Oraby et al. 2007; Sticklen 2008). This approach is seen as a potential solution to reduce the recalcitrance of the tissue to deconstruction (Taylor et al. 2008). Cellulases in vegetative tissues contribute to more digestible cellulose, although processing conditions for the biomass must be tailored to prevent enzyme inactivation during pretreatment.

An alternative approach is to produce cellulases in a concentrated tissue that allows inexpensive growth, harvest, storage, and processing (Howard and Hood 2005). In this way the enzymes can be scaled to meet market demands and stably stored until required for biomass processing. One such system is producing enzymes in seed of commodity crops such as corn, wheat or soybeans. Production and processing of maize grain is a well-developed industry (Alexander 1994). Two cellulase genes have been successfully expressed in corn and production lines developed. (Hood et al. 2007). High levels of protein have been recovered from the grain after breeding of the original transformants into elite germplasm or high oil germplasm (Clough et al. 2006; Hood and Howard 2008; Hood et al. 2002). This method has been particularly effective for increasing cellulase accumulation in maize seed (Hood et al. 2011). In fact, protein levels for cellulase E1 (from *Acidothermus cellulolyticus*) in elite maize grain is as high as 0.2% of dry weight and for CBH I (from *Trichoderma reesei*) as high as 0.45% of dry weight. Because the enzyme is concentrated entirely in the germ (embryo) fraction of the grain, the grain can be fractionated through established methods (wet or dry milling) to physically concentrate the enzyme. In traditional corn processing, the germ is a low value co-product of the starch from the endosperm and is commonly sold as animal feed. The germ can be defatted, also through established protocols to recover the corn oil, leaving the defatted germ as a co-product. In this case, the defatted germ is the most valuable part of the grain, as it contains the cellulases (Howard et al. 2011).

Most of the transgenic plants containing cellulases have focused on using one or another of the various endo- or exo-cellulases that are active on biomass. Each of these enzymes by itself is quite ineffective at cellulose deconstruction because the enzymes evolved to work in concert. Thus, a potentially more effective approach would be to stack the genes together into a single plant to allow all the enzymes to work together. This can be accomplished through breeding plants with individual enzymes in the tissues, or through putting the genes together on mini chromosomes (Carlson et al. 2007; Yu et al. 2007). Each of these technologies has its own challenges and is several years from deployment.

8.3 Conclusions

Plant-based bioproduction of proteins for industrial applications has been developing for several years. The first product sold was bovine trypsin (Trypzean, T3568, Sigma Chemical Co.). The current market is relatively limited, although potential applications of this enzyme in pharmaceutical manufacturing are lucrative. This enzyme is used to mature pro-insulin to insulin produced from bacteria and to free pharmaceutical-producing cells from their substrate in order to passage them. The first large-scale product from the technology has just entered the marketplace—Enogen amylase from Syngenta. According to a recent press release, this excellent technology will be deployed at large scale in 2012 (http://advancedbiofuelsusa.info/usda-approves-corn-amylase-trait-for-enogentm). The Enogen example paves the way for future industrial products from the plant production system.

The use of plants for biofuels and biomaterials, though it has environmentally positive impacts, is currently of concern among some groups because of its potential negative impact on land use (increased cultivation) or loss of food-production acreage. These issues must be addressed, and one way to do so is to increase productivity of all crops whether for food, biofuel/bioproduct or industrial use. Numerous technologies have been applied to agricultural production over the years including domestication, mechanization, breeding, fertilizer application, and genetic engineering (Fedoroff 2010). Each technology has contributed significantly to improvements in productivity. If investments continue to be made in these technologies, all food and non-food uses of plants will likely be feasible on current arable land. Considering the current climate surrounding the use of genetically engineered (GE) plants for food, feed, fiber and fuel, delays in deployment of new crops using this technology exist and are likely to continue. Certainly crop safety is an important issue to address for these GE crops. However, the risk of not using all the tools available to us to achieve increased crop productivity is much greater—crop failure leads to starvation. "Thus, the challenges to agriculture in the twenty-first century are profound" (Fedoroff 2010).

The global market for industrial enzymes is continuously growing and is expected to reach anywhere from US$3.74 billion to US$7 billion by 2015 (http://www.prweb.com/releases/industrial_enzymes/proteases_carbohydrases/prweb8121185.htm; http://www.reportlinker.com/p0148002/World-Enzymes-Market.html#ixzz1FZu2b3Bq). Continued strong demand for specialty enzymes has fueled the growing interest for new technologies to produce them. This is in part due to the cost efficiency of their use but also to the increasing interest in the substitution of petrochemical products.

The development of novel and superior enzymes is linked to the advances in the production technology and biotechnology has changed the enzyme production industry. The use of genetically modified plants will satisfy a need for more efficient, environmentally friendly and low cost production system.

References

Alexander RJ (1994) In: Watson SA, Ramstad PE (eds) Corn dry milling: Processes, products and applications. American Association of Cereal Chemists, Inc., St. Paul, pp 351–376

Austin S, Bingham ET, Koegel RG, Mathews DE, Shahan MN, Straub RJ, Burgess RR (1994) An overview of a feasibility study for the production of industrial enzymes in transgenic alfalfa. Ann N Y Acad Sci 721:234–244

Austin S, Bingham ET, Mathews DE, Shahan MN, Will J, Burgess RR (1995) Production and field performance of transgenic alfalfa (Medicago sativa L.) expressing alpha-amylase and manganese-dependent lignin peroxidase. Euphytica 85:381–393. doi:10.1007/BF00023971

Bae H-J, Kim HJ, Kim YS (2008) Production of a recombinant xylanase in plants and its potential for pulp biobleaching applications. Bioresour Technol 99:3513–3519. doi:10.1016/j.biortech.2007.07.064

Bailey MR, Woodard SL, Callaway E, Beifuss K, Magallanes-Lundback M, Lane JR, Horn ME, Mallubhotla H, Delaney DD, Ward M, Van Gastel F, Howard JA, Hood EE (2004) Improved recovery of active recombinant laccase from maize seed. Appl Microbiol Biotechnol 63:390–397

Baker J, Ehrman C, Adney W, Thomas S, Himmel M (1998) Hydrolysis of cellulose using ternary mixtures of purified celluloses. Appl Biochem Biotechnol 70–72:395–403

Banci L, Bartalesi I, Ciofi-Baffoni S, Ming T (2003) Unfolding and pH studies on manganese peroxidase: role of heme and calcium on secondary structure stability. Biopolymers 72:38–47

Banerjee G, Scott-Craig JS, Walton JD (2010) Improving enzymes for biomass conversion: a basic research perspective. BioEnergy Res 3:82–92. doi:10.1007/s12155-009-9067-5

Beg QK, Kapoor M, Mahajan L, Hoondal GS (2001) Microbial xylanases and their industrial applications: a review. Appl Microbiol Biotechnol 56:326–338. doi:10.1007/s002530100704

Belanger FC, Kriz AL (1991) Molecular basis for allelic polymorphism of the maize Globulin-1 gene. Genetics 129:863–872

Bergquist P, Te'o V, Gibbs M, Cziferszky A, de Faria F, Azevedo M, Nevalainen H (2002) Expression of xylanase enzymes from thermophilic microoroganisms in fungal hosts. Extremophiles 6:177–184

Bhat MK (2002) Cellulases and related enzymes in biotechnology. Biotechnol Adv 18:355–383

Bourbonnais R, Paice M (1990) Oxidation of non-phenolic substratesAn expanded role for laccase in lignin biodegradation. FEBS Lett 267:99–102. doi:10.1016/0014-5793(90)80298-W

Brock R (2011) USDA approves corn designed for ethanol production. Corn and soybean digest

Camacho NA, Aguilar OG (2002) Production, purification, and characterization of a low-molecular-mass xylanase from Aspergillus sp. and its application in baking. Appl Biochem Biotechnol 104:159–171. doi:10.1385/ABAB:104:3:159

Cantarel BL, Coutinho PM, Rancurel C, Bernard T, Lombard V, Henrissat B (2009) The Carbohydrate-Active EnZymes database (CAZy): an expert resource for Glycogenomics. Nucleic Acids Res 37:D233–D238

Cardona CA, Sánchez OJ (2007) Fuel ethanol production: process design trends and integration opportunities. Bioresour Technol 98:2415–2457. doi:10.1016/j.biortech.2007.01.002

Carlson SR, Rudgers GW, Zieler H, Mach JM, Luo S, Grunden E, Krol C, Copenhaver GP, Preuss D (2007) Meiotic transmission of an in vitro-assembled autonomous maize minichromosome. PLoS Genet 3:1965–1974. doi:10.1371/journal.pgen.0030179

Clough R, Pappu K, Thompson K, Beifuss K, Lane J, Delaney D, Harkey R, Drees C, Howard J, Hood EE (2006) Manganese peroxidase from the white-rot fungus Phanerochaete chrysosporium is enzymatically active and accumulates to high levels in transgenic maize seed. Plant Biotechnol J 4:53–62

Crabb WD, Mitchinson C (1997) Enzymes involved in the processing of starch to sugars. Trends Biotechnol 15:349–352. doi:10.1016/S0167-7799(97)01082-2

Economist (2009) The parable of the sower. The Economist, pp 71–73

Elegir G, Bussini D, Antonsson S, Lindstrom M, Zoia L (2007) Laccase-initiated cross-linking of lignocellulose fibres using a ultra-filtered lignin isolated from kraft black liquor. Appl Microbiol Biotechnol 77:809–817

Fedoroff N (2010) The past, present and future of crop genetic modification. New Biotechnol 27:461–465

Gray BN, Ahner BA, Hanson MR (2009) High-level bacterial cellulase accumulation in chloroplast-transformed tobacco mediated by downstream box fusions. Biotechnol Bioeng 102:1045–1054

Gupta R, Beg Q, Lorenz P (2002) Bacterial alkaline proteases: molecular approaches and industrial applications. Appl Biochem Biotechnol 59:15–32

Harbak L, Thygesen HV (2002) Safety evaluation of a xylanase expressed in Bacillus subtilis. Food Chem Toxicol 40:1–8

Hayes TL, Zimmerman N, Hackle A (2007) World enzymes; Industry study 2229, Cleveland, OH

Heinzelman P, Snow CD, Smith MA, Yu X, Kannan A, Boulware K, Villalobos A, Govindarajan S, Minshull J, Arnold FH (2009) SCHEMA recombination of a fungal cellulase uncovers a single mutation that contributes markedly to stability. J Biol Chem 284:26229–26233

Holladay JE, White JF, Bozell JJ, Johnson D (2007) Top value-added chemicals from biomass volume II — results of screening for potential candidates from biorefinery lignin. Pacific Northwest National Laboratory, Richland

Hood E, Howard J, Delaney D (2002) Method of Increasing Heterologous Protein Expression in Plants. US patent # 7, 541, 515

Hood E, Howard J (2008) Over-expression of novel proteins in maize. In: Kriz A, Larkins B (eds) Molecular genetic approaches to maize improvement. Springer, Berlin/Heidelberg, pp 91–105

Hood E, Vicuna Requesens D (2011) Recombinant protein production in plants: challenges and solutions. In: Lorence A (ed) Methods in molecular biology: recombinant gene expression. Humana Press, New York

Hood E, Mr B, Beifuss K, Magallanes-Lundback M, Horn M, Callaway E, Drees C, Delaney D, Clough R, Howard J (2003) Criteria for high-level expression of a fungal laccase gene in transgenic maize. Plant Biotechnol J 1:129–140. doi:10.1046/j.1467-7652.2003.00014.x

Hood E, Love R, Lane J et al (2007) Subcellular targeting is a key condition for high-level accumulation of cellulase protein in transgenic maize seed. Plant Biotechnol J 5:709–719

Howard JA, Hood E (2005) Bioindustrial and biopharmaceutical products produced in plants. Adv. Agron. 85:91–124

Howard J, Hood E (2007) Methods for growing nonfood products in transgenic plants. Crop Sci 47:1255. doi:10.2135/cropsci2006.09.0594

Howard J, Nikolov Z, Hood E (2011) Enzyme production systems for biomass conversion. In: Hood E, Nelson P, Powell R (eds) Plant biomass conversion. Wiley Press, Ames, pp 227–253

Hood EE, Devaiah SP, Fake G, Egelkrout E, Teoh K, Vicuna Requesens D, Hayden C, Hood KR, Pappu K, Carroll J and Howard JA (2011) Manipulating corn germplasm to increase recombinant protein accumulation. Plant Biotechnology Journal. Online DOI: 10.1111/j.1467-7652.2011.00627.x

Hyunjong B, Lee D-S, Hwang I (2006) Dual targeting of xylanase to chloroplasts and peroxisomes as a means to increase protein accumulation in plant cells. J Exp Bot 57:161–169. doi:10.1093/jxb/erj019

Jamai L, Ettayebi K, El Yamani J, Ettayebi M (2007) Production of ethanol from starch by free and immobilized Candida tropicalis in the presence of alpha-amylase. Bioresour Technol 98:2765–2770. doi:10.1016/j.biortech.2006.09.057

Jin R, Richter S, Zhong R, Lamppa GK (2003) Expression and import of an active cellulase from a thermophilic bacterium into the chloroplast both in vitro and in vivo. Plant Mol Biol 51:493–507

Jørgensen H, Kristensen JB, Felby C (2007) Enzymatic conversion of lignocellulose into fermentable sugars: challenges and opportunities. Biofpr 1:119–134

Kabir Kazi F, Fortman J, Anex R, Kothandaraman G, Hsu D, Aden A, Dutta A (2010) Techno-Economic Analysis of Biochemical Scenarios for Production of Cellulosic Ethanol. Technical Report. NREL/TP-6A2-46588

Kim J, Kavas M, Fouad W, Nong G, Preston J, Altpeter F (2010) Production of heperthermostable GH10 xylanase Xyl10B from Thermotoga maritima in transplastomic plants enables complete hydrolysis of methylglucuronoxylan to fermentable sugars for biofuels production. Plant Mol Biol. doi:10.1007/s11103-010-9712-6

Koutinas AA, Wang R, Webb C (2004) Restructuring upstream bioprocessing: technological and economical aspects for production of a generic microbial feedstock from wheat. Biotechnol Bioeng 85:524–538

Krishnan M (2000) Economic analysis of fuel ethanol production from corn starch using fluidized-bed bioreactors. Bioresour Technol 75:99–105. doi:10.1016/S0960-8524(00)00047-X

Kuan IC, Tien M (1993) Stimulation of Mn peroxidase activity: a possible role for oxalate in lignin biodegradation. Proc Natl Acad Sci USA 90:1242–1246

Kulkarni N, Shendye A, Rao M (1999) Molecular and biotechnological aspects of xylanases. FEMS Microbiol Rev 23:411–456

Kumagai M (2000) Rapid, high-level expression of glycosylated rice α-amylase in transfected plants by an RNA viral vector. Gene 245:169–174. doi:10.1016/S0378-1119(00)00015-9

Lagrimini LM, Bradford S, Rothstein S (1990) Peroxidase-induced wilting in transgenic tobacco plants. Plant Cell 2:7–18. doi:10.1105/tpc.2.1.7

Lau OS, Sun SSM (2009) Plant seeds as bioreactors for recombinant protein production. Biotechnol Adv 27:1015–1022. doi:10.1016/j.biotechadv.2009.05.005

Leelavathi S, Gupta N, Maiti S, Ghosh A, Reddy VS (2003) Overproduction of an alkali- and thermo-stable xylanase in tobacco chloroplasts and efficient recovery of the enzyme. Mol Breed 11:59–67

Lehninger AL, Nelson DL, Cox MM (2005) Lehninger principles of biochemistry, vol 1. W.H. Freeman, New York

Leonowicz A, Cho N, Luterek J, Wilkolazka A, Wojtas-Wasilewska M, Matuszewska A, Hofrichter M, Wesenberg D, Rogalski J (2001) Fungal laccase: properties and activity on lignin. J Basic Microbiol 41:185–227. doi:10.1002/1521-4028(200107)41:3/4<185::AID-JOBM185>3.0.CO;2-T

Li X-L, Skory CD, Ximenes EA, Jordan DB, Dien BS, Hughes SR, Cotta MA (2007) Expression of an AT-rich xylanase gene from the anaerobic fungus Orpinomyces sp. strain PC-2 in and secretion of the heterologous enzyme by Hypocrea jecorina. Appl Microbiol Biotechnol 74:1264–1275. doi:10.1007/s00253-006-0787-6

Liu J-H, Selinger LB, Cheng K-J, Beauchemin KA, Moloney MM (1997) Plant seed oil-bodies as an immobilization matrix for a recombinant xylanase from the rumen fungus Neocallimastix patriciarum. Biochem J 3:463–470

Loera Corral O, Pérez Pérez MCI, Barbosa Rodríguez JR, Villaseñor Ortega F, Guevara-González RG, Torres-Pacheco I (2006) Laccases. In: Ramón Gerardo Guevara-González and Irineo Torres-Pacheco (eds) Advances in Agricultural and Food Biotechnology. Research Signpost, Kerala, India, pp 323–340

Ma JK-C, Drake PMW, Christou P (2003) The production of recombinant pharmaceutical proteins in plants. Nat Rev Genet 4:794–805. doi:10.1038/nrg1177

Mandavilli S (2000) Performance characteristics of an immobilized enzyme reactor producing ethanol from starch. J Chem Eng Japan 33:886–890

Mattinen M-L, Suortti T, Gosselink R, Argyropoulos DS, Evtuguin D, Suurnakki A, de Jong E, Tamminen T (2008) Polymerization of different lignins by laccase. BioResources 3:549–565

McElroy D (2003) Sustaining agbiotechnology through lean times. Nat Biotechnol 21:996–1002

Mei C, Park S-H, Sabzikar R, Callista Ransom CQ, Mariam S (2009) Green tissue-specific production of a microbial endo-cellulase in maize (Zea mays L.) endoplasmic-reticulum and mitochondria converts cellulose into fermentable sugars. J Chem Technol Biotechnol 84:689–695

Merino ST, Cherry J (2007) Progress and challenges in enzyme development for biomass utilization. Adv Biochem Eng Biotechnol 108:95–120

Mitsui T, Itoh K (1997) The alpha-amylase multigene family. Trends Plant Sci 2:255–261

Moharrery A, Hvelplund T, Weisbjerg MR (2009) Effect of forage type, harvesting time and exogenous enzyme application on degradation characteristics measured using in vitro technique. Anim Feed Sci Technol 153:178–192. doi:10.1016/j.anifeedsci.2009.06.001

Mukherjee A, Borah M, Rai S (2009) To study the influence of different components of fermentable substrates on induction of extracellular α-amylase synthesis by Bacillus subtilis DM-03 in solid-state fermentation and exploration of feasibility for inclusion of α-amylase in laundry detergen. Biochem Eng J 43:149–156. doi:10.1016/j.bej.2008.09.011

Nieves RA, Ehrman CI, Adney WS, Elander RT, Himmel ME (1998) Technical communication: survey and analysis of commercial cellulase preparations suitable for biomass conversion to ethanol. World J Microb Biotechnol 14:301–304

Oraby H, Venkatesh B, Dale B, Ahmad R, Ransom C, Oehmke J, Mariam S (2007) Enhanced conversion of plant biomass into glucose using transgenic rice-produced endoglucanase for cellulosic ethanol. Transgenic Res 16:739–749

Pandey A, Nigam P, Soccol C, Soccol V, Singh D, Mohan R (2000) Advances in microbial amylases. Biotechnol Appl Biochem 31:135–152

Patel M, Johnson JS, Brettell RIS, Jacobsen J, G-ping X (2000) Transgenic barley expressing a fungal xylanase gene in the endosperm of the developing grains. Mol Breed 6:113–123

Pen J, van den Ooyen A, Elzen P, Rietveld K, Hoekema A (1992) Direct screening for high-level expression of an introduced alpha-amylase gene in plants. Plant Mol Biol 18:1133–1139

Polizeli MLTM, Rizzatti ACS, Monti R, Terenzi HF, Jorge JA, Amorim DS (2005) Xylanases from fungi: properties and industrial applications. Appl Microbiol Biotechnol 67:577–591. doi:10.1007/s00253-005-1904-7

Raab RM (2010) Transgenic plants expressing CIVPS or intein modified proteins and related method. US Patent # 20110138502

Rodrigues M, Pinto P, Bezerra R, Dias A, Guedes C, Cardoso V, Cone J, Ferreira L, Colaco J, Sequeira C (2008) Effect of enzyme extracts isolated from white-rot fungi on chemical composition and in vitro digestibility of wheat straw. Anim Feed Sci Technol 141:326–338. doi:10.1016/j.anifeedsci.2007.06.015

Schülein M (2000) Protein engineering of cellulases. Biochim Biophys Acta 1543:239–252

Senior D, Hamilton J, Taiplus P, Torvinin J (1999) Enzyme use can lower bleaching costs, aid ECF conversions. Pulp and Paper 73(7):59–65

Silveira M, Jonas R (2002) The biotechnological production of sorbitol. Appl Microbiol Biotechnol 59:400–408

Smith AM (1999) Making starch. Curr Opin Plant Biol 2:223–229. doi:10.1016/S1369-5266(99)80039-9

Sticklen Mb (2008) Plant genetic engineering for biofuel production: towards affordable cellulosic ethanol. Nat Rev Genet 9:433–443

Streatfield SJ (2007) Approaches to achieve high-level heterologous protein production in plants. Plant Biotechnol J 5:2–15. doi:10.1111/j.1467-7652.2006.00216.x

Taherzadeh MJ, Karimi K (2007) Enzyme-based hydrolysis processes for ethanol from lignocellulosic materials: a review. BioResources 2:707–738

Taylor LE II, Dai Z, Decker SR, Brunecky R, Adney William S, Ding S-Y, Himmel Michael E (2008) Heterologous expression of glycosyl hydrolases in planta: a new departure for biofuels. Trends Biotechnol 26:413–424

Urbanchuk JM, Kowalski DJ, Dale BE, Kim S (2009) Corn amylase: improving the efficiency and environmental footprint of corn to ethanol through plant biotechnology. AgBioforum 12:149–154

Uthandi S, Saad B, Humbard MA, Maupin-Furlow JA (2010) LccA, an archaeal laccase secreted as a highly stable glycoprotein into the extracellular medium by Haloferax volcanii. Appl Environ Microbiol 76:733–743. doi:10.1128/AEM.01757-09

van der Maarel M (2002) Properties and applications of starch-converting enzymes of the α-amylase family. J Biotechnol 94:137–155. doi:10.1016/S0168-1656(01)00407-2

Wolt J, Karaman S (2007) Estimated environmental loads of alpha-amylase from transgenic high-amylase maize. Biomass Bioenerg 31:831–835. doi:10.1016/j.biombioe.2007.04.003

Woodard SL, Mayor JM, Bailey MR, Barker DK, Love RT, Lane JR, Delaney DE, McComas-Wagner JM, Mallubhotla HD, Hood EE, Dangott LJ, Tichy SE, Howard JA (2003) Maize-derived bovine trypsin: characterization of the first large-scale, commercial protein product from transgenic plants. Biotechnol Appl Biochem 38:123–130

Yang P, Wang Y, Bai Y, Meng K, Luo H, Yuan T, Fan Y, Yao B (2007) Expression of xylanase with high specific activity from Streptomyces olivaceoviridis A1 in transgenic potato plants (Solanum tuberosum L.). Biotechnol Lett 29:659–667. doi:10.1007/s10529-006-9280-7

Yu W, Han F, Gao Z, Vega JM, Birchler JA (2007) Construction and behavior of engineered minichromosomes in maize. Proc Natl Acad Sci USA 104:8924–8929

Zhao J, Li X, Qu Y (2006) Application of enzymes in producing bleached pulp from wheat straw. Bioresour Technol 97:1470–1476. doi:10.1016/j.biortech.2005.07.012

Chapter 9
Transient Expression Using Agroinfiltration and Its Applications in Molecular Farming

Rima Menassa, Adil Ahmad, and Jussi J. Joensuu

Abstract Transient expression via agroinfiltration and/or viral vectors has quickly emerged as the preferred expression system for plant-made recombinant proteins. Transient expression can serve as a valuable research tool for finding optimal expression parameters before tedious and time-consuming production of stable transgenic plants or it can be scaled up to commercial production scale with vacuum infiltration. This technology is poised to compete with conventional production systems, large-scale production facilities are currently available and others are in the process of development. This chapter will introduce background and rationale in development of transient expression systems in plants and summarize the latest developments and examples in this area.

9.1 Introduction

To satisfy the increasing rigor of industrial production, a recombinant protein production system must fulfill several criteria: scalability, simple and inexpensive purification methods, short generation and production timelines, high production rates, capability to carry out co- and post-translational modifications, biosafety and product reproducibility.

R. Menassa (✉) • A. Ahmad
Southern Crop Protection and Food Research Centre, Agriculture and Agri-Food Canada, 1391 Sandford St., London, ON N5V 4T3, Canada
e-mail: Rima.Menassa@agr.gc.ca

J.J. Joensuu
VTT Biotechnology, Technical Research Centre of Finland, 02044 VTT Espoo, Finland

A. Wang and S. Ma (eds.), *Molecular Farming in Plants: Recent Advances and Future Prospects*, DOI 10.1007/978-94-007-2217-0_9,

Currently, no classical recombinant expression system satisfies all of these requirements. For example, complex recombinant proteins produced in microbial cell culture are not always properly processed or folded resulting in a protein incapable of biological activity. Accordingly, prokaryotic expression systems have been utilized for the production of simple recombinant proteins that do not require extensive post-translational modifications such as insulin, interferon and human growth hormone (Walsh and Jefferis 2006). Due to the limitations of microbial production systems, focus has been placed towards optimization of competing eukaryotic expression systems, yeast, fungi and mammalian cultures. These production systems possess disadvantages such as hyperglycosylation for yeast and fungi, and the potential of harboring human pathogens coupled with scalability and ethical concerns for mammalian cell culture and transgenic animals. These limitations combined with growing demand for therapeutic and industrial proteins have contributed to the emergence of plants as a much safer, low-cost alternative for the production of biologically active recombinant proteins. Plant expression systems can perform post-translational modifications and produce a variety of functional mammalian proteins and industrial enzymes. The production of correctly folded and assembled multi-subunit proteins such as antibodies in plant cells is a good illustration that plants possess the ability to produce and assemble complex mammalian proteins (Nuttall et al. 2005).

9.2 Plant Expression Systems

Stable transformation of plants has been until recently the preferred method of overexpression of recombinant proteins. In stably transformed plant tissue, *Agrobacterium* strains introduce recombinant genes into the plant nuclear genome through virulence factors coded from the Ti-plasmid. The portion of the Ti plasmid which contains the gene of interest is delineated by left and right T-DNA border sequences and is randomly integrated into the plant nuclear genome (Zambryski 1988). This method is largely utilized by most plant scientists; however it has proven more successful for dicotyledonous plants. A major advantage of stable transgenic plants is that the heterologous protein production trait is heritable, resulting in a permanent resource, allowing for simple and rapid scale-up and almost unlimited and sustainable production capacity only requiring planting of seeds in a large area and harvesting it (Gray et al. 2009). Disadvantages of using transgenic plants as production platform include the long developmental phase required to regenerate and analyse the transformants, unpredictable expression due to chromosomal position effects associated with random gene insertion, recombinant protein stability issues leading to low accumulation levels, and physiological effects on the host plant such as toxicity of the recombinant protein (Hobbs et al. 1990; Krysan et al. 2002).

Since the first idea of expressing vaccines in edible plant organs such as bananas and potatoes, much has been accomplished in the area of producing recombinant proteins in plants. It has become clear that food crops cannot be used for the production of pharmaceutical proteins due to the risk of contaminating the food supply. As well, the idea of oral administration of plant material has evolved to oral administration of purified and quality-controlled products because of the need for accurate dosing which would not be possible with raw plant materials (Rybicki 2010). This is due to the variability in expression levels between plants, within the same plant, and depending on the physiological condition of each plant. Therefore, even though vaccines such as the hepatitis B surface antigen (HBsAg) had been successfully produced in potato and had shown to be effective in human clinical trials (Thanavala et al. 2005), attention was redirected to non-food plants and expression in transient systems that allow faster timelines and higher expression levels than stable transgenic plants. Several breakthroughs in transient expression have been reported in recent years that allow for unprecedented expression levels within 1–2 weeks. Such systems rely on two classes of plant pathogens, plant viruses and *Agrobacterium*, a bacterial soil pathogen.

9.2.1 Agrobacterium Infiltration

In agroinfiltration, recombinant *Agrobacterium tumefaciens* bacteria harbouring a binary expression vector are introduced directly into plant leaves using vacuum infiltration or direct syringe injection. Following infection, single-stranded T-DNA is transferred from the *Agrobacterium* to the plant cells. Once moved into the plant cell by bacterial and plant encoded proteins, this T-DNA is trafficked to the nucleus with the aid of chaperones. Only a small percentage is integrated into the host chromosomes leading to stable transformed cells that can subsequently be regenerated into transgenic plants (Zambryski 1988). The long-term fate of the T-DNAs that do not integrate into chromosomes is unclear; however, it appears that the free T-DNA molecules are transcriptionally competent, thus providing an opportunity for a short-lived burst of recombinant protein production and harvest (Voinnet et al. 2003).

The significantly reduced production timeline and convenience of agroinfiltration technique yields results within 2–5 days, making transient plant-based expression an attractive option for the production of proteins. Thus, agroinfiltration can be used to rapidly evaluate the activity of expression constructs and to produce small amounts of recombinant protein for functional analysis (Wroblewski et al. 2005). These timelines compare very favourably to the time- and resource-intensive process of generating stable transgenic plants, which usually takes 3–6 months. The transient expression system is also flexible, as it allows for the expression of multiple genes simultaneously, and provides a reliable and reproducible indicator of expression construct performance, since it avoids the positional effects normally associated with stable transgenic plants (Kapila et al. 1997).

9.2.2 Virus-Based Expression

Virus-based expression of recombinant proteins is the subject of another chapter in this book (Wang, Chap. 10), thus we will very briefly describe virus-based systems and focus on recent developments where deconstructed viruses are Agro-infiltrated and result in very high expression levels.

The first virus-based expression vectors were simple gene replacement vectors, in which a foreign gene of interest replaced the capsid protein gene of a virus. These vectors were limited in the expression of these genes and although they could move from cell to cell, they could not move systemically in plants. Eventually, plant RNA viruses were constructed to express a foreign gene in addition to all required viral genes, so after inoculation, the plant virus would systemically infect all cells of the plants and generate multiple transcripts of the transgene (Pogue et al. 2002).

More recently, a new system relying on agroinfiltration of deconstructed viral vectors has shown the highest levels of expression achieved in any system. This system, called magnifection, was developed by the German biotechnology company Icon Genetics (now a subsidiary of Bayer Innovation). Magnifection combines the high transfection efficiency of *Agrobacterium* with high expression yield of deconstructed viral vectors, leading to accumulation levels up to 5 g recombinant protein per kg of fresh leaf weight (FLW), equivalent to about 50% TSP (Gleba et al. 2005, 2007). The process consists of an infiltration of whole mature plants with a diluted suspension of *Agrobacterium* carrying T-DNAs encoding viral replicons. The bacteria carry on the function of primary infection while the virus provides cell-to-cell spread, amplification and high-level expression. This system allows the expression of heteromeric recombinant proteins such as monoclonal antibodies (mAbs). For this, the magnICON® system uses two non-competitive viral vectors based on turnip vein clearing tobamovirus (TVCV) and potato virus X (PVX). One vector carries the heavy chain while the other vector carries the light chain. Both vectors are co-infiltrated into *N. benthamiana* where both chains are expressed and assembled into a fully functional mAb (Giritch et al. 2006; Hiatt and Pauly 2006).

The scale up for this technique is essentially the same as for agroinfiltration requiring an infiltration apparatus. Because of very high expression levels in this system, proteins can be produced in a contained facility minimizing biosafety-related risk. The company Kentucky BioProcessing (KBP), in collaboration with Bayer Innovation and Icon Genetics, has adapted the MagnICON system to infiltrate kilograms of plants per hour, allowing the production of 25–75 g of antibody to be produced per greenhouse lot in about 2 weeks (Pogue et al. 2010).

9.2.3 Silencing and Its Suppression

In transgenic plants, post-transcriptional gene silencing (PTGS) is observed as the reduction in steady-state levels of transcript being coded from a foreign DNA sequence (Voinnet et al. 1999). This reduction is caused by an increased turnover of target RNA, with the transcription level of corresponding genes remaining unaffected.

RNA silencing triggered by the presence of transgenes requires an RNA-dependant RNA polymerase (RdRp)-like protein to catalyze synthesis of RNA complementary to the target species (Dalmay et al. 2000). Double-stranded RNA is then recognized by a specific nuclease and cleaved to produce 21–23 nucleotide RNA species (Zamore et al. 2000). These small RNAs are proposed to associate with nuclease-like proteins and serve as guides for sequence specific cleavage of target RNA transcripts (Voinnet et al. 1999).

PTGS can be avoided by expressing simultaneously the gene of interest and a suppressor of silencing. Individual plant viruses seem to produce their own suppressor of silencing and the characterization of a large number of suppressors is currently in progress (Lienard et al. 2007). The best characterized suppressor of silencing is the p19 protein, encoded by tomato bushy stunt virus (TBSV). Through agroinfiltration methods, recombinant protein expression has been greatly improved when co-infiltrated in the presence of p19 (Voinnet et al. 2003), up to 50-fold. Use of this method provides a simple route for increasing recombinant protein production in any plant species which is amenable to agroinfiltration.

9.2.4 Syringe vs Vacuum Infiltration

Initially, transient expression by agroinfiltration was developed for assessing the ability of various constructs to induce the expression of recombinant proteins. The best performing constructs would then be used to produce stable transgenic lines. For this, an *Agrobacterium* suspension is infiltrated into the abaxial side of the leaves, the plant most frequently used for these experiments is *N. benthamiana*, but *Nicotiana tabacum* can also be used, as well as other plants such as *Arabidopsis thaliana*, tomato or lettuce (Wroblewski et al. 2005). In addition to the choice of plant species, the developmental stage and physiological status of infiltrated plants can have a major effect on expression levels of target proteins. When co-infiltrated with a suppressor of PTGS such as HcPro from potato virus Y, expression levels can be as high as 1.5 g recombinant protein/kg of LFW (Vézina et al. 2009). Because of the high expression levels and the speed of protein production using *Agrobacterium* syringe infiltration, this method has been scaled up for vacuum infiltration of kilogram amounts of tissue. About half the amount of recombinant protein was obtained using vacuum infiltration in a side by side comparison with syringe infiltration (Vézina et al. 2009). Nevertheless, vacuum infiltration has now been adopted by several groups including Medicago Inc. for the production of influenza vaccines (D'Aoust et al. 2010) and automated by Kentucky BioProcessing, LLC for the magnifection system to infiltrate kilograms of plants/hour (Pogue et al. 2010).

9.3 Applications in Molecular Farming

This section will briefly describe advances in plant-made pharmaceuticals developed for treatment of human diseases, with a focus on transient expression, which seems to be method of choice for fast high level production of complex proteins.

Detailed reviews are available for each of these categories, and not all produced proteins will be discussed, rather a sample of the most successful developed examples will be described.

9.3.1 Vaccines

9.3.1.1 Hepatitis B Virus

Hepatitis B surface antigen (HBsAg) has been produced in transgenic potatoes and shown to be immunogenic in a human clinical trial as a booster in previously immunized human volunteers (Thanavala et al. 2005). However, low expression levels and concerns about using food crops for the production of pharmaceutical proteins triggered the investigation of alternative plant hosts and ways to improve expression levels. *Agrobacterium* infiltration for transient expression of HBsAg was first reported in 2004 by Huang and Mason (Huang and Mason 2004) who used this method for the evaluation of antigen conformation with and without a fusion partner. To optimize the transient expression system, the same authors used the deconstructed MagnICON viral vectors, and showed that HbsAg properly assembled into dimers and virus-like particles and accumulated to 300 mg/kg LFW. Further, immunization of mice with partially purified HBsAg elicited HBsAg-specific antibodies (Huang et al. 2008). In another attempt of developing a hepatitis B vaccine, the MagnIcon vectors were used for expressing the hepatitis B core antigen (HBc) in *N. benthamiana.* HBc accumulated to 2.38 g/kg LFW, assembled into virus-like particles which were immunogenic in mice (Huang et al. 2006). In an attempt at developing other expression vectors, the same group developed a deconstructed viral vector based on the geminivirus bean yellow dwarf virus (BeYDV) to transiently produce HBc to levels up to 1 g/kg LFW (Chen et al. 2011).

9.3.1.2 Human Papilloma Virus

Human papilloma viruses are responsible for causing cervical cancers in women, and are implicated in anogenital and head and neck tumors in both men and women (Bosch et al. 2002). Vaccination is the most efficient way for fighting HPV infections, and two prophylactic vaccines are now available, Gardasil produced in yeast by Merck, and Cervarix produced in insect cells by GlaxoSmithKline. Both vaccines protect against the two high risk HPV types 16 and 18 which cause 70% of all cervical cancers (Bosch et al. 2008). Several groups have focused their efforts at making HPV vaccines in plants as a way to significantly reduce costs of production. The most successful attempt achieved expression levels of 0.5 g/kg LFW (17% TSP) in a transient agroinfiltration system, co-expressing the tomato spotted wilt virus non-structural small silencing suppressor protein, and targeting the HPV-16 L1 protein to the chloroplast (Maclean et al. 2007). Virus-like particles were formed and parenteral administration of concentrated plants extracts elicited high titer antisera in the same range as those reported for human trial subjects injected with commercial vaccines (Giorgi et al. 2010). A further 50% increase in expression levels was achieved by the same group upon the use of replicating geminivirus sequences in the agro-infiltrated binary vector (Regnard et al. 2010).

9.3.1.3 HIV

The type I human immunodeficiency virus (HIV) is responsible for the acquired immunodeficiency syndrome (AIDS) and currently infects more than 40 million people worldwide, and is continuing to spread, mainly in sub-Saharan Africa. The development of an effective vaccine is essential for controlling the epidemic, and treatments such as neutralizing antibodies for treating infected patients are needed. In either case, large amounts of recombinant proteins are required, and plants can be an inexpensive system for their production, provided accumulation levels are high enough (Rybicki 2010). Several efforts have focused on the production of HIV antigens as chimeric proteins in plants, including structural proteins Gag (and its component proteins p24, p17, and p17/p24) and Env and regulatory proteins Tat and Nef (De Virgilio et al. 2008; Karasev et al. 2005; Meyers et al. 2008; Yusibov et al. 1997; Zhou et al. 2008). Only two of these antigens accumulated to high enough levels, Nef when fused to zeolin accumulated to 1.5% TSP and p24-Nef when expressed from the chloroplast genome accumulated to 40% TSP in petite Havana, a small laboratory tobacco cultivar, and to 6% TSP in a high biomass tobacco cultivar (Marusic et al. 2009; Zhou et al. 2008). Similarly, the Gag-derived p17/p24 fusion protein could be expressed to 5 mg/kg FW by agroinfiltration when targeted to the chloroplast (Meyers et al. 2008).

Another approach for controlling the transmission of HIV consists in producing griffithsin, a potent viral entry inhibitor from the red alga Griffithsia (Mori et al. 2005). This protein is a lectin that targets the high-mannose glycans displayed on the surface of HIV envelope glycoproteins, and inactivates the virus on contact (Emau et al. 2007; Ziółkowska et al. 2006). Griffithsin was produced in *N. benthamiana* through infection with a tobacco mosaic virus-based vector at extremely high levels of more than 1 g/kg of LFW in just 12 days. Plant-made griffithsin was shown to be active, directly virucidal, and capable of blocking cell to cell HIV transmission (O'Keefe et al. 2009).

9.3.1.4 Influenza

Influenza viruses evolve rapidly and require the development of a new vaccine every year for the seasonal influenza season. The probability of occurrence of a pandemic influenza and the identity of the virus causing it are unknown, and therefore the ability of organizations to prepare and stockpile such vaccines is very limited. Once the virus is identified, it usually takes about 6 months to produce the vaccine product which consists of inactivated viruses grown in eggs (D'Aoust et al. 2010). These concerns have triggered interest in developing influenza vaccines in plants using the quick agroinfiltration technology. Antigenic domains of H5 from strain A/Vietnam/04 (H5N1) and of H3 from A/Wyoming/3/03 (H3N2), both fused to a carrier protein were produced by agroinfiltration in *N. benthamiana*. The immunogenicity of H3 was demonstrated in a ferret study and showed that a single dose in combination with the neuraminidase antigen induced a strong immune response (Mett et al. 2008; Musiychuk et al. 2007). However, the dose of 200 μg required to induce the immune response was much higher than industry standards. Agroinfiltration of the hemagglutinin (HA) domain spanning the outside of the viral envelope from the human seasonal influenza virus A/Wyoming/03/03 (H3N2) and several highly pathogenic H5N1 avian strains was also reported (Shoji et al. 2008, 2009a, b). These studies showed that specific expressed domains induced a significant immune response in mice with the addition of an adjuvant. However, (Shoji et al. 2009a) showed that three high doses were necessary for protecting ferrets against a lethal challenge with the homologous strain. A more successful strategy was developed by the Canadian biotechnology company Medicago Inc., which involves the expression of the entire HA protein from H1N1 strain A/New Caledonia/20/99 and H5N1 strain A/Indonesia/5/05. They found that virus-like particles are produced, bud off the plasma membrane and accumulate between the plasma membrane and the cell wall (D'Aoust et al. 2008). They also showed that the VLPs are more immunogenic than the HA protein that had not assembled into VLPs (D'Aoust et al. 2009). Further, all mice injected with two doses of 0.5 μg of H5-VLPs were protected against a lethal challenge of a different H5N1 isolate (D'Aoust et al. 2008). Medicago Inc. further adapted the agroinfiltration method for scaling up their production capacity and are able to agro-infiltrate batches of 1200–1500 plants weekly from which 25 kg of leaf biomass can be harvested and the VLPs are subsequently purified (D'Aoust et al. 2010). This system was tested for speed of production during the outbreak of the H1N1 pandemic in the spring of 2010, and this group showed that they were capable of obtaining VLPs only 3 weeks after the sequence of the novel A/H1N1 strain A/California/04/09 became available. Further, they showed that the produced VLPs were highly efficacious as a vaccine in a mouse study that lasted another 6 weeks. Therefore, a pandemic VLP vaccine can be produced in plants much faster than by conventional vaccine manufacturing (D'Aoust et al. 2010). These positive results, and the unpreparedness for quickly facing an influenza pandemic have influenced the U.S. Defence Department to invest $21 million US in Medicago Inc, which is now building a large manufacturing facility in North Carolina to grow tobacco plants and produce about 40 million doses of seasonal flu vaccine per year, or 120 million doses of pandemic flu vaccine for the U.S. market (http://www.cbc.ca/news/health/story/2010/11/24/flu-vaccine-tobacco-plants-medicago.html).

9.3.2 Antibodies

Among the biopharmaceutical drugs in development, monoclonal antibodies constitute the fastest growing group because of their outstanding specificities (Aggarwal 2009).

Demand for therapeutic antibodies far exceeds their production capacity in current mammalian expression systems, and plants provide an attractive production platform because they can correctly produce, fold and assemble complex multimeric proteins such as antibodies (Ma et al. 1995). Transient expression in plants using agroinfiltration either with traditional binary vectors (Vézina et al. 2009), with MagnICON vectors (Giritch et al. 2006) or with the use of cowpea mosaic vector hypertranslatable deleted RNA2 (Sainsbury and Lomonossoff 2008) has so far provided the highest expression levels of antibody production of any other plant system; however, glycosylation patterns of secreted monoclonal antibodies are different in plants and can lead to immunogenicity in humans. As well, the approach of retaining the antibodies in the ER which prevents plant specific and immunogenic glycan addition causes their rapid clearance from the blood stream, and causes lower complement-dependent cytotoxicity (Jefferis 2009). Therefore, close attention has been paid to engineering glycosylation pathways in plants to produce "humanized" glycosylation patterns. This has been done by knocking-out endogenous glycosyltransferases and/or knocking-in human glycosyltransferases.

N. benthamiana plants lacking immunogenic β1,2-xylose and core α1,3-fucose were generated by RNAi and were used to transiently produce the 2G12 HIV antibody (Strasser et al. 2008). The antibody was found to effectively contain homogeneous N-glycans without detectable β1,2-xylose and α1,3-fucose residues. The functional properties of the plant produced antibody were similar to those of a Chinese hamster ovary (CHO)-produced 2G12 antibody. For further humanizing the glycosylation patterns of antibodies produced in plants, a human galactosyltransferase was introduced into the previously produced RNAi line lacking α1,3-fucosyl- and β1,2-xylosyl- transferase (Fuc-T and Xyl-T) activities. Two HIV antibodies, 2G12 and 4E10 were subsequently expressed in the resulting plants and were shown to be fully galactosylated and to possess much more homogeneous glycans than CHO cells were able to produce (Strasser et al. 2009). The resulting antibodies displayed improved virus neutralization potency when compared with other glycoforms produced in plants and Chinese hamster ovary cells.

A similar result was achieved in a very elegant experiment where instead of knocking-down Fuc-T and Xyl-T, Vezina et al. (2009) expressed a chimeric form of the human β1,4-galactosyltransferase (Gal-T) together with a diagnostic antibody, C5-1. The chimeric protein consisted in the fusion of the N-terminus of GNTI to the catalytic domain of human Gal-T. GNTI is typically expressed in the ER and cis-Golgi, upstream from the Fuc-T and Xyl-T in the secretory pathway. This allowed the production of C5-1 completely lacking plant-specific β1,2-xylose and α1,3-fucose glycans presumably because the hybrid Gal-T acted at the earliest stage of complex N-glycan synthesis and inhibited further transfer of plant specific xylose and fucose to the core oligosaccharide (Vézina et al. 2009).

Therefore it is clear that complex antibodies can be transiently produced in plants with appropriate post-translational modifications at high levels and with high activity.

9.3.3 Fusions for Improved Accumulation and Purification

Recently, the use of fusion tags has gained popularity in plant-production platforms. Most common fusion tags in use were originally conceived as aids for isolation and downstream purification such as Arg-tag, His-tag, FLAG-tag, c-myc-tag, GST-tag etc.… (Lichty et al. 2005; Terpe 2003). As well, the use of protein-stabilizing proteins as fusions to "difficult to express" proteins has been successful in improving expression levels such as ubiquitin (Mishra et al. 2006), β-glucuronidase (Dus Santos et al. 2002), cholera toxin B subunit (Molina et al. 2004), and human immunoglobulin α-chains (Obregon et al. 2006). However, three fusion partners have emerged that have shown very significant improvements in accumulation levels and that result in the accumulation of recombinant proteins inside round structures that are reminiscent of seed protein bodies (PBs); these are γ-zein from maize, elastin-like polypeptides of mammalian origin, and hydrophobins from fungi. All three partners can also be used in protein purification either by density centrifugation (γ-zein), inverse transition cycling (ELP) or aqueous two phase separation system (hydrophobins). As well, all three fusion partners were shown to give the best results in plants when expressed transiently by agroinfiltration into *N. benthamiana*.

9.3.3.1 Zera Fusions

γ-zein, a major constituent of maize storage proteins is able to induce the formation of PBs in seed and vegetative organs of transgenic dicots in the absence of the other zein subunits α-zein and β-zein (Geli et al. 1994). The N-terminal domain of γ-zein, termed Zera® (Era Biotech, Barcelona, Spain), was shown to be sufficient for the formation of PBs and for the increase in accumulation levels of several proteins; for example, epidermal growth factor accumulated to 100-fold higher levels up to 0.5 g/kg LFW, and human growth hormone accumulation increased by 13-fold to 3.2 g/kg LFW (Torrent et al. 2009). A detailed study of the N-terminal γ – zein domains showed that the two N-terminal Cys residues are critical for oligomerization, the first step toward PB formation in *N. benthamiana* (Llop-Tous et al. 2010). Zera has been very recently used in transient agroinfiltration of *N. benthamiana* leaves for the expression of a zera-xylanase fusion. Expression levels were up to 9% TSP, corresponding to 1.6 g of the fusion/kg LFW. The fused protein was shown to accumulate as biologically active insoluble aggregates inside PBs (Llop-Tous et al. 2011).

9.3.3.2 ELP Fusions

Elastin-like polypeptides (ELP) are synthetic proteins composed of the pentapeptide repeat VPGVG that occur in mammalian elastins (Raju and Anwar 1987; Urry 1988). Various sizes of ELP tags are used by various groups for expression of recombinant protein fusions in plants; for example, (Scheller et al. 2006) showed that a 100-mer ELP tag increased accumulation of a recombinant antibody fragment up to 40 times in tobacco seeds, while (Patel et al. 2007) showed that a 28-mer ELP tag increased accumulation of interleukin-4 (IL-4) and IL-10 by 85 and 90-fold, respectively, in tobacco leaves. Therefore, (Conley et al. 2009a) produced an ELP size library starting at 5 repeats up to 240 repeats and tested this library with four different proteins by agroinfiltration. In this study the size of the ELP tag was shown to have a significant impact on the accumulation and the purification of recombinant proteins by inverse transition cycling. Smaller tags were more beneficial for protein accumulation, leading to the accumulation of IL-10 to 4.5% TSP, a 1000-fold improvement from the best expressing IL-10 transgenic line previously reported (Menassa et al. 2007), while larger tags allowed a higher recovery rate of the protein fusion by ITC. An ELP size of 30–40 pentapeptide repeats was found to be a good compromise for both accumulation and purification (Conley et al. 2009a). Furthermore, the fusion protein was found to accumulate in novel spherical, membrane bound structures reminiscent of Zera PBs. It is thought that these PBs allow higher accumulation levels by protecting the recombinant protein from the degradative machinery of the cell (Conley et al. 2009b).

9.3.3.3 Hydrophobin Fusions

Hydrophobins are a class of small surface active proteins produced by filamentous fungi, and are thought to be involved in the adaptation of fungi to their environment (Talbot 1999). Hydrophobins coat fungal surfaces and have been proposed to protect against desiccation and wetting and to contribute to spore and conidial dissemination (Linder et al. 2005). These small proteins (ca. 10 kDa) contain a large proportion of hydrophobic amino acids and eight cysteines all of which involved in four intramolecular disulfide bonds (Hakanpaa et al. 2004). The unique surface-active properties of hydrophobins can be transferred to their fusion partners, a property that has been exploited for a rapid and inexpensive surfactant-based aqueous two phase purification system, ATPS (Linder et al. 2004). Recently, we have shown by agroinfiltration of *N. benthamiana* that hydrophobin fusions allow for extremely high accumulation levels of GFP, up to 3.7 g/kg LFW. Similar high expression levels were obtained with glucose oxidase, an enzyme that did not express well in other conventional systems (Bankar et al. 2009; Joensuu et al. 2010). In the same study we also found that GFP-hydrophobin fusion accumulates in PBs, and that infiltrated leaves survived longer when the fusion rather than GFP alone was used in the infiltration. This indicates that PBs not only protect the recombinant protein from degradation as observed with ELP, they also protect cells from the toxicity of highly expressed proteins (Conley et al. 2011; Joensuu et al. 2010).

9.4 Conclusions

Agroinfiltration enhances the value of plant-based expression systems, serves as a very useful tool to boost research in plant biology and opens new possibilities for studying protein overexpression in plants. With the fast production timeline, good yield and well defined co- and post-translational modifications, agroinfiltration shows promise for plant-derived bioproducts and can now compete with classical eukaryotic expression systems.

Acknowledgement This work was supported by a grant from an A-base grant from Agriculture and Agri-Food Canada and the Agricultural Bioproducts Innovation Program.

References

Aggarwal S (2009) What's fueling the biotech engine – 2008. Nat Biotechnol 27:987–993

Bankar SB, Bule MV, Singhal RS, Ananthanarayan L (2009) Glucose oxidase – an overview. Biotechnol Adv 27:489–501

Bosch FX, Lorincz A, Munoz N, Meijer CJ, Shah KV (2002) The causal relation between human papillomavirus and cervical cancer. J Clin Pathol 55:244–265

Bosch FX, Burchell AN, Schiffman M, Giuliano AR, de Sanjose S, Bruni L, Tortolero-Luna G, Kjaer SK, Muñoz N (2008) Epidemiology and natural history of human papillomavirus infections and type-specific implications in cervical neoplasia. Vaccine 26:K1–K16

Chen Q, He J, Phoolcharoen W, Mason HS (2011) Geminiviral vectors based on bean yellow dwarf virus for production of vaccine antigens and monoclonal antibodies in plants. Hum Vaccin 7:331–338

Conley AJ, Joensuu JJ, Jevnikar AM, Menassa R, Brandle JE (2009a) Optimization of elastin-like polypeptide fusions for expression and purification of recombinant proteins in plants. Biotechnol Bioeng 103:562–573

Conley AJ, Joensuu JJ, Menassa R, Brandle JE (2009b) Induction of protein body formation in plant leaves by elastin-like polypeptide fusions. BMC Biol 7:48

Conley AJ, Zhu H, Le LC, Jevnikar AM, Lee BH, Brandle JE, Menassa R (2011) Recombinant protein production in a variety of Nicotiana hosts: a comparative analysis. Plant Biotechnol J 9:434–444

D'Aoust MA, Lavoie PO, Couture MMJ, Trépanier S, Guay JM, Dargis M, Mongrand S, Landry N, Ward BJ, Vézina LP (2008) Influenza virus-like particles produced by transient expression in Nicotiana benthamiana induce a protective immune response against a lethal viral challenge in mice. Plant Biotechnol J 6:930–940

D'Aoust M-A, Couture M, Ors F, Trepanier S, Lavoie P-O, Dargis M, Vézina L-P, Landry N (2009) Recombinant influenza virus-like particles (VLPs) produced in transgenic plants expressing hemagglutinin. International patent application PCT/CA2009/000032

D'Aoust M, Couture MM, Charland N, Trépanier S, Landry N, Ors F, Vézina L (2010) The production of hemagglutinin-based virus-like particles in plants: a rapid, efficient and safe response to pandemic influenza. Plant Biotechnol J 8:607–619

Dalmay T, Hamilton A, Mueller E, Baulcombe DC (2000) Potato virus X amplicons in Arabidopsis mediate genetic and epigenetic gene silencing. Plant Cell 12:369–379

De Virgilio M, De Marchis F, Bellucci M, Mainieri D, Rossi M, Benvenuto E, Arcioni S, Vitale A (2008) The human immunodeficiency virus antigen Nef forms protein bodies in leaves of transgenic tobacco when fused to zeolin. J Exp Bot 59:2815–2829

Dus Santos MJ, Wigdorovitz A, Trono K, Rios RD, Franzone PM, Gil F, Moreno J, Carrillo C, Escribano JM, Borca MV (2002) A novel methodology to develop a foot and mouth disease virus (FMDV) peptide-based vaccine in transgenic plants. Vaccine 20:1141–1147

Emau P, Tian B, O'Keefe BR, Mori T, McMahon JB, Palmer KE, Jiang Y, Bekele G, Tsai CC (2007) Griffithsin, a potent HIV entry inhibitor, is an excellent candidate for anti-HIV microbicide. J Med Primatol 36:244–253

Geli MI, Torrent M, Ludevid D (1994) Two structural domains mediate two sequential events in gamma-zein targeting: protein endoplasmic reticulum retention and protein body formation. Plant Cell 6:1911–1922

Giorgi C, Franconi R, Rybicki EP (2010) Human papillomavirus vaccines in plants. Expert Rev Vaccines 9:913–924

Giritch A, Marillonnet S, Engler C, Van Eldik G, Botterman J, Klimyuk V, Gleba Y (2006) Rapid high-yield expression of full-size IgG antibodies in plants coinfected with noncompeting viral vectros. Proc Natl Acad Sci USA 103:14701–14706

Gleba Y, Klimyuk V, Marillonnet S (2005) Magnifection – a new platform for expressing recombinant vaccines in plants. Vaccine 23:2042–2048

Gleba Y, Klimyuk V, Marillonnet S (2007) Viral vectors for the expression of proteins in plants. Curr Opin Biotechnol 18:134–141

Gray BN, Ahner BA, Hanson MR (2009) High-level bacterial cellulase accumulation in chloroplast-transformed tobacco mediated by downstream box fusions. Biotechnol Bioeng 102:1045–1054

Hakanpaa J, Paananen A, Askolin S, Nakari-Setala T, Parkkinen T, Penttila M, Linder MB, Rouvinen J (2004) Atomic resolution structure of the HFBII hydrophobin, a self-assembling amphiphile. J Biol Chem 279:534–539

Hiatt A, Pauly M (2006) Monoclonal antibodies from plants: a new speed record. Proc Natl Acad Sci USA 103:14645–14646

Hobbs SLA, Kpodar P, DeLong CMO (1990) The effect of T-DNA copy number, position and methylation on reporter gene expression in tobacco transformants. Plant Mol Biol 15:851–864

Huang Z, Mason HS (2004) Conformational analysis of hepatitis B surface antigen fusions in an Agrobacterium-mediated transient expression system. Plant Biotechnol J 2:241–249

Huang Z, Santi L, LePore K, Kilbourne J, Arntzen CJ, Mason HS (2006) Rapid, high-level production of hepatitis B core antigen in plant leaf and its immunogenicity in mice. Vaccine 24:2506–2513

Huang Z, LePore K, Elkin G, Thanavala Y, Mason HS (2008) High-yield rapid production of hepatitis B surface antigen in plant leaf by a viral expression system. Plant Biotechnol J 6:202–209

Jefferis R (2009) Recombinant antibody therapeutics: the impact of glycosylation on mechanisms of action. Trends Pharmacol Sci 30:356–362

Joensuu JJ, Conley AJ, Lienemann M, Brandle JE, Linder MB, Menassa R (2010) Hydrophobin fusions for high-level transient protein expression and purification in Nicotiana benthamiana. Plant Physiol 152:622–633

Kapila J, De Rycke R, Van Montagu M, Angenon G (1997) An Agrobacterium-mediated transient gene expression system for intact leaves. Plant Sci 122:101–108

Karasev AV, Foulke S, Wellens C, Rich A, Shon KJ, Zwierzynski I, Hone D, Koprowski H, Reitz M (2005) Plant based HIV-1 vaccine candidate: Tat protein produced in spinach. Vaccine 23:1875–1880

Krysan PJ, Young JC, Jester PJ, Monson S, Copenhaver G, Preuss D, Sussman MR (2002) Characterization of T-DNA insertion sites in Arabidopsis thaliana and the implications for saturation mutagenesis. Omics 6:163–174

Lichty JJ, Malecki JL, Agnew HD, Michelson-Horowitz DJ, Tan S (2005) Comparison of affinity tags for protein purification. Protein Expr Purif 41:98–105

Lienard D, Sourrouille C, Gomord V, Faye L (2007) Pharming and transgenic plants. Biotechnol Annu Rev 13:115–147

Linder MB, Qiao M, Laumen F, Selber K, Hyytia T, Nakari-Setala T, Penttila ME (2004) Efficient purification of recombinant proteins using hydrophobins as tags in surfactant-based two-phase systems. Biochemistry 43:11873–11882

Linder MB, Szilvay GR, Nakari-Setala T, Penttila ME (2005) Hydrophobins: the protein-amphiphiles of filamentous fungi. FEMS Microbiol Rev 29:877–896

Llop-Tous I, Madurga S, Giralt E, Marzabal P, Torrent M, Ludevid MD (2010) Relevant elements of a Maize γ-zein domain involved in protein body biogenesis. J Biol Chem 285:35633–35644

Llop-Tous I, Ortiz M, Torrent M, Ludevid MD (2011) The expression of a xylanase targeted to ER-protein bodies provides a simple strategy to produce active insoluble enzyme polymers in tobacco plants. PLoS One 6:e19474

Ma JK, Hiatt A, Hein M, Vine ND, Wang F, Stabila P, van Dolleweerd C, Mostov K, Lehner T (1995) Generation and assembly of secretory antibodies in plants. Science 268:716–719

Maclean J, Koekemoer M, Olivier AJ, Stewart D, Hitzeroth II, Rademacher T, Fischer R, Williamson AL, Rybicki EP (2007) Optimization of human papillomavirus type 16 (HPV-16) L1 expression in plants: comparison of the suitability of different HPV-16 L1 gene variants and different cell-compartment localization. J Gen Virol 88:1460–1469

Marusic C, Vitale A, Pedrazzini E, Donini M, Frigerio L, Bock R, Dix PJ, McCabe MS, Bellucci M, Benvenuto E (2009) Plant-based strategies aimed at expressing HIV antigens and neutralizing antibodies at high levels. Nef as a case study. Transgenic Res 18:499–512

Menassa R, Du C, Yin ZQ, Ma S, Poussier P, Brandle J, Jevnikar AM (2007) Therapeutic effectiveness of orally administered transgenic low-alkaloid tobacco expressing human interleukin-10 in a mouse model of colitis. Plant Biotechnol J 5:50–59

Mett V, Musiychuk K, Bi H, Farrance CE, Horsey A, Ugulava N, Shoji Y, de la Rosa P, Palmer GA, Rabindran S et al (2008) A plant-produced influenza subunit vaccine protects ferrets against virus challenge. Influenza Other Respir Viruses 2:33–40

Meyers A, Chakauya E, Shephard E, Tanzer FL, Maclean J, Lynch A, Williamson AL, Rybicki EP (2008) Expression of HIV-1 antigens in plants as potential subunit vaccines. BMC Biotechnol 8:53

Mishra S, Yadav DK, Tuli R (2006) Ubiquitin fusion enhances cholera toxin B subunit expression in transgenic plants and the plant-expressed protein binds GM1 receptors more efficiently. J Biotechnol 127:95–108

Molina A, Hervas-Stubbs S, Daniell H, Mingo-Castel AM, Veramendi J (2004) High-yield expression of a viral peptide animal vaccine in transgenic tobacco chloroplasts. Plant Biotechnol J 2:141–153

Mori T, O'Keefe BR, Sowder Ii RC, Bringans S, Gardella R, Berg S, Cochran P, Turpin JA, Buckheit RW Jr, McMahon JB et al (2005) Isolation and characterization of Griffithsin, a novel HIV-inactivating protein, from the red alga Griffithsia sp. J Biol Chem 280:9345–9353

Musiychuk K, Stephenson N, Bi H, Farrance CE, Orozovic G, Brodelius M, Brodelius P, Horsey A, Ugulava N, Shamloul AM et al (2007) A launch vector for the production of vaccine antigens in plants. Influenza Other Respir Viruses 1:19–25

Nuttall J, Ma JKC, Frigerio L (2005) A functional antibody lacking N-linked glycans is efficiently folded, assembled and secreted by tobacco mesophyll protoplasts. Plant Biotechnol J 3:497–504

O'Keefe BR, Vojdani F, Buffa V, Shattock RJ, Montefiori DC, Bakke J, Mirsalis J, D''Andrea AL, Hume SD, Bratcher B et al (2009) Scaleable manufacture of HIV-1 entry inhibitor griffithsin and validation of its safety and efficacy as a topical microbicide component. Proc Natl Acad Sci USA 106:6099–6104

Obregon P, Chargelegue D, Drake PM, Prada A, Nuttall J, Frigerio L, Ma JK (2006) HIV-1 p24-immunoglobulin fusion molecule: a new strategy for plant-based protein production. Plant Biotechnol J 4:195–207

Patel J, Zhu H, Menassa R, Gyenis L, Richman A, Brandle J (2007) Elastin-like polypeptide fusions enhance the accumulation of recombinant proteins in tobacco leaves. Transgenic Res 16:239–249

Pogue GP, Lindbo JA, Garger SJ, Fitzmaurice WP (2002) Making an ally from an enemy: plant virology and the new agriculture. Annu Rev Phytopathol 40:45–74

Pogue GP, Vojdani F, Palmer KE, Hiatt E, Hume S, Phelps J, Long L, Bohorova N, Kim D, Pauly M et al (2010) Production of pharmaceutical-grade recombinant aprotinin and a monoclonal antibody product using plant-based transient expression systems. Plant Biotechnol J 8:638–654

Raju K, Anwar RA (1987) A comparative analysis of the amino acid and cDNA sequences of bovine elastin a and chick elastin. Biochem Cell Biol 65:842–845

Regnard GL, Halley-Stott RP, Tanzer FL, Hitzeroth II, Rybicki EP (2010) High level protein expression in plants through the use of a novel autonomously replicating geminivirus shuttle vector. Plant Biotechnol J 8:38–46

Rybicki EP (2010) Plant-made vaccines for humans and animals. Plant Biotechnol J 8:620–637

Sainsbury F, Lomonossoff GP (2008) Extremely high-level and rapid transient protein production in plants without the use of viral replication. Plant Physiol 148:1212–1218

Scheller J, Leps M, Conrad U (2006) Forcing single-chain variable fragment production in tobacco seeds by fusion to elastin-like polypeptides. Plant Biotechnol J 4:243–249

Shoji Y, Chichester JA, Bi H, Musiychuk K, de la Rosa P, Goldschmidt L, Horsey A, Ugulava N, Palmer GA, Mett V et al (2008) Plant-expressed HA as a seasonal influenza vaccine candidate. Vaccine 26:2930–2934

Shoji Y, Bi H, Musiychuk K, Rhee A, Horsey A, Roy G, Green B, Shamloul M, Farrance CE, Taggart B et al (2009a) Plant-derived hemagglutinin protects ferrets against challenge infection with the A/Indonesia/05/05 strain of avian influenza. Vaccine 27:1087–1092

Shoji Y, Farrance CE, Bi H, Shamloul M, Green B, Manceva S, Rhee A, Ugulava N, Roy G, Musiychuk K et al (2009b) Immunogenicity of hemagglutinin from a/Bar-headed Goose/Qinghai/1A/05 and a/Anhui/1/05 strains of H5N1 influenza viruses produced in nicotiana benthamiana plants. Vaccine 27:3467–3470

Strasser R, Stadlmann J, Schähs M, Stiegler G, Quendler H, Mach L, Glössl J, Weterings K, Pabst M, Steinkellner H (2008) Generation of glyco-engineered Nicotiana benthamiana for the production of monoclonal antibodies with a homogeneous human-like N-glycan structure. Plant Biotechnol J 6:392–402

Strasser R, Castilho A, Stadlmann J, Kunert R, Quendler H, Gattinger P, Jez J, Rademacher T, Altmann F, Mach L et al (2009) Improved virus neutralization by plant-produced anti-HIV antibodies with a homogeneous beta1,4-galactosylated N-glycan profile. J Biol Chem 284:20479–20485

Talbot NJ (1999) Fungal biology. Coming up for air and sporulation. Nature 398:295–296

Terpe K (2003) Overview of tag protein fusions: from molecular and biochemical fundamentals to commercial systems. Appl Microbiol Biotechnol 60:523–533

Thanavala Y, Mahoney M, Pal S, Scott A, Richter L, Natarajan N, Goodwin P, Arntzen CJ, Mason HS (2005) Immunogenicity in humans of an edible vaccine for hepatitis B. Proc Natl Acad Sci USA 102:3378–3382

Torrent M, Llompart B, Lasserre-Ramassamy S, Llop-Tous I, Bastida M, Marzabal P, Westerholm-Parvinen A, Saloheimo M, Heifetz PB, Ludevid MD (2009) Eukaryotic protein production in designed storage organelles. BMC Biol 7:5

Urry DW (1988) Entropic elastic processes in protein mechanisms. I. Elastic structure due to an inverse temperature transition and elasticity due to internal chain dynamics. J Protein Chem 7:1–34

Vézina LP, Faye L, Lerouge P, D'Aoust MA, Marquet-Blouin E, Burel C, Lavoie PO, Bardor M, Gomord V (2009) Transient co-expression for fast and high-yield production of antibodies with human-like N-glycans in plants. Plant Biotechnol J 7:442–455

Voinnet O, Pinto YM, Baulcombe DC (1999) Suppression of gene silencing: a general strategy used by diverse DNA and RNA viruses of plants. Proc Natl Acad Sci USA 96:14147–14152

Voinnet O, Rivas S, Mestre P, Baulcombe D (2003) An enhanced transient expression system in plants based on suppression of gene silencing by the p19 protein of tomato bushy stunt virus. Plant J 33:949–956

Walsh G, Jefferis R (2006) Post-translational modifications in the context of therapeutic proteins. Nat Biotechnol 24:1241–1252

Wroblewski T, Tomczak A, Michelmore R (2005) Optimization of Agrobacterium-mediated transient assays of gene expression in lettuce, tomato and Arabidopsis. Plant Biotechnol J 3:259–273

Yusibov V, Modelska A, Steplewski K, Agadjanyan M, Weiner D, Hooper DC, Koprowski H (1997) Antigens produced in plants by infection with chimeric plant viruses immunize against rabies virus and HIV-1. Proc Natl Acad Sci USA 94:5784–5788

Zambryski P (1988) Basic processes underlying Agrobacterium-mediated DNA transfer to plant cells. Annu Rev Genet 22:1–30

Zamore PD, Tuschl T, Sharp PA, Bartel DP (2000) RNAi: double-stranded RNA directs the ATP-dependent cleavage of mRNA at 21 to 23 nucleotide intervals. Cell 101:25–33

Zhou F, Badillo-Corona JA, Karcher D, Gonzalez-Rabade N, Piepenburg K, Borchers AMI, Maloney AP, Kavanagh TA, Gray JC, Bock R (2008) High-level expression of human immunodeficiency virus antigens from the tobacco and tomato plastid genomes. Plant Biotechnol J 6:897–913

Ziółkowska NE, O'Keefe BR, Mori T, Zhu C, Giomarelli B, Vojdani F, Palmer KE, McMahon JB, Wlodawer A (2006) Domain-swapped structure of the potent antiviral protein griffithsin and its mode of carbohydrate binding. Structure 14:1127–1135

Chapter 10
Plant Virus-Mediated Expression in Molecular Farming

Aiming Wang

Abstract Plant viruses have contrasting abilities. On one hand, they can induce gene silencing, termed as virus-induced gene silencing. On the other hand, they have evolved mechanisms that suppress gene silencing and allow the accumulation of very high levels of viral proteins in infected plants. The latter is the driving force for the manipulation of plant viruses for molecular farming in plants. In comparison to the transgenic approach which is often associated with low levels of expression and the requirement of a time-consuming and labour-intensive genetic transformation process, the plant virus-mediated expression approach has several advantages such as easy manipulation, high yield and fast manufacturing. This approach uses plant virus-based expression vectors as a vehicle to produce therapeutic proteins such as antibodies, enzymes, vaccines, and other recombinant proteins of interest in plants. Over the last two decades, a number of plant viruses have been developed and optimized for expression of a variety of pharmaceutical proteins. Some of these recombinant proteins are currently under pre-clinical or clinical trials. In this chapter, I will summarize recent progress, current challenges and future prospects of plant virus-mediated expression in molecular farming.

10.1 Introduction

Since the discovery of the first plant virus, *Tobacco mosaic virus* (TMV), in 1898, virology has become a subject of science (Levine 2001). Over the last century, particularly since the 1980s, plant virology has contributed enormously to the

---

A. Wang (✉)
Southern Crop Protection and Food Research Centre, Agriculture and Agri-Food Canada,
1391 Sandford St., London, ON N5V 4T3, Canada
e-mail: Aiming.Wang@AGR.GC.CA

A. Wang and S. Ma (eds.), *Molecular Farming in Plants: Recent Advances and Future Prospects*, DOI 10.1007/978-94-007-2217-0_10,

understanding of the fundamental concepts in modern biology as well as the advancement of modern plant biotechnology. Plant viral elements such as promoters, terminators, translational enhancers, and gene silencing suppressors have been extensively studied and widely used in plant biotechnology (Hull 2002). Over the last two decades, plant viruses or their modified versions have been directly employed to drive the transient expression of recombinant proteins in plants (Lomonossoff and Porta 2001; Gleba et al. 2007; Lindbo 2007; Sainsbury et al. 2010a).

As a powerful and versatile platform technology for the expression of recombinant proteins in plants, plant virus-mediated expression systems are superior to the transgenic approach. The foreign proteins encoded by transgenes usually do not accumulate to high levels *in planta*, which bottlenecks the application of the transgenic approach in molecular farming (Doran 2006). However, plant viruses have evolved mechanisms that overcome the plant innate anti-foreign RNA system, e.g., posttranscriptional or virus-induced gene silencing, to overpower host cells to produce large quantities of viral proteins, and to prevent protein degradation with a yet unknown mechanism, allowing for accumulation of large amounts of the viral genome and the viral proteins (Baulcombe 2000). Thus, plant viruses may be manipulated to produce and accumulate large amounts of recombinant proteins within a short period of time. This is significantly superior to the lengthy process of generation and characterization of transgenic plants. Since the turnaround time for expression test is short, the plant viral vector system may be optimized through screening for high expression levels, suitable crop species, and proper protein antigenic sites. Toxic proteins are also less problematic for this system because healthy plants at proper growth stages can be selected for inoculation. Therefore, this system is fast, easy to manipulate, free from chromosomal position effects, highly efficient, and flexible with target proteins and plant species. At early stages, the "full virus" vector strategy was employed to construct the first-generation viral vectors (Gleba et al. 2004, 2007). Essentially the viral vector is an infectious clone that contains the full-length cDNA of the wild-type virus. The second-genera vectors are engineered by the "deconstructed virus" vector strategy. As the deletion versions of the full-length viral vectors, these vectors eliminate undesired or limiting viral genes but retain speed and high productivity (Gleba et al. 2004, 2007).

To date, a number of plant RNA viruses have been developed as powerful and versatile expression vectors for the production of a wide range of heterologous proteins in plants. The most commonly used RNA viruses include TMV, *Potato virus X* (PVX), *Cowpea mosaic virus* (CPMV), and a number of potyviruses. In fact, most of the success in virus-mediated expression in molecular farming has been made using RNA viruses, although breakthroughs in the use of DNA viruses have also been documented recently. As several aspects of this research area have been extensively discussed in a few excellent recent reviews (Lico et al. 2008; De Muynck et al. 2010; Pogue et al. 2010; Rybicki 2010; Gleba and Giritch 2011; other chapters in this book), I will outline in this chapter the most commonly studied plant viruses as expression systems in molecular farming and further briefly summarize recent progress, current challenges and future prospects of plant virus-mediated expression.

10.2 DNA Virus-Mediated Expression

10.2.1 Cauliflower Mosaic Virus

The earliest plant virus expression system for the production of foreign proteins in plants was developed not based on plant RNA viruses but on a DNA virus, *Cauliflower mosaic virus* (CaMV) (Brisson et al. 1984; Fütterer et al. 1990). As the first plant virus was found using DNA instead of RNA as genetic material, CaMV has played an essential and unique role in the fundamental research of plant molecular biology and biotechnology. Its 35S promoter has been widely used as a strong, constitutive expression promoter for various plant research projects as well as commercial applications (Scholthof et al. 1996; Haas et al. 2002). CaMV is the type member of the genus *Caulimovirus* in the family *Caulimoviridae*. The icosahedral virions are non-enveloped isometric particles with 420 coat protein (CP) subunits. The viral genome is a double-stranded circular DNA molecule of approximately 8 kb. It has two intergenic regions of regulation and six major open reading frames (ORFs). Its genome replication is through reverse transcription of a pregenomic RNA. Brisson et al. (1984) replaced an aphid transmission factor domain in the ORF II with a bacterial dihydrofolate reductase (DHFR) that confers resistance to methotrexate. The resulting vector was used to inoculate turnip leaves and the recombinant virus spread systemically in the inoculated plants. Typical CaMV symptoms were observed. The plant-derived DHFR directed by CaMV was biologically active (Brisson et al. 1984). The CaMV vector was also successfully used for the expression of a Chinese hamster metallothionein (CHMT II) and a human interferon in turnip plants (Lefebvre et al. 1987; de Zoeten et al. 1989). CaMV showed some potential as the recombinant protein could reach 0.5% of the total soluble leaf protein (Lefebvre et al. 1987). It was quickly found out that the use of the CaMV vector is hampered by the narrow range of plants (limited to the *Cruciferae* family and a few species in the family *Solanaceae*) infected by CaMV, and by practical limitations on inserting foreign DNA that are imposed by the biology of CaMV (Fütterer et al. 1990; Scholthof et al. 1996; Haas et al. 2002).

10.2.2 Geminiviruses

Viruses in another DNA virus family *Geminiviridae* have also been developed as vectors for the production of proteins of interest in plants. The viral genome of this family consists of one or two circular single-stranded DNA (ssDNA) molecules of 2.5–3.0 kb in length (Gutirrez 1999, 2000). Based on genomic organization, host range and insect vectors, viruses in this family are classified into four genera, e.g., *Mastrevirus*, *Begomovirus*, *Curtovirus*, and *Topocuvirus*. Although viruses in all the four genera have been studied as potential expression vectors, excellent progress has been made from several viruses in the first two genera.

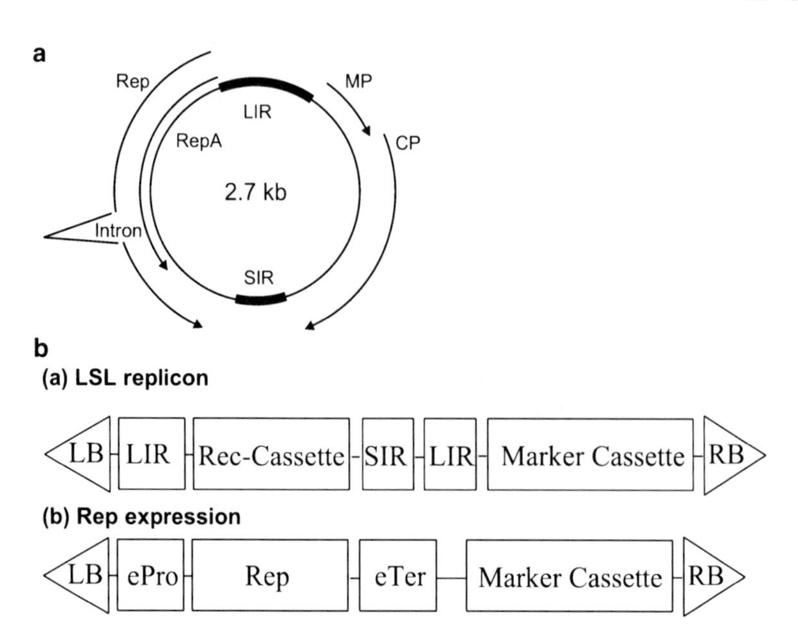

Fig. 10.1 Geminivirus-mediated expression in molecular farming. (**a**) Genomic organization of mastreviruses including *Bean yellow dwarf virus* (BeYDV). (**b**) The BeYDV-based dual-vector system for the high-level expression of recombinant proteins in plants. *CP* coat protein, *ePro* plant eukaryotic promoter, *eTer* plant eukaryotic terminator, *IR* intergenic region, *LB* left boarder, *LIR* large intergenic region, *Marker cassette* marker gene expression cassette, *MP* movement protein, *RB* right boarder, *Rec-Cassette* recombinant protein expression cassette, *Rep* replicase, *RepA* resulting from translation of the differentially spliced transcript

The leafhopper-transmitted mastreviruses including *Bean yellow dwarf virus* (BeYDV), *Maize streak virus* (MSV), *Wheat dwarf virus* (WDV), and *Tobacco yellow dwarf virus* (TYDV) have a single genome component of about 2.7 kb (Needham et al. 1998; Hefferon and Fan 2004; Hefferon et al. 2004; Huang et al. 2009; Regnard et al. 2010). Most mastreviruses are confined to monocotyledonous plants, but some of them such as BeYDV and TYDV infect dicots. The mastreviral genome has a long intergenic region (LIR) that consists of transcriptional promoters and the viral origin of replication, and a short intergenic region (SIR) that contains transcription termination signals and the DNA primer binding site for complementary strand DNA synthesis (Fig. 10.1) (Regnard et al. 2010). Similar to other geminiviruses, mastreviruses carry out their genome replication in the nuclei of the infected cells using a rolling circle mechanism (Gutierrez 1999, 2000). During genome replication, numerous double-stranded DNA replicative form intermediates are produced for both replication and transcription. Mastreviruses have only three genes: the viral sense genes *V1* and *V2* encoding the movement protein (MP) and the coat protein (CP), respectively and the complementary sense gene *Rep* coding for two replicase proteins, Rep and RepA (resulting from an alternative splicing) (Fig. 10.1). Early endeavors were concentrated on MSV and WDV (Laufs et al. 1990;

Matzeit et al. 1991; Ugaki et al. 1991; Timmermans et al. 1992, 1994). In those MSV or WDV-derived vectors, MP or CP was replaced by foreign proteins, and infection and replication were limited within protoplasts or primarily infected cells. An improved MSV-derived vector where foreign genes were inserted into the non-coding region of the viral genome was shown to be able to systematically infect maize plants with an increased level of recombinant protein accumulation (Shen and Hohn 1994, 1995). Unfortunately, the overall foreign protein yield was not satisfactory.

Recent focus has shifted on dicot-infecting mastreviruses such as BeYDV (Mor et al. 2003; Zhang et al. 2006; Huang et al. 2009; Regnard et al. 2010). A BeYDV-derived LSL replicon from a plasmid or a chromosome-integrated transgene could be rescued by the presence of the replication initiation protein, Rep (Fig. 10.1). High-level expression of GUS was found in tobacco NT1 cell suspensions co-transfected with the LSL vector carrying a GUS reporter and a Rep-supplying vector (Mor et al. 2003). In this system, Rep induced release of the BeYDV replicon and episomal replication to high copy number (Mor et al. 2003). This system was tested with two transgenes. One of them was the BeYDV replicon with an expression cassette (allowing for expression of the genes of interest) flanked by *cis*-acting DNA elements of BeYDV. The other was to express Rep under an alcohol-inducible promoter. Both of them were transformed into the tobacco NT1 cells and potato plants. After ethanol treatment, transgene mRNA and protein levels in the NT1 cells and the leaves of whole potato plants increased by 80 and 10 times, respectively (Zhang et al. 2006). To avoid genetic transformation, co-delivery of the BeYDV-derived vector and the Rep/RepA supplying vector was attempted by agroinfiltration of *Nicotinana benthamiana* (Huang et al. 2009) (Fig. 10.1). The improved system largely enhanced recombinant protein accumulation (Huang et al. 2009). More recently, Regnard et al. (2010) developed a new BeYDV vector, pRIC. This vector differs from other BeYDV- and other geminivirus-derived vectors published previously. In this vector system, the BeYDV replicase proteins were included *in cis*. High level protein expression in plants by using this autonomously replicating shuttle vector was reported (Regnard et al. 2010).

In addition to mastreviruses, the whitefly-transmitted begomoviruses, such as *Bean golden mosaic virus* (BGMV), *Tomato gold mosaic virus* (TGMV), *Tomato yellow leaf curl virus* (TYLCV) and *Ageratum yellow vein virus* (AYVV), have also been investigated for their suitability in molecular farming. This group of geminiviruses exclusively infects dicotyledonous plants. Most begomoviruses have a bipartite genome with two ssDNA molecules (DNA A and DNA B) of approximately 2.6 kb each. As early as in 1988, Hayes et al. successfully developed a TGMV-based vector. This vector was used to express bacterial neomycin phosphotransferase (NPT) (gene *neo*) and bacterial β-glucuronidase (GUS) (gene *uidA*) in tobacco (Hayes et al. 1988, 1989). However, this vector seemed unstable as deletion was evident (Hays et al. 1989). A few members of begomoviruses such as AYVV and TYLCV that have a monopartite genome have also been tested as viral expression vectors (Tamilselvi et al. 2004; Perez et al. 2007). Although not much progress has been made so far, such monopartite begomoviruses are predicted to be of great potential in molecular farming as they are dicotyledonous plant viruses and share the similar genomic structure with BeYDV-like mastreviruses.

10.3 RNA Virus-Mediated Expression

Numerous plant RNA viruses have been explored for their uses in molecular farming. In most cases, the foreign gene is either to replace or to be fused in-frame with CP and the viral cDNA containing the foreign gene is placed under the control of a strong bacteriophage promoter such as SP6, T3 and T7 in an *in vitro* transcription vector or under a eukaryotic constitutive expression promoter such as the CaMV 35S promoter in a binary vector (Fig. 10.2). For the former, infectious transcripts are obtained by *in vitro* transcription and introduced into plants via mechanical inoculation. For the latter, transient expression is achieved by mechanical inoculation, biolistic delivery or agroinfection. Here I will cover the most commonly used plant RNA virus-mediated expression systems.

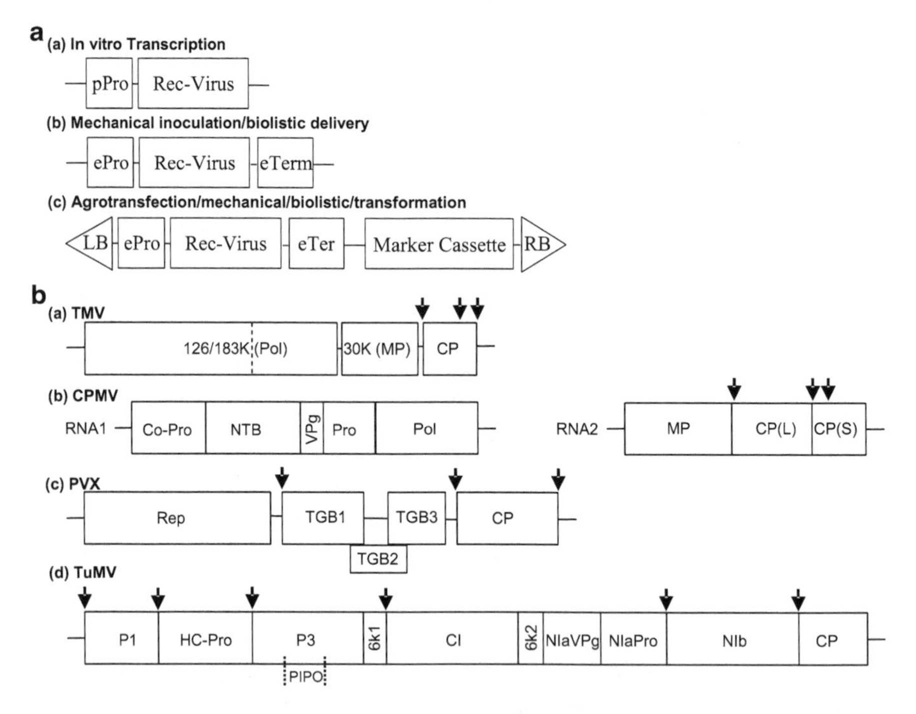

Fig. 10.2 RNA virus-mediated expression in molecular farming. (**a**) Commonly used strategies for vector construction. *pPro* prokaryotic promoter (such as SP6, T3 and T7), *Rec-virus* recombinant virus, *ePro* plant eukaryotic promoter, *eTerm* plant eukaryotic terminator. (**b**) Construction of genomic organization of plant RNA viruses that have been extensively used to express peptides and proteins of interest in plants. *Arrows* point to the positions where foreign proteins are inserted. *TMV* Tobacco mosaic virus, *CPMV* Cowpea mosaic virus, *PVX* Potato virus X, and *TuMV* Turnip mosaic virus

10.3.1 TMV

TMV is the type member of the genus *Tobamovirus*. As the first named virus, TMV is considered the best studied RNA virus. The TMV-based viral vector is also the most commonly used one in molecular farming. TMV has a simple viral genome, which is a 6.4 kb single-stranded positive sense RNA molecule encoding four viral proteins (Fig. 10.2). The first ORF encodes two replicase proteins, 126- and 183-kDa proteins, which are required for viral RNA replication. The 183-kDa replicase results from read-through of the amber stop codon for the 126-kDa protein. The 30-kDa MP and 17-kDa CP are translated via two subgenomic RNAs from the 3'-proximal ORFs. As the 17-kDa CP is the most abundant protein, CP fusion or replacement of CP with the foreign gene to be expressed was the primary strategy for the expression of foreign proteins (Takamatsu et al. 1987). However, this kind of recombinant TMVs was unable to infect systemically due to the lack or disturbance of the functional CP. This problem was solved by cloning the target gene under the control of the TMV CP subgenomic promoter and inserting this expression unit between MP and CP of the viral vector (Kumagai et al. 1993). This strategy was successfully used to express, at high levels, a number of recombinant proteins such as α-trichosanthin (Kumagai et al. 1993), human papillomavirus CP L1 (Varsani et al. 2006), GFP (Lindbo 2007), and human growth hormone (Skarjinskaia et al. 2008) in plants.

10.3.2 CPMV

CPMV, the type member of the genus *Comovirus*, is a bipartite virus consisting of two positive-sense single-stranded RNAs, RNA1 and RNA2 as its genome. Each RNA has a 3'-poly(A) tail and a viral protein genome-linked (VPg) covalently linked at the 5' terminus (Lomonossoff and Porta 2001). RNA1 is about 5.9 kb and RNA2 ~3.5 kb in length, each encoding a single polyprotein that is processed by the RNA1-encoded 24-kDa protease domain (Fig. 10.2). RNA1-encoded proteins are required for viral genome replication and polyprotein processing; RNA2 encodes MP and two CPs, i.e., CP(L) and CP(S). As RNA1 is essential for replication, RNA2 is the target of manipulation. Early work was focused on fusion or replacement of the CP(S) for the expression of epitopes such as an epitope of VP1 of *Foot-and-mouth disease virus* (FDMV) (Usha et al. 1993), epitopes derived from human rhinovirus and human immunodeficiency virus type 1 (HIV-1) (Porta et al. 1994) and several other epitopes (McLain et al. 1995; Dalsgaard et al. 1997; Brennan et al. 1999; Taylor et al. 2000). In addition to epitopes, CPMV has also been used for the high-level expression of a variety of recombinant proteins. Cañizares et al. (2006) reported an improved CPMV expression system. In this system, the chimeric gene coding for the 2A protease of FMDV and the target protein was in frame fused to the C-terminus of the CP(S) of RNA2. Upon translation, the target protein can be released by the 2A protease. Transgenic plants containing the recombinant CPMV RNA2 cDNA controlled by the *CaMV 35S* promoter was generated. When the transgenic plants derived from this vector were agroinoculated with a plant expression vector to produce RNA1-encoded proteins or genetically crossed with another transgenic plant expressing RNA1, active recombinant CPMV replication would occur, leading to the production of large amounts of recombinant protein. Cañizares et al. (2006) reported a deletion version of RNA2 where the recombinant RNA2 cDNA was precisely deleted to keep the 5' UTR and 3' UTR under the control of the 35S promoter. The target gene can be inserted between the 5' and 3' UTR for expression. More recently, Sainsbury et al. (2008, 2010b) have further modified the CPMV expression system to produce high quality functional anti-HIV antibodies.

10.3.3 PVX

PVX, the type member of the genus *Potexvirus*, is a monopartite virus with a single, positive-stranded RNA molecule of approximately 6.5 kb as its genome (Chapman et al. 1992). The 5' capped, 3' polyadenylated genomic RNA has five ORFs encoding a replicase (Rep), a set of three movement proteins (triple gene block: TGB1, TGB2 and TGB3) and CP (Batten et al. 2003; Avesani et al. 2007) (Fig. 10.2). PVX-mediated expression has been extensively studied for expression of heterologous proteins in plants (Chapman et al. 1992; Baulcombe et al. 1995; Hammond-Kosack et al. 1995; Sablowski et al. 1995; Angell and Baulcombe 1997; Santa et al. 1996; O'Brien et al. 2000; Ziegler et al. 2000; Marusic et al. 2001; Toth et al. 2001; Franconi et al. 2002; Avesani et al. 2003; Čeřovská et al. 2004; Manske et al. 2005; Uhde et al. 2005; Komorova et al. 2006; Ravin et al. 2008). As the recombinant PVX lacking CP (replaced by the target gene) failed to infect plants systemically resulting in poor accumulation of recombinant proteins, this gene replacement strategy seemed ineffective (Chapman et al. 1992). A new strategy, i.e., CP fusions, was proposed and tested for the production of vaccine antigenic sites (Santa et al. 1996; O'Brien et al. 2000; Marusic et al. 2001; Uhde et al. 2005). However, this system requires that the CP-fusion does not compromise viral particle assembly and the systemic infectivity of the recombinant virus. Therefore, it is only suitable for the expression of small protein tags or antigen epitopes. The gene insertion strategy was further developed to overcome this drawback. This system allows for insertion of the target gene into the viral genome coupled to a duplicated copy of the CP promoter (Baulcombe et al. 1995). This improved version infected systemically and directed high-levels of expression of a variety of recombinant proteins such as GFP and the human islet autoantigen glutamic acid decarboxylase (hGAD65) (Baulcombe et al. 1995; Avesani et al. 2003). Recent studies have shown that a minimal PVX vector in which the triple block and CP were removed can allow for high-level expression (Komorova et al. 2006; Ravin et al. 2008). The target gene was inserted downstream of the first viral subgenomic promoter and transcription of the recombinant PVX was controlled by the 35S promoter. In agroinfiltrated plant leaves, the recombinant protein accumulated up to 2% of total soluble proteins (Ravin et al. 2008).

10.3.4 Potyviruses

Potyviruses represent the largest and most agriculturally important plant virus group (Urcuqui-Inchima et al. 2001). The monopartite virus has a single-stranded positive-sense RNA molecular of ~10 kb as its genome. Similar to CPMV, the genomic RNA has a 3'-poly(A) tail and a viral protein genome-linked (VPg) covalently linked at the 5' terminus. The virus adopts a polyprotein strategy. The only long ORF encodes a large polyprotein of ~350 kDa, and also a shorter polyprotein as a result of translational frameshift in the P3 coding region (Fig. 10.2). The two polyproteins are processed by virus-encoded proteases to release 11 mature proteins, from the N-terminus, P1, HC-Pro, P3, P3N-PIPO, 6 K1, CI, 6 K2, NIa-VPg, NIa-Pro, NIb, and CP (Chung et al. 2008). Since recombinant potyviruses can allow for simultaneous equimolecular expression of multiple foreign genes and the lengths of the foreign genes are flexible, a number of potyviruses have been developed as a viral expression vector, including *Turnip mosaic virus* (TuMV), *Tobacco etch virus* (TEV), *Plum pox virus* (PPV), *Lettuce mosaic virus* (LMV), *Clover yellow vein virus* (ClYVV), *Pea seed-borne mosaic virus* (PSbMV), *Potato virus A* (PVA), *Zucchini yellow mosaic virus* (ZYMV) and *Soybean mosaic virus* (SMV) (Dolja et al. 1992; Verchot et al. 1995; Guo et al. 1998; Whitham et al. 1999; German-Retana et al. 2000; Masuta et al. 2000; Johansen et al. 2001; Hsu et al. 2004; Beauchemin et al. 2005; Kelloniemi et al. 2008; Wang et al. 2008). TEV was the first potyvriusderived vector and the foreign gene was in-frame inserted at the junction of P1 and HC-Pro for expression (Dolja et al. 1992). In addition to the P1/HC-Pro site, other junctions such as P1/HC-Pro, NIa-Pro/NIb and NIb/CP have also been found suitable for the simultaneous expression of several heterologous proteins (Chen et al. 2007; Kelloniemi et al. 2008; Bedoya et al. 2010).

10.4 Virus-Mediated Expression of Antibodies

Antibodies play an important role in several aspects of medical science such as research, therapy and diagnostics. Virus-based expression has been explored for the production of recombinant antibodies in plants to satisfy the growing demand. Verch et al. (1998) pioneered such research by using a TMV-based vector to express monoclonal antibody (mAb) CO17-1A, directed to a colon cancer antigen, in *N. benthamiana*. Two recombinant TMV clones were engineered to express heavy and light chains of this antibody. Plants co-transfected with both recombinant viral constructs expressed the heavy and light chains that were assembled into a biologically active full-length antibody (Verch et al. 1998). To enhance the recombinant antibody yield, Giritch et al. (2006) used two noncompeting viral vectors derived from TMV and PVX, each expressing one different chain of the human tumor-specific mAb A5 independently. The two viral vectors effectively coexpressed the heavy and light chains in the same cell throughout the plant with yields of up to 0.5 g of assembled mAbs per kg of fresh-leaf biomass (Giritch et al. 2006).

The same two-noncompeting viral vector system was successfully used to produce the humanized murine mAb, Hu-E16 (Lai et al. 2010). The plant-derive mAb had therapeutic activity as effective as the mammalian-cell-produced HuE16 against *West Nile virus* (Lai et al. 2010).

10.5 Virus-Mediated Expression of Vaccines

A number of plant-derived candidate vaccines through virus-mediated expression have been produced against the causal pathogen such as a virus, bacterium or parasite (termed prophylactic vaccine) and a disease such as cancer (termed therapeutic vaccine) in humans and animals. About 30 representative plant-based antigens expressed through viral vectors before 2007 were listed in an excellent review by Lico et al. (2008). Based on the nature of the recombinant proteins, these plant-based vaccines can be classified into two types: free proteins (vaccine subunits) and peptide or epitope fusions to viral CP (which forms empty viral particles) or to other proteins (Pogue et al. 2002; Rybicki 2010). Unlike the traditional vaccines that are often injected intraperitoneally for vaccination, plant-derived vaccines either through virus-mediated expression or other expression systems can be administered into the body orally, intranasally or by needle injection, as most of plants are edible (Awram et al. 2002). Oral or nasal administration that can induce specific mucosal immune response may be the best approach against pathogens as the vast majority of pathogens enter the body through the mucosal surface.

10.5.1 Prophylactic Vaccines

The TMV-based vector was used to express the malarial epitope-CP fusion protein, one of the first plant-based vaccines (Turpen et al. 1995). The recombinant TMV technology was modified to express numerous prophylactic and therapeutic vaccines. For instance, the 5B19 epitope of the spike protein of *Murine hepatitis virus* (MHV) was fused to the C-terminus of the CP of a TMV vector. The mice immunized through either subcutaneous or intranasal routes survived challenge with a lethal dose of MHV strain JHM (Koo et al. 1999). In the case of free vaccine subunits, the structural proteins VP1 of FMDV carrying critical epitopes responsible for the induction of neutralizing antibodies was expressed by a TMV-based vector and needle injection of mice with foliar extracts containing VP1 induced immune protection against a lethal FMDV infection (Wigdorovitz et al. 1999). Recently, a new version of the TMV-based vector that was delivered into plants via agroinfiltration has been successfully used to express several antigens against both viral and bacterial pathogens (Musiychuk et al. 2007; Mett et al. 2008; Chichester et al. 2009).

Plant-based vaccines were expressed by other viral vectors too. For instance, a PPV-based vector was employed to express the entire VP60 of *Rabbit hemorrhagic disease virus* (RHDV) by insertion into the junction of NIb/CP. Immunization of rabbits via the subcutaneous route with protein extracts containing VP60 protected the immunized rabbits against a lethal challenge with RHDV (Fernádez-Fernádez et al. 2001). A PVX-based vector was used to express the CP fusion with a highly conserved ELDKWA epitope from glycoprotein (gp) 41 of HIV-1. Normal or immunodeficient mice were immunized intraperitoneally or intranasally with the purified chimeric particles from *N. benthamiana* leaves inoculated with the recombinant PVX. High levels of HIV-1-specific immune response were found in these mice. Sera from either normal or immunodeficient mice immunized with the plant-derived CP fusions showed an anti-HIV-1 neutralizing activity (Marusic et al. 2001).

10.5.2 Therapeutic Vaccines

The recombinant TMV technology was also used to express therapeutic vaccines against challenging chronic diseases. An excellent example was to express therapeutic vaccines against cancers. B cell tumours express a unique cell surface Ig which is a tumor-specific marker and vaccination of patients with this Ig often achieves a superior clinical outcome (McCormick et al. 1999). An ideotype-specific single-chain Fv fragment (scFv) of the immunoglobulin from the 38 C13 mouse B cell lymphoma was cloned into a TMV vector and expressed in *N. benthamiana*. Mice immunized with the plant-made scFV were protected from challenge by a lethal dose of the syngenetic 38 C13 tumor cells (McCormick et al. 1999). Recently, the same viral expression system has been used to express patient-specific scFVs from individual patient's tumour. Results from a Phase I clinical study by immunization of patients with their own individual therapeutic antigen produced in plants suggest that the ideotype vaccines produced through virus-mediated expression are safe to administer and offer follicular lymphoma patients with a viable option for ideotype-specific immune therapy (McCormick et al. 2008).

10.6 Virus-Mediated Expression of Recombinant Proteins with Other Functions

Plant-produced proteins via virus-mediated expression may have diverse functions such as insecticides and industrial enzymes. For example, a TMV vector was used to produce rice α-amylase in *N. Benthamiana* (Kumagai et al. 2000). The plant-produced enzyme was moderately glycosylated and was accumulated to levels of at least 5% of total soluble protein. A potential larvicide was also produced through TMV-mediated expression in plants (Borovsky et al. 2006). A mosquito decapeptide hormone, the *Aedes aegypti* trypsin-modulating oostatic factor (TMOF) was fused with CP. In *N. tabacum* infected by the recombinant TMV, the TMOF fusion could reach to levels of 1.3% of total soluble protein. TMOF fusion-expressing tobacco discs effectively inhibited the growth of *Heliothis virescens*. Purified CP-TMOF virions fed to mosquito larvae arrested larval growth and caused death (Borovsky et al. 2006).

10.7 Conclusions

Like other plant production systems, the virus-based expression system has its strengths and weaknesses. There are several challenges that may prevent the application of this system for commercial uses. First, each particular virus-mediated expression system is limited to certain plant species due to the host range of plant viruses. Second, plant virus-mediated expression is associated with the potential high error frequency and rapid recombination (van Vloten-Doting et al. 1985; Drake and Holland 1999). The relatively high mutation rate of the viral RNA-dependent RNA polymerase due to the lack of proofreading ability may result in the production of mixed recombinant proteins containing undesirable proteins. Rapid recombination may cause instability and deletion of the foreign genes (Scholthof et al. 1996; García-Arenal et al. 2003). Third, there are bio- and environmental concerns about the impact of recombinant viruses (Pogue et al. 2002). Although the risk of human and animal infection by exposure in the field or in food products to plant viruses is ruled out due to their non-infectivity to human beings and animals, recombinant plant viruses can spread to weeds or other crops in their host range. Containment measures through physical barriers or biotechnology (mutations in the genes responsible for insect transmission to prevent insect transmission in the field) should be assessed. In addition, other practical issues, such as expression stability, the biological activity of the product, and downstream processing, must be addressed before large-scale commercial uses of the plant viral vector system.

In view of the growing demand of the market for recombinant proteins with diverse functions, virus-mediated expression in plants has proved to be a promising alternative means to satisfy this need. In the past several years, several new vector systems revolutionized the area with improved viral vector stability, high protein yields and low bio- and environmental risks (Sanchez-Navarro et al. 2001; Perez et al. 2007; Huang et al. 2009; Regnard et al. 2010; Sainsbury et al. 2010a). Some systems require just a few weeks from obtaining the DNA sequence of a viral pathogen to the production of candidate vaccine subunits against it. This very short turn-around time also makes "personalized therapeutic proteins" such as antibodies possible and practical. Along with the improvement of virus-mediated expression technologies, the number of virus-derived recombinant proteins is expected to increase significantly. As a result, more and more plant-derived recombinant proteins via virus-mediated expression will move through clinic trials into commercialization in the coming years. This exciting vision will certainly stimulate new investments on research and development of next generation virus-mediated expression systems that will circumvent current technological drawbacks and new challenges emerging during the course of research, production and commercialization.

Acknowledgement This work was supported in part by an A-base grant from Agriculture and Agri-Food Canada.

References

Angell SM, Baulcombe DC (1997) Consistent gene silencing in transgenic plants expressing a replicating potato virus X RNA. EMBO J 16:3675–3684

Avesani L, Falorni A, Tornielli GB, Marusic C, Porceddu A, Polverari A, Faleri C, Calcinaro F, Pezzotti M (2003) Improved in planta expression of human islet autoantigen glutamic acid decarboxylase (GAD65). Transgenic Res 12:203–212

Avesani L, Marconi G, Morandini F, Albertini E, Bruschetta M, Bortesi L, Pezzotti M, Porceddu A (2007) Stability of *Potato virus X* expression vectors is related to insert size: implications for replication models and risk assessment. Transgenic Res 16:587–597

Awram P, Gardner RC, Forster RL, Bellamy AR (2002) The potential of plant viral vectors and transgenic plants for subunit vaccine production. Adv Virus Res 58:81–123

Batten JS, Yoshinari S, Hemenway C (2003) Potato virus X: a model system for virus replication, movement and gene expression. Mol Plant Pathol 4:125–131

Baulcombe DC (2000) Unwinding RNA silencing. Science 290:1108–1109

Baulcombe DC, Chapman S, Santa Cruz S (1995) Jellyfish green fluorescent protein as a reporter for virus infections. Plant J 7:1045–1053

Beauchemin C, Bougie V, Laliberté J-F (2005) Simultaneous production of two foreign proteins from a potyvirus-based vector. Virus Res 112:1–8

Bedoya L, Martínez F, Rubio L, Daròs J-A (2010) Simultaneous equimolecular expression of multiple proteins in plants from a disarmed potyvirus vector. J Biotechnol 150:268–275

Borovsky D, Rabindran S, Dawson WO, Powell CA, Iannotti DA, Morris TJ, Shabanowitz J, Hunt DF, DeBondt H, DeLoof A (2006) Expression of *Aedes* trypsin-modulating oostatic factor on the virion of TMV: a potential larvicide. Proc Natl Acad Sci USA 103:18963–18968

Brennan FR, Jones TD, Gilleland LB, Bellaby T, Xu F, North PC, Thompson A, Staczek J, Lin T, Johnson JE, Hamilton WDO, Gilleland HE (1999) *Pseudomonas aeroginosa* outer-membrane protein F epitopes are highly immunogenic when expressed on a plant virus. Microbiology 145:211–220

Brisson N, Paszkowski J, Penswick JR, Gronenborn B, Potrykus I, Hohn T (1984) Expression of a bacterial gene in plants by suing a viral vector. Nature 310:511–514

Cañizares MC, Liu L, Perrin Y, Tsakiris E, Lomonossoff GP (2006) A bipartite system for the constitutive and inducible expression of high levels of foreign proteins in plants. Plant Biotechnol J 4:183–193

Čeřovská N, Pečenková T, Tomáš M, Velemínský J (2004) Transient expression of heterologous model gene in plants using *Potato virus X*-based vector. Plant Cell Tissue Org Cult 79:147–152

Chapman S, Kavanagh T, Baulcombe D (1992) Potato virus x as a vector for gene expression in plants. Plant J 2:549–557

Chen C-C, Chen T-C, Raja JAJ, Chang C-A, Chen L-W, Lin S-S, Yeh S-D (2007) Effectiveness and stability of heterologous proteins expressed in plants by Turnip mosaic virus vector at five different insertion sites. Virus Res 130:210–227

Chichester JA, Musiychuk K, Farrance CE, Mett V, Lyons J, Mett V, Yusibov V (2009) A single component two-valent LcrV-F1 vaccine protects non-human primates against pneumonic plaque. Vaccine 27:3471–3474

Chung BY, Miller WA, Atkins JF, Firth AE (2008) An overlapping essential gene in the *Potyviridae*. Proc Natl Acad Sci USA 105:5897–5902

Dalsgaard K, Uttenthal Å, Jones TD, Xu F, Merryweather A, Hamilton WDO, Langeveld JPM, Boshuizen RS, Kamstrup S, Lomonossoff GP, Porta C, Vela C, Casal JI, Meloen RH, Rodgers PB (1997) Plant-derived vaccine protects target animals against a virus disease. Nat Biotechnol 15:248–252

De Muynck B, Navarre C, Boutry M (2010) Production of antibodies: status after twenty years. Plant Biotechnol J 8:529–563
De Zoeten GA, Penswick JR, Horisberger MA, Ahl P, Schulze M, Hohn T (1989) The expression, localization, and effect of a human interferon in plants. Virology 172:213–222
Dolja VV, McBride HJ, Carrington JC (1992) Tagging of plant potyvirus replication and movement by insertion of beta-glucuronidase into the viral polyprotein. Proc Natl Acad Sci USA 89:10208–10212
Doran P (2006) Foreign protein degradation and instability in plants and plant cultures. Trends Biotechnol 24:426–432
Drake JW, Holland JJ (1999) Mutation rates among RNA viruses. Proc Natl Acad Sci USA 96:13910–13913
Fernández-Fernández MR, Mouriño M, Rivera J, Rodriguez F, Plana-Durán P, Garcia JA (2001) Protection of rabbits against *Rabbit hemorrhagic disease virus* by immunization with the VP60 protein expressed in plants with a potyvirus-based vector. Vaccine 280:283–291
Franconi R, Di Bonito P, Dibello F (2002) Plant derived-human papillomavirus 16 E7 oncoprotein induces immune response and specific tumor protection. Cancer Res 62:3654–3658
Fütterer J, Bonneville JM, Hohn T (1990) Cauliflower mosaic virus as a gene expression vector for plants. Physiol Plant 79:154–157
García-Arenal F, Fraile A, Malpica JM (2003) Variation and evolution of plant virus populations. Int Microbiol 6:225–232
German-Retana S, Candresse T, Alias E, Delbos R-P, Le Gall O (2000) Effects of green fluorescent protein and β-glucuronidase tagging on the accumulation and pathogenicity of a resistance-breaking *Lettuce mosaic virus* isolate in susceptible and resistant lettuce cultivars. Mol Plant Microbe Interact 13:316–324
Giritch A, Marillonnet S, Engler C, van Eldik G, Botterman J, Klimyuk V, Gleba Y (2006) Rapid high-yield expression of full-size IgG antibodies in plants coinfected with noncompeting viral vectors. Proc Natl Acad Sci USA 103:14701–14706
Gleba Y, Giritch A (2011) Plant viral vectors for protein expression. In: Caranta C, Aranda MA, Tepfer M, Lopez-Moya JJ (eds) Recent advances in ant virology. Caister Academic Press, Norfolk, pp 387–412
Gleba Y, Marillonnet S, Klimyuk V (2004) Engineering viral expression vectors for plants: the 'full virus' and the 'deconstructed virus' strategies. Curr Opin Plant Biol 7:182–188
Gleba Y, Klimyuk V, Marillonnet S (2007) Viral vectors for the expression of protein in plants. Curr Opin Biotechnol 18:134–141
Guo HS, López-Moya JJ, Garcia JA (1998) Susceptibility to recombinant rearrangements of a chimeric plum pox potyvirus genome after insertion of a foreign gene. Virus Res 57:183–195
Gutierrez C (1999) Geminivirus DNA replication. Cell Mol Life Sci 56:313–329
Gutierrez C (2000) DNA replication and cell cycle in plants: learning from geminiviruses. EMBO J 19:792–799
Haas M, Bureau M, Geldreich A, Yot P, Keller M (2002) Cauliflower mosaic virus: still in the news. Mol Plant Pathol 3:419–429
Hammond-Kosack KE, Staskawicz BJ, Jones JDG, Baulcombe DC (1995) Functional expression of a fungal avirulence gene from a modified potato virus X genome. Mol Plant Microbe Interact 8:181–185
Hayes RJ, Petty ITD, Coutts RHA, Buck KW (1988) Gene amplification and expression in plants by a replicating geminivirus vector. Nature 334:179–182
Hayes RJ, Coutts RHA, Buck KW (1989) Stability and expression of bacterial genes in replicating geminivirus vectors. Nucleic Acids Res 17:2391–2403
Hefferon KL, Fan Y (2004) Expression of a vaccine protein in a plant cell line using a geminivirus-based replicon system. Vaccine 23:404–410
Hefferon KL, Kipp P, Moon YS (2004) Expression and purification of heterologous proteins in plant tissue using a geminivirus vector system. J Mol Microbiol Biotechnol 7:109–114

Hsu C-H, Lin S-S, Liu F-L, Su W-C, Yeh S-D (2004) Oral administration of a mite allergen expressed by zucchini yellow mosaic virus in cucurbit species downregulate allergen-induced airway inflammation and IgE synthesis. J Allergy Clin Immunol 113:1079–1085

Huang Z, Chen Q, Hjelm B, Arntzen C, Mason H (2009) A DNA replicon system for rapid high-level production of virus-like particles in plants. Biotechnol Bioeng 103:706–714

Hull R (2002) Control and uses of plant viruses. In: Hull R (ed) Matthews' plant virology. Academic, London, pp 675–741

Johansen IE, Lund OS, Hjulsager CK, Laursen J (2001) Recessive resistance in *Pisum sativum* and potyvirus pathotype resolved in a gene-for-cistron correspondence between host and virus. J Virol 75:6609–6614

Kelloniemi J, Mäkinen K, Valkonen JPT (2008) Three heterologous proteins simultaneously expressed from a chimeric potyvirus: infectivity, stability and the correlation of genome and virion lengths. Virus Res 135:282–291

Komorova TV, Skulachev MV, Zvereva AS, Schwartz AM, Dorokhov YL, Atabekov JG (2006) New viral vector for efficient production of target proteins in plants. Biochemistry (Moscow) 71:846–850

Koo M, Bendahmane M, Lettiri GA, Paoletti AD, Lane TE, Fitchen JH, Buchmeier MJ, Beachy RN (1999) Protective immunity against murine hepatitis virus (MHV) induced by intranasal or subcutaneous administration of hybrids of tobacco mosaic virus that carries an MHV epitope. Proc Natl Acad Sci USA 96:7774–7779

Kumagai MH, Turpen TH, Weinzettl N, Della-Cioppa G, Turpen AM, Donson J, Hilf ME, Grantham GL, Dawson WO, Chow TP, Piatak M Jr, Grill LK (1993) Rapid, high-level expression of biologically active α-trichosanthin in transfected plants by an RNA viral vector. Proc Natl Acad Sci USA 90:427–430

Kumagai MH, Donson J, della-Cioppa G, Grill LK (2000) Rapid, high-level expression of glycosylated rice α-amylase in transfected plants by an RNA viral vector. Gene 245:169–174

Lai H, Engle M, Fuchs A, Keller T, Johnson S, Gorlatov S, Diamond MS, Chen Q (2010) Monoclonal antibody produced in plants efficiently treats West Nile virus infection in mice. Proc Natl Acad Sci USA 107:2419–2424

Laufs J, Wirtz U, Kammann M, Matzeit V, Schaefer S, Schell J, Chernilofsky AP, Baker B, Gronenborn B (1990) Wheat dwarf AC/DS vectors: expression and excision of transposable elements introduced into various cereals by a viral replicon. Proc Natl Acad Sci USA 87:7752–7756

Lefebvre DD, Miki BL, Laliberte JF (1987) Mammalian metallothionein functions in plants. Biotechnology 5:1053–1056

Levine AJ (2001) The origins of virology. In: Knipe DM, Howley PM (eds) Fields virology (volume 1). Lippincott Williams & Wilkins, Philadelphia, pp 3–18

Lico C, Chen Q, Santi L (2008) Viral vectors for production of recombinant proteins in plants. J Cell Physiol 216:366–377

Lindbo JA (2007) TRBO: a high-efficiency *Tobacco mosaic virus* RNA-based overexpression vector. Plant Physiol 145:1232–1240

Lomonossoff GP, Porta C (2001) Cowpea mosaic virus as a versatile system for the expression of foreign peptides and proteins in legumes. In: Toutant JP, Balázs E (eds) Molecular farming. INRA Editions, Paris, pp 151–160

Manske U, Schiemann J (2005) Development and assessment of a *Potato virus X*-based expression system with improved biosafety. Environ Biosafety Res 4:45–57

Marusic C, Rizza P, Lattanzi L, Mancini C, Spada M, Belardelli F, Benvenuto E, Capone I (2001) Chimeric plant virus particles as immunogens for inducing murine and human immune responses against human immunodeficiency virus type 1. J Virol 75:8434–8439

Masuta C, Yamana T, Tacahashi Y, Uyeda I, Sato M, Ueda S, Matsumura T (2000) Development of clover yellow vein virus as an efficient, stable gene expression system for legume species. Plant J 23:539–545

Matzeit V, Schaefer S, Kammann M, Schalk HJ, Schell J, Gronenborn B (1991) Wheat dwarf virus vectors replicate and express foreign genes in cells of monocotyledonous plants. Plant Cell 3:247–258

McCormick AA, Kumagai MH, Hanley K, Turpen TH, Hakim I, Grill LK, Tusé D, Levy S, Levy R (1999) Rapid production of specific vaccine for lymphoma by expression of the tumor-derived single-chain Fv epitopes in tobacco plants. Proc Natl Acad Sci USA 96:703–708

McCormick AA, Reddy S, Reinl SJ, Cameron TI, Czerwinkski DK, Vojdani F, Hanley KM, Garger SJ, White EL, Novak J, Barrett J, Holtz RB, Tusé D, Levy R (2008) Plant-produced idiotype vaccines for the treatment of non-Hodgkin's lymphoma: safety and immunogenicity in a phase I clinical study. Proc Natl Acad Sci USA 105:10131–10136

McLain L, Porta C, Lomonossoff GP, Durrani Z, Dimmock NJ (1995) Human immunodeficiency virus type 1 neutralizing antibodies raised to a gp41 peptide expressed on the surface of a plant virus. AIDS Res Hum Retroviruses 11:327–334

Mett V, Musiychuk K, Bi H, Farrance CE, Horsey A, Ugulava N, Shoji Y, de la RP, Palmer GA, Rabindran S, Streatfield SJ, Boyers A, Russell M, Mann A, Lambkin R, Oxford JS, Schild GC, Yusibov V (2008) A plant-produced influenza subunit vaccine protects ferrets against virus challenge. Influenza Other Respi Viruses 2:33–40

Mor TS, Moon YS, Palmer KE, Mason HS (2003) Geminivirus vectors for high-level expression of foreign proteins in plant cells. Biotechnol Bioeng 81:430–437

Musiychuk K, Stephenson N, Bi H, Farrance CE, Orozovic G, Brodelius M, Brodelius P, Horsey A, Ugulava N, Shamloul AM, Mett V, Rabindran S, Streatfield SJ, Yusibov V (2007) A launch vector for the production of vaccine antigens in plants. Influenza Other Respi Viruses 1:19–25

Needham PD, Atkinson RG, Morris BAM, Gardner RC, Gleave AP (1998) GUS expression patterns from a tobacco yellow dwarf virus-based episomal vector. Plant Cell Rep 17:631–639

O'Brien GJ, Bryant CJ, Voogd C, Greenberg HB, Gardner RC, Bellamy AR (2000) Rotarvirus VP6 expressed by PVX vectors in Nicotiana benthamiana coats PVX rods and also assembles into virus like particles. Virology 270:444–453

Perez Y, Mozes-Koch R, Akad F, Tanne E, Czosnek H, Sela I (2007) A universal expression/silencing vector in plants. Plant Physiol 145:1251–1263

Pogue GP, Lindbo JA, Garger SJ, Fitmaurice WP (2002) Making an ally from an enemy: plant virology and the agriculture. Annu Rev Phytopathol 40:45–74

Pogue GP, Vojdani F, Palmer KE, Hiatt E, Hume S, Phelps J, Long L, Bohorova N, Kim D, Velasco J, Whaley K, Zeitlin L, Garger SJ, White E, Bai Y, Haydon H, Bratcher B (2010) Production of pharmaceutical-grade recombinant aprotinin and a monoclonal antibody product using plant-based transient expression systems. Plant Biotechnol J 8:638–654

Porta C, Spall VE, Loveland J, Johnson JE, Barker PJ, Lomonossoff GP (1994) Development of cowpea mosaic virus as a high-yielding system for the presentation of foreign peptides. Virology 202:949–955

Ravin NV, Kuprianov VV, Zamchuk LA, Kochetov AV, Dorokhov YL, Atabekov JG, Skryabin KG (2008) High efficient expression of *Escherichia coli* heat-labile enterotoxin B subunit in plants suing *Potato virus X*-based vector. Biochemistry (Moscow) 73:1108–1113

Regnard GL, Halley-Stott RP, Tanzer FL, Hitzeroth II, Rybicki EP (2010) High level protein expression in plants through the use of a novel autonomously replicating geminivirus shuttle vector. Plant Biotechnol J 8:38–46

Rybicki EP (2010) Plant-made vaccines for humans and animals. Plant Biotechnol J 8:620–637

Sablowski RWM, Baulcombe DC, Bevan M (1995) Expression of a flower-specific Myb protein in leaf cells using a viral vector causes ectopic activation of a target promoter. Proc Natl Acad Sci USA 92:6901–6905

Sainsbury F, Lomonossoff GP (2008) Extremely high-level and rapid transient protein production in plants without the use of viral replication. Plant Physiol 148:1212–1218

Sainsbury F, Canizares MC, Lomonossoff GP (2010a) Cowpea mosaic virus: the plant virus-based biotechnology workhorse. Annu Rev Phytopathol 48:437–455

Sainsbury F, Sack M, Stadlmann J, Quendler H, Fischer R, Lomonossoff GP (2010b) Rapid transient production in plants by replicating and non-replicating vectors yields high quality functional anti-HIV antibodies. PLoS One 5:e13976

Sanchez-Navarro J, Miglino R, Ragozzino A, Bol JF (2001) Engineering of *Alfalfa mosaic virus* RNA3 into an expression vector. Arch Virol 146:923–939

Santa Cruz S, Chapman S, Roberts AG, Roberts IM, Prior DAM, Oparka KJ (1996) Assembly and movement of a plant virus carrying a green fluorescent protein overcoat. Proc Natl Acad Sci USA 93:6286–6290

Scholthof HB, Scholthof K-BG (1996) Plant virus gene vectors for transient expression of foreign proteins in plants. Annu Rev Phytopathol 34:299–323

Shen WB, Hohn B (1994) Amplification and expression of the β-glucuronidase gene in maize plants by vectors based on maize streak virus. Plant J 5:227–236

Shen WB, Hohn B (1995) Vectors based on maize streak virus can replicate to high copy numbers in maize plants. J Gen Virol 76:965–969

Skarjinskaia M, Karl J, Aurajo A, Ruby K, Rabindran S, Streatfield SJ, Yusibov V (2008) Production of recombinant proteins in clonal root culture using episomal expression vectors. Biotechnol Bioeng 100:814–819

Takamatsu N, Ishikawa M, Meshi T, Okada Y (1987) Expression of bacterial chloramphenicol acetyltransferase gene in tobacco plants mediated by TMV-RNA. EMBO J 6:307–311

Tamilselvi D, Anand G, Swarup S (2004) A geminivirus AYVV-derived shuttle vector for tobacco BY2 cells. Plant Cell Rep 23:81–90

Taylor KM, Lin T, Porta C, Mosser AG, Giesing HA, Lomonossoff GP, Johnson JE (2000) Influence of three-dimensional structure on the immunogenicity of a peptide expressed on the surface of a plant virus. J Mol Recognit 13:71–82

Timmermans MCP, Das OP, Messing J (1992) *Trans* replication and high copy numbers of wheat dwarf virus vectors in maize cells. Nucleic Acids Res 20:4047–4054

Timmermans MCP, Das OP, Messing J (1994) Geminiviruses and their uses as extrachromosomal vectors. Annu Rev Plant Physiol Plant Mol Biol 45:79–112

Toth RL, Chapman S, Carr F, Santa Cruz S (2001) A novel strategy for the expression of foreign genes from plant virus vectors. FEBS Lett 489:215–219

Turpen TH, Reinl SJ, Charoenvit Y, Hoffman SL, Fallarme V, Grill LK (1995) Malarial epitopes expressed on the surface of recombinant tobacco mosaic virus. Biotechnology (NY) 13:53–57

Ugaki M, Ueda T, Timmermans MC, Vieira J, Elliston KO, Messing J (1991) Replication of a geminivirus derived shuttle vector in maize endosperm cells. Nucleic Acids Res 19:371–377

Uhde K, Fischer R, Commandeur U (2005) Expression of multiple foreign epitopes presented as synthetic antigens on the surface of *Potato virus X* particles. Arch Virol 150:327–340

Urcuqui-Inchima S, Haenni AL, Bernardi F (2001) Potyvirus proteins: a wealth of functions. Virus Res 74:157–175

Usha R, Rohll JB, Spall VE, Shanks M, Maule AJ, Johnson JE, Lomonossoff GP (1993) Expression of an animal virus antigenic site on the surface of a plant virus particle. Virology 197:366–374

van Vloten-Doting L, Bol JF, Cornelissen B (1985) Plant-virus-based vectors for gene transfer will be of limited use because of the high error frequency during viral RNA synthesis. Plant Mol Biol 4:323–326

Varsani A, Williamson AL, Stewart D, Rybicki EP (2006) Transient expression of human papillomavirus type 16L1 protein in *Nicotiana benthamiana* using an infectious tobamovirus vector. Virus Res 120:91–96

Verch T, Yusibov V, Koprowski H (1998) Expression and assembly of a full-length monoclonal antibody in plants using a plant virus vector. J Immunol Methods 220:69–75

Verchot J, Carrington JC (1995) Evidence that the potyvirus P1 proteinase functions in trans as an accessory factor for genome amplification. J Virol 69:567–576

Wang A, Chowda Reddy RN, Chen H (2008) Development of a plant-based vaccine against *Porcine reproductive and respiratory syndrome virus*. J Biotechnol 36S:S232–S233

Whitham SA, Yamamoto ML, Carrington JC (1999) Selectable viruses and altered susceptibility mutants in *Arabidopsis thaliana*. Proc Natl Acad Sci USA 96:772–777

Wigdorovitz A, Pérez DM, Robertson N, Carrillo C, Sadir AM, Morris TJ, Borca MV (1999) Protection of mice against challenges with *Foot and mouth disease virus* (FDMV) by immunization with foliar extracts from plants infected with recombinant tobacco mosaic virus expressing the FDMV structural protein VP1. Vaccine 264:85–91

Zhang X, Mason HS (2006) Bean yellow dwarf virus replicons for high-level transgene expression in transgenic plants and cell cultures. Biotechnol Bioeng 93:271–279

Ziegler A, Cowan GH, Torrance L, Ross HA, Davies HV (2000) Facile assessment of cDNA constructs for expression of functional antibodies in plants using the Potato virus X vector. Mol Breed 6:327–335

Chapter 11
Downstream Processing of Transgenic Plant Systems: Protein Recovery and Purification Strategies

Lisa R. Wilken and Zivko L. Nikolov

Abstract Plant-derived recombinant proteins provide alternatives to proteins produced by mammalian, microbial, and insect cell cultures due to significant upstream technological achievements over the past 5–10 years. Plants offer flexibility in both growth methods (open-field, greenhouse, and bioreactor) and host expression systems (seeds, leaves, and cell culture). The diversity of plant production systems provides for numerous commercial applications of plant-derived recombinant proteins but economic viability must be ensured through high expression levels and scalable manufacturing processes. Initial research efforts in plant biotechnology were focused on expression strategies and, thus, upstream production achievements have not be matched by downstream processing advancements. However, case-by-case extraction studies for numerous recombinant proteins led to the development of plant system-based approaches. Other progress includes the development of pre-treatment strategies to improve purification efficiency and to reduce downstream processing costs for purification from green tissue homogenates that contain chlorophyll, phenolics, and active enzymes. In spite of all the progress and positive developments made in the last 10 years, continual research and technological breakthroughs in downstream processing are needed to capitalize on the lower production cost of transgenic biomass. This chapter describes general advantages and disadvantages of seed-, leaf-, and bioreactor-based plant systems and strategies used for primary recovery and purification of recombinant proteins.

L.R. Wilken • Z.L. Nikolov (✉)
Biological & Agricultural Engineering, Texas A&M University,
College Station, TX 77843, USA
e-mail: znikolov@tamu.edu

A. Wang and S. Ma (eds.), *Molecular Farming in Plants: Recent Advances and Future Prospects*, DOI 10.1007/978-94-007-2217-0_11,

11.1 Introduction

Plant-derived recombinant proteins provide an alternative to proteins produced by mammalian, microbial, and insect cell cultures due to significant upstream technological achievements over the past 5–10 years. Plant systems have been evaluated for production of recombinant human therapeutics (Giddings et al. 2000; Stöger et al. 2005), nutraceuticals (Adkins and Lönnerdal 2004), antibodies (Arntzen et al. 2005; De Muynck et al. 2010; Fischer et al. 2009; Nikolov et al. 2009; Stöger et al. 2002), industrial enzymes (Hood et al. 2007; Howard et al. 2011), and vaccine antigens (Ling et al. 2010; Sala et al. 2003).

The first transgenic plants were generated in 1983 (Fraley et al. 1983; Zambryski et al. 1983) and early efforts of recombinant proteins production in plants focused primarily on the expression, localization, and bioactivity of pharmaceutical proteins, such as α-interferon (De Zoeten et al. 1989; Edelbaum et al. 1992; Zhu et al. 1994) and antibodies (Hiatt et al. 1989; Ma et al. 1994, 1995; Vaquero et al. 1999). The first pharmaceutical protein expressed in a plant system was human growth hormone, which was expressed as a fusion protein in tobacco callus tissue and reported in 1986 (Barta et al. 1986). Three years later, the first full-length serum monoclonal antibody was expressed in tobacco (Hiatt et al. 1989) followed by the first vaccine candidate, hepatitis B virus surface antigen, and the first industrial enzyme, *Bacillus licheniformis* α-amylase, in 1992 (Mason et al. 1992). Nearly all of these proteins were expressed in tobacco as the first pharmaceutical protein was not expressed rice until 1994 (Zhu et al. 1994), soybeans until 1998 (Zeitlin et al. 1998), and algae in 2003 (Mayfield et al. 2003). Other expression systems used over the past 20 years for the production of vaccines, therapeutic proteins, and industrial enzymes include canola, alfalfa, lettuce, *Lemna minor* (duckweed), potato, carrot cell culture, and hairy root culture (Daniell et al. 2009a; Franconi et al. 2010; Karg and Kallio 2009; Sharma and Sharma 2009). The diversity of products and plant production systems provides for numerous potential prospects and applications, but the economic viability of each plant system must be ensured through high expression levels and scalable manufacturing processes (Davies 2010).

Plants have been used as a source of natural products for thousands of years, thus, processing methods for recovering food and feed products from both seed and leafy tissues are readily available. However, the developed methods are suited for crude separations using conditions that may not be compatible with recombinant proteins, which usually requires retention of activity. The processing knowledge developed for nontransgenic plants has been applied to transgenic plants and served as the basis for the designing methods for separation of native plant components from recombinant proteins. The first reports of downstream processing (extraction, purification) for transgenic plants in a systematic manner (Evangelista et al. 1998; Hood et al. 1997; Kusnadi et al. 1998), rather than for protein characterization, were published almost 10 years after it was demonstrated that fully functional recombinant proteins could be produced in plants. The first large-scale commercial production of a pharmaceutical protein (trypsin) from transgenic plants was reported in 2003 (Woodard et al. 2003).

Some of disadvantages of using plant systems, cited before 2000, include concerns about the development time for biopharmaceuticals, glycosylation differences between plants and mammalian systems, regulatory uncertainties for open-field production and manufacturing of transgenic tissue, and expression levels. Many of these concerns have been successfully addressed in certain plant systems. For example, the use of transient expression in tobacco drastically reduced the product development time from >24 months to 14–20 days (Hiatt and Pauly 2006), glycosylation differences were resolved by inhibiting plant-specific modifications in *Lemna* (Cox et al. 2006), and expression levels reached 1% seed weight in rice (Zhang et al. 2010) and 0.15% fresh weight in tobacco (Bendandi et al. 2010). By addressing these concerns, research efforts are beginning to shift to downstream processing, as lack of efficient processing methods has been highlighted as a major barrier to cost-effective recombinant protein production (Conley et al. 2010).

Plant systems are typically grouped into three categories: seed-based systems, leafy-based systems, and bioreactor-based systems. In the following sections, we describe general advantages and disadvantages of each system followed by strategies used for the recovery and purification of recombinant proteins.

11.1.1 Seed-Based Systems

Using seed crops to produce vaccines, therapeutic proteins, and industrial enzymes is advantageous compared to other production systems because of the established production infrastructure (Kusnadi et al. 1997; Nikolov and Hammes 2002), relatively high protein content, and low risk of contamination (Ma et al. 2003). Accumulation of recombinant proteins within naturally desiccant seed storage organelles (Nikolov and Hammes 2002) that are devoid of proteases and other enzymatic activities (Fischer et al. 2009) allows for the separation of the production fields, storage facilities, and processing operations (Boothe et al. 2010). Endogenous protease inhibitors protect seed-expressed proteins throughout storage, extraction, and purification (Menkhaus et al. 2004a). Transgenic seeds expressing a single chain antibody remained stable for at least 5 months (Stöger et al. 2000) and more than a year (Ramírez et al. 2001) at room temperature without significant loss of activity.

Seed-based systems, such as rice and corn, have generally regarded as safe (GRAS) status and can potentially be used for producing oral vaccines, nutraceuticals, and industrial enzymes with minimal processing and purification (Hood and Howard 2009; Nandi et al. 2002; Yang et al. 2008). Seed crops typically have lower biomass yields per unit surface area compared to leaf-based systems, but for open-field production, the economy of scale and stability of seed-expressed proteins outweighs the biomass yield disadvantage (Nikolov and Hammes 2002; Schillberg et al. 2005). For large-scale manufacturing (>1 t protein/year), open-field production is the only feasible option. Several plant biotechnology companies have recently received USDA approval to grow transgenic corn, rice, safflower, and barley in open fields for the production of in-process reagents (human lysozyme, lactoferrin, and human serum albumin), vaccines (hepatitis B virus surface antigen), and industrial enzymes (cellulases).

11.1.2 Leaf-Based Systems

The first plant-produced monoclonal antibody was expressed in tobacco (Hiatt et al. 1989), which still remains the leading leaf–based system for commercial production of recombinant proteins (Tremblay et al. 2010; Twyman et al. 2003). Advantages of using leafy tissue include high biomass yield, possibility for multiple growth cycles per year, and established agricultural infrastructure. The main disadvantages of leafy tissue are high water content, storage stability of harvested biomass, and related to these, recombinant protein stability which does not allow decoupling of upstream and downstream processing. Leaf-expressed proteins are synthesized in an aqueous environment and can degrade during transport or storage and should either be processed immediately after harvesting or dried or frozen to reduce metabolic activities. Other advantages and disadvantages of using tobacco and alfalfa systems for producing therapeutic and diagnostic antibodies have been recently reviewed by Fischer et al. (2009).

Significant advancements have been achieved using transient expression in tobacco (*N. benthamiana*), which reduced the development time needed to produce gram-level quantities of therapeutic products from >24 months to 14–20 days (Conley et al. 2010; D'Aoust et al. 2010; Joensuu et al. 2010). Transient expression in leafy tissues provides a distinct advantage for greenhouse-produced tobacco, compared to other plant-based systems, when rapid production of pandemic vaccines is required (D'Aoust et al. 2010). Transient expression alleviates transgenic regulatory concerns but is currently only suited for low volume protein production such as personalized therapeutics, seasonal and pandemic vaccines, and other specialized markets (Fischer et al. 2009; Pogue et al. 2010).

11.1.3 Bioreactor-Based Systems

Bioreactor-based production systems include liquid-medium supported systems, such as hairy root culture, aquatic plants, plant cell culture, moss, and algae. These contained systems have been evaluated primarily for the production of injectable biopharmaceuticals and other highly regulated recombinant proteins. Tobacco, mainly *N. benthamiana*, has been used for the production of numerous proteins including a human anti-rabies antibody (Girard et al. 2006), human growth hormone (Xu et al. 2010), and interferon α2b (Xu et al. 2007). Other notable products include the intracellular product, glucocerebrosidase, produced in carrot cell culture (Shaaltiel et al. 2007), intracellular, IgG1 *Lemna minor* (Cox et al. 2006; Nikolov et al. 2009), and the secreted protein, human serum albumin, in rice cell suspension (Huang et al. 2005). Bioreactor-based systems are advantageous from a regulatory perspective due to controlled growth conditions with complete containment, product yield consistency, quality, and homogeneity, and production speed (Fischer et al. 2009; Huang et al. 2009). Disadvantages include relatively low concentrations (10–250 mg/L) of secreted products (Weathers et al. 2010; Xu et al. 2011), higher capital investment costs compared to open-field and greenhouse production, and bioreactor scale-up limitations similar to those faced with mammalian cell cultures. From a downstream processing perspective, the potential advantage of a secreted product is a simpler and less costly purification (Doran 2006; Hellwig et al. 2004). However, the secretion efficiency may be limited by protein size, hydrophobicity, and/or charge (Fischer et al. 2009). For proteins that are not secreted, plant cell homogenates have similar complexities to those of leafy plants and have similar processing requirements. Thus, intracellular recombinant proteins produced by bioreactor-based systems may be less commercially attractive compared to other heterologous protein expression systems since potential processing savings (compared to secreted products) are lost (Doran 2006).

11.2 Downstream Processing

Downstream processing of plant systems generally consists of product extraction, solid–liquid separation, extract pretreatment and/or conditioning, and purification (capture, intermediate purification, and polishing) (Fig. 11.1). For seed-based systems, tissue fractionation is often the first step in downstream processing, but is rarely used with leaf-based systems. The first three steps of downstream processing for leaf- and seed-based systems, consisting of product release, solid–liquid separation, and pretreatment and/or conditioning, are categorized as primary recovery steps (Lee 1989). Extracellular proteins produced by plant cell-culture and aquatic plants (e.g. *Lemna minor*) have only two recovery steps because there is no need for product release, a fact considered an advantage in terms of potential process cost savings. Irrespective of the number and type of primary recovery steps, the common objective of protein recovery is to maximize the product titer, reduce process volume, and to prepare a clarified feed stream for purification (Fig. 11.1).

Downstream processing for bioreactor-based systems starts with cell tissue harvesting by centrifugation, membrane filtration, or by pumping the liquid media out of the bioreactor vessel. Because intracellular proteins require disintegration of harvested biomass for product release, the primary recovery after biomass harvesting is the same as that for leaf- and seed-based systems. For media-secreted proteins, the culture media is typically concentrated, clarified, and conditioned prior to the first chromatography step. For seeds with recombinant protein expression targeted to a particular organelle, fractionation prior to grinding reduces extraction volumes. After grinding of the whole seed or seed fraction, the flour is added to water or buffer for extraction. The crude extract is clarified by separating the clarified extract from

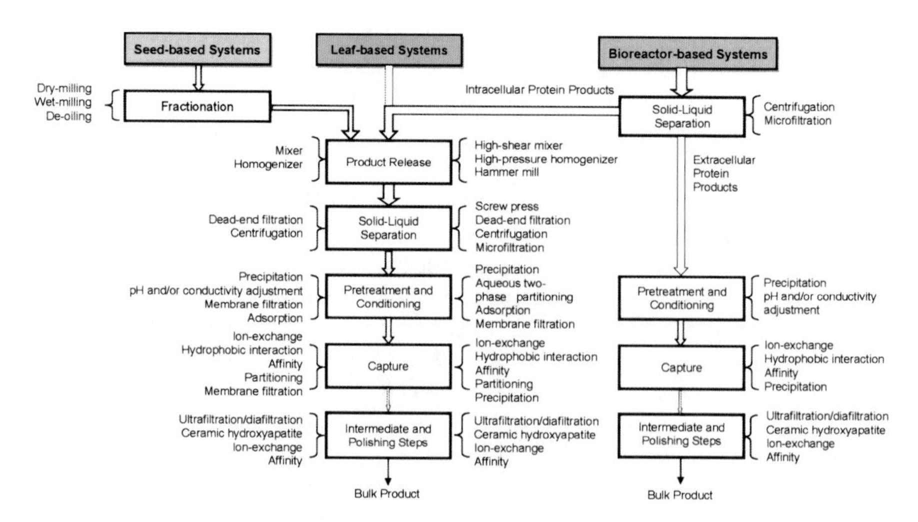

Fig. 11.1 Downstream processing flow diagrams with unit operations for recombinant protein purification from seed-based, leaf-based, and bioreactor-based systems

the spent solids by centrifugation or dead-end filtration. The recombinant protein is then purified from the clarified extract by one or more steps using methods such as chromatography, two-phase partitioning, membrane filtration, and/or precipitation. Leafy tissues are processed in a similar manner, with the exception of product release, which can be accomplished by screw pressing without the addition of extraction buffer. Exceptions from these generalized process schemes include protein products formulated as crude protein extracts or plant tissue fractions intended for direct delivery of protein sweeteners, industrial enzymes, and vaccine products.

11.2.1 Fractionation

Fractionation prior to extraction provides an opportunity to reduce the amount of biomass for processing, which can enrich the recombinant protein concentration in the tissue fivefold to tenfold for seed systems. In addition, seed fractionation using established processing methods, such as dry milling and wet milling, allows for generation of co-product revenues (Moeller et al. 2010; Paraman et al. 2010a, 2010b; Zhang et al. 2009a) when the recombinant protein is targeted to a particular organelle or compartment.

Corn fractionation using dry milling results in germ-, endosperm-, and bran-rich fractions for extraction and purification or for use in a partially purified form (Moeller et al. 2010; Shepherd et al. 2008a, 2008b; Zhang et al. 2009a). In the case of edible vaccines where high purity is not required, fractionation can significantly reduce the required dosage volume (Lamphear et al. 2002; Moeller et al. 2010).

Oil removal after degermination for a germ-expressed product or for oil-rich seeds (canola, safflower, soybeans) further enhances recombinant protein concentration, reduces protein losses due to protein-oil emulsification (Bai and Nikolov 2001), and improves purification efficiency (Zhang et al. 2009a). De-oiling is also required for direct delivery of germ-expressed industrial enzymes or oral vaccines to reduce product rancidity due to lipid oxidation. De-oiling of transgenic seeds is typically achieved using organic solvents, such as hexane, but at lower temperatures than for non-transgenic seeds. An exception to this de-oiling requirement is oleosin-fusion proteins, which are designed to remain attached to the surface of seed oilbodies during aqueous extraction. Engineered oilbodies with attached recombinant protein are extracted with water by homogenization and then recovered in the oil phase by centrifugation (Boothe et al. 2010).

Wet milling methods provide better seed fractionation and thus, higher purity fractions for co-product use (starch) and higher recombinant protein enrichment compared to dry milling. Recent studies indicate that traditional wet milling conditions need to be modified to retain recombinant protein quality and/or activity during fractionation (Paraman et al. 2010a; Zhang et al. 2009a). Variations in wet milling processing, like the modified quick-germ methods, provide an alternative for traditional fractionation methods and can reduce recombinant protein losses due to proteolysis or leaching (Paraman et al. 2010a).

The main concerns with using transgneic corn for ethanol production are the effect of recombinant extraction on ethanol yield and presence of residual recombinant activity in the unfermented corn residue that will be used for animal feed. Paraman et al. (2010b) evaluated the possibility of integration of recombinant protein recovery process into a dry fractionation, corn-to-ethanol plant. The study determined that degermination and extraction of an endosperm-expressed subunit vaccine of *Escherichia coli* enterotoxin or germ-expressed human collagen did not adversely affect ethanol production yields from the starch-rich fractions.

Although there are no systematic studies published for fractionation of leaf-based systems, harvesting of only plant leaves would be desirable because recombinant protein expression per unit biomass is significantly lower in stalks and stems compared to leaves.

11.2.2 Extraction Methods

Extraction requires efficient homogenization of plant tissue and disruption of plant cell walls, which is a critical recovery step that dictates the total extract volume, recombinant protein concentration and purity, and the type and quantity of impurities that have to be removed during purification (Hassan et al. 2008; Nikolov et al. 2009). Extraction conditions that maximize the concentration of recombinant protein and minimize the amount of soluble native protein and other impurities in the extract are desirable and are important considerations for reducing purification costs (Azzoni et al. 2002; Menkhaus et al. 2004a; Nikolov and Woodard 2004).

Typically, optimal extraction conditions are determined using small-scale studies to screen factors such as tissue disruption technique and particle size, extraction time, buffer composition, plant tissue-to-buffer ratio, and expression compartment (Azzoni et al. 2005; Farinas et al. 2005a; Hassan et al. 2008; Menkhaus et al. 2004a; Wilken and Nikolov 2006; Woodard et al. 2009; Zhang et al. 2010). Modifying these conditions can enhance the extraction of a recombinant protein by reducing its interactions with plant debris and organelles and by reducing extracted impurities. The extraction methods summary for seed-, leaf-, and bioreactor-based systems (Table 11.1) shows that extraction conditions are typically screened using small amounts of plant tissue (1–20 g). Ideally, tissue homogenization and extraction should be conducted using kilogram quantities of transgenic tissue because mixing pattern, shear rate, heat generation, and time required for maximal product release are volume (vessel size) dependent and not linearly scalable.

11.2.2.1 Leaf-Based Systems

Physical disruption of high-water content biomass, such as *Lemna* and algae, can be accomplished by using high-shear mixers or high-pressure homogenizers with buffer at a tissue-to-buffer ratios ranging from 1:1.5 to 1:8. Leafy tissues that contain a significant amount of stem and stalk fiber can be screw-pressed with or without previous hammer milling (Bratcher et al. 2005; D'Aoust et al. 2004) to produce an extract with chlorophyll and other pigments that is referred to as "green juice". Screw-presses are used on a pilot and manufacturing scale whereas fruit juice presses are an adequate option for bench-scale processing. Screw pressing minimizes buffer usage and, thus, minimizes the overall extract volume. Algae, plants cell cultures, and small-leaf plants like *Lemna minor* are too small in size for screw pressing and are processed by homogenization. Milder extraction procedures include vacuum infiltration (Turpen et al. 2001) and enzymatic digestion of the plant cell wall (Fischer et al. 1999). These methods are attractive but their efficiencies and economic viabilities for large-scale manufacturing have not yet been demonstrated.

Leafy plants require careful attention because of the instability of expressed proteins in metabolically active tissues and their extracts (De Muynck et al. 2010; Doran 2006) and should be processed immediately or frozen or dried after harvesting for future processing. Grinding, pressing, or homogenization of leaf tissue releases additional components, such as proteases, phenol oxidases, and plant phenolics that can degrade or modify the recombinant protein. Carefully controlled extraction conditions and the inclusion of protease inhibitors, non-reducing agents, and anti-chelating agents can improve recombinant protein stability during and after extraction. Extraction buffers used for leafy tissue homogenization typically contain buffer additives such as β-2-mercaptoethanol (β-ME), dithiothreitol (DTT), polyvinyl polypyrrolidone (PVPP), ascorbic acid, and sodium metabisulphite to reduce phenolics interference on adsorption processes, phenolics-protein interactions, or phenolics oxidation (Holler et al. 2007; Holler and Zhang 2008; Peckham et al. 2006). Since plant homogenates also contain active proteases, cocktails of protease

Table 11.1 Selected examples of recombinant protein extraction methods and results (Modified from Wilken and Nikolov 2011)

Platform	Recombinant protein	pI of recombinant protein	Application	Amount	Method	Tissue to buffer ratio	Buffer	Product concentration	Reference
Seeds									
Corn seed	Aprotinin	10.5	Protease inhibitor	20 g	Mixing	1 to 5	200 mM NaCl, pH 3.0	0.7 μg/mL	Azzoni et al. (2002)
Corn germ	Aprotinin	10.5	Protease inhibitor	30 g	Mixing	1 to 5	Deionized water, pH 3.0	NR	Zhong et al. (2007)
Corn seed (defatted germ)	Collagen	8.3–8.9	NR	5 g	Mixing	1 to 5 (2 times)	0.1 M phosphoric acid, 0.15 M NaCl, pH 1.8	120 mg/kg germ fraction	Zhang et al. (2009b)
Corn meal	IgG (pI > 8.0)		NR	50 g	Mixing	1 to 10	150 mM NaCl	0.52 mg/g (0.052 mg/mL)	Lee and Forciniti (2010)
Corn seed	IgG (2 G12)	8.0	anti-HIV	5 g	Mixing	1 to 4	20 mM NaPhosphate, pH 6.0	> 75 mg/kg DW	Ramessar et al. (2008)
Rapeseed	Chymosin	4.6	Industrial/food enzyme	40	Homogenization	1 to 10	Water with 250 mM NaCl	4.2% TSP (47 mg/kg)	Van Rooijen et al. (2008)
Rice seed	Human lysozyme	10.2	Gastrointestinal infections, formula and cell culture additive	20 g	Mixing	1 to 5	50 mM NaAcetate, 50 mM NaCl, pH 4.5	4.4 mg/g (3.3 mg/g for selected extraction conditions)	Wilken and Nikolov (2006)
Rice seed	Human lactoferrin	8.2	Gastrointestinal infections, formula and cell culture additive	1 g to 2 kg	Mixing	1 to 10	0.02 M NaPhosphate, 0.3 M NaCl, pH 7.0	6 mg/g (4.2 mg/g for selected extraction conditions)	Nandi et al. (2005)
Rice seed	Transferrin	6.3	Cell culture media additive for regulation; drug carrier	NR	Mixing	1 to 10	25 mM Tris–HCl, pH 7.5	NR, 10 g/kg expression level	Zhang et al. (2010)

(continued)

Table 11.1 (continued)

Platform	Recombinant protein	pI of recombinant protein	Application	Amount	Method	Tissue to buffer ratio	Buffer	Product concentration	Reference
Soybean seed (defatted flour)	β-Glucuronidase (GUS)	~5.5	Model protein	5 g	Mixing	1 to 20	50 mM citrate buffer, pH 5.3	NR, 3 mg/mL TSP	Robić et al. (2006)
Leafy crops									
Tobacco leaves	Aprotinin	10.5	Protease inhibitor	NR	Homogenization	NR	NR	~750 mg/kg (greenhouse), ~ 300 mg/kg (open field)	Pogue et al. (2010)
Tobacco leaves	β-Glucuronidase	5.5	Model protein	NR	Homogenization	1 to 5	50 mM NaPhosphate, 10 mM BME, 1 mM EDTA, pH 7.0	NR, 1.4 mg/mL TSP	Holler and Zhang (2008)
Tobacco leaves	IgG (patient 69)	>8.0	Idiotype vaccines for non-hodgkin's lymphoma	5 kg	Homogenization	NR	NR	1.5 g/kg FW	Bendandi et al. (2010)
Tobacco leaves	IgG	>8.0	anti-WNV	10 g to 5 kg	Homogenization	NR	PBS, 1 mM EDTA, 0.3 mg/mL PMSF, 10 mg/mL NaAscorbate, 10 ug/mL leupeptin	0.8 g/kg FW	Lai et al. (2010)
Tobacco leaves	IgG (H10)	>8.0	anti-tenascin-C	250 g	Homogenization	1 to 3	PBS, 5 mM EDTA, protease inhibitor cocktail (1%v/v), pH 7.4	600 mg/kg	Lombardi et al. (2010)
Tobacco leaves	IgG	>8.0	Antibody to hepatitis B surface antigen	10 to 600 kg	Hammer milling	NR (wetted leaves)	150 mM PBS, 0.56 mM ascorbic acid	22–46 mg IgG/kg biomass (lowest for 600 kg process)	Padilla et al. (2009)

Tobacco leaves	IgG (4E10) and IgG (2 G12)	>8.0	anti-HIV	5 g	Homogenization and mixing	1 to 3	50 mM NaPO4, pH 5.0	0.5–1.0% of TSP	Platis et al. (2008), Platis and Labrou (2009)
Tobacco leaves	IgG (αCCR5)	>8.0	anti-HIV activity	40 to 60 kg	Homogenization and screw pressing	NR	NR	250 mg/kg after purification	Pogue et al. (2010)
Tobacco leaves	IgG (C5-1)	8.4–9.7	Diagnostic	100 to 150 g	Homogenization	1 to 3	20 mM NaPhosphate, 150 mM NaCl, 2 mM sodium metabisulphite, pH 5.8	570 mg/kg FW	Vézina et al. (2009)
Bioreactor-based systems									
Carrot suspension cell culture	Glucocerebrosidase (intracellular)	7.4	Gaucher's disease	NR	Homogenization	NR	20 mM NaPO4, 20 mM EDTA, 0.1 mM PMSF; 1% Triton X-100; 20 mM Ascorbic acid; 0.1 mM DL-dithiothreitol, pH 7.2	NR	Shaaltiel et al. (2007)
Lemna fronds	IgG1 (anti-CD 30)	>8.0	Non-Hodgkin's lymphoma and anaplastic large-cell lymphoma	NR	High-shear homogenization	1 to 5	50 mM NaPhosphate, 300 mM NaCl, 10 mM EDTA, pH 7.2	2% of TSP	Cox et al. (2006), Nikolov et al. (2009)
Lemna fronds	IgG1 (CNTD)	>8.0	NA	50 g	High-shear homogenization	1 to 8	100 mM NaAcetate, 300 mM NaCl, 10 mM EDTA, pH 4.5	50% of TSP @ pH 4.5; 10% TSP at pH 7.5	Woodard et al. (2009)

(continued)

Table 11.1 (continued)

Platform	Recombinant protein	pI of recombinant protein	Application	Amount	Method	Tissue to buffer ratio	Buffer	Product concentration	Reference
Tobacco suspension cell culture	Anti-rabies Mab (intracellular)	>8.0	Human anti-rabies virus Mab for treatment	1 g	Mixer mill then homogenization	NR	40 mM Tris–HCl, 20% glycerol, 2% SDS, 4% β-mercaptoethanol, protease inhibitors, pH 8	3–10 μg/g leaf FW	Girard et al. (2006)
Tobacco suspension cell culture	Anti-rabies Mab (intracellular)	>8.0	Human anti-rabies virus Mab for treatment	1.45 kg	French press	1 to 10	20 mM NaPhosphate, pH 7.2	fourfold less than that with 1 g extraction	Girard et al. (2006)
Tobacco suspension cell culture	GFP-fusion protein (intracellular)	~5.0	Reporter tag	20 g	Microson ultrasonic cell disruptor	1 to 2	0.25 M Boric acid, 0.25 M NaCl, 1% caffeine, 1%ascorbic acid, 0.1% Triton X-100, 1 mM DTT, 1 mM PMSF, 1% protease cocktail, pH 8	10.7 mg TSP	Peckham et al. (2006)

NR not reported, *TSP* total soluble protein

inhibitors are often added to extraction buffers (Table 11.1). For membrane-associated proteins and chloroplast-expressed proteins, the addition of detergents is required to reduce hydrophobic interactions with cell membranes (Boyhan and Daniell 2010; Daniell et al. 2009b; Tran et al. 2009).

Extraction buffer ionic strength and pH can be modified to reduce interactions between the recombinant protein and plant matrix and maximize recombinant protein release. For example, 300 mM NaCl was required for maximal extraction of recombinant plasmin (pI=6.3) from *Lemna minor* at pH 3.5, whereas at pH 6.5 and 9.0, 150 mM NaCl was sufficient. Furthermore, using 50 mM NaCl recovered 75% of the highest plasmin concentration at pHs 6.5 and 9 but extracted only 5% of the maximum amount at pH 3.5 (Wilken and Nikolov, unpublished data). Extraction of IgG (mAb) from *Lemna minor* by homogenization at pH 4.5 and 7.5 required at least 300 mM NaCl for maximal release of mAb in the homogenate. At pH 4.5, the solubility of ribulose bisphosphate carboxylase/oxygenase (rubisco) and other acidic *Lemna* proteins was reduced, resulting in a fivefold enhancement of mAb concentration (expressed as percent of total soluble protein) compared to pH 7.5 (Woodard et al. 2009). A more detailed study of the extraction of three mAb variants from transgenic tobacco at constant ionic strength of 0.1 M was conducted by Hassan et al. (2008) and included the effect of grinding methods, extraction pH and temperature, and presence of detergent. This study concluded that temperature and grinding methods did not have a significant effect on the extraction yield of mAb, whereas the buffer pH was a critical variable. Extractions at pH 5 and 7, below the mAb isoelectric point (pI) of 8–9.5, were optimal while extraction at a pH close to the pI reduced the yield twofold due to low mAb solubility. Addition of detergent (Triton X-100) increased the extraction yield threefold for membrane-bound mAb but did not improve extractability of the same antibody targeted to the apoplast or endoplasmic reticulum (Hassan et al. 2008). Thus, at a constant ionic strength buffer, pH had the dominant effect on extraction of the apoplast-expressed mAb. Since we did not observe a lower solubility effect with plasmin or mAb near their respective pIs, one could assume that both plant matrix and recombinant protein properties determine the optimal extraction conditions.

11.2.2.2 Seed-Based Systems

Ground seed is extracted with low-shear mixers using tissue-to-buffer ratios ranging from 1:4 to 1:20 (Table 11.1). Wet grinding can also be used with seed-based systems with concomitant protein extraction, although the preferred method for low-oil-containing seeds is dry grinding followed by low-shear mixing with an aqueous buffer. Extraction of oil bodies for oleosin-fusion technology requires 1:12–1:20 water-to-tissue ratios (Boothe et al. 2010; Nykiforuk et al. 2006, 2011). Using lower ratios reduces process volume but may also decrease the percentage of recombinant protein extracted. An extraction time of 30 min is usually sufficient for solubilizing proteins from soybean (Robić et al. 2006, 2010), canola (Zhang and Glatz 1999), corn (Kusnadi et al. 1998), and rice (Zhang et al. 2010) if diffusion limitations were minimized by grinding the seed to particle sizes less than 1 mm. In some cases, the addition of sodium chloride may be necessary to increase recombinant protein solubility by reducing electrostatic interactions with insoluble components.

Several studies have been dedicated to investigating the effect of particle size, pH, and ionic strength on protein extractability from seed systems (Azzoni et al. 2002, 2005; Bai and Nikolov 2001; Bai et al. 2002; Farinas et al. 2005b; Menkhaus et al. 2004b; Zhang et al. 2005, 2010). Optimization of extraction conditions with respect to process integration and/or cost analysis has been addressed for select seed systems (Azzoni et al. 2002; Kusnadi et al. 1998; Menkhaus et al. 2002, 2004b; Nandi et al. 2005; Wilken and Nikolov 2006, 2010). Buffer pH and ionic strength have a strong effect on recombinant protein and native protein extractability that appear to be protein specific rather than seed specific (Azzoni et al. 2002; Farinas et al. 2005b; Robić et al. 2010; Wilken and Nikolov 2006, 2010; Zhong et al. 2007). In the pH range below the pI of the recombinant protein, extractability was increased moderately by increasing the ionic strength (Wilken and Nikolov 2006, 2010; Zhong et al. 2007). Near or above the protein pI, the ionic strength did not affect extractability of human lactoferrin (Nandi et al. 2005), transferrin (Zhang et al. 2010), or lysozyme (Wilken and Nikolov 2010) from transgenic rice and β-glucuronidase from transgenic soybeans (Robić et al. 2010). In all cases, increasing ionic strength increased the extractability of native protein and thus, decreases recombinant protein purity in the extract expressed as percent total soluble protein or specific activity (Robić et al. 2010; Wilken and Nikolov 2006). Most of the published studies with seed systems optimized extraction parameters by measuring only total protein and recombinant protein concentrations. Although these two outputs are of paramount importance for downstream processing, information on other extractable components such as phytic acid, phenolics, lipids, and reducing sugars could be critical for purification. Farinas et al. (2005a) investigated the effect of pH and ionic strength on the extraction of phenolic compounds, lipids, and sugars from transgenic corn seed and Robić et al. (2010) from transgenic soybeans. The consensus reached in both studies was that low pH and low ionic strength buffers minimize the extraction of reducing sugars, phenolics, and native protein and that the effect of buffer pH on these compounds was more pronounced than ionic strength. The presence of phytic acid, in addition to phenolic compounds, could affect both recombinant protein extractability and purification efficiency (Wilken and Nikolov 2010). Since acidic pH increases extractability of phytic acid, and basic pH enhances phenolics extraction, there is no single approach to simultaneously minimizing their concentrations by manipulating pH and/or ionic strength. The adverse effect of these two compounds and potential remedies will be addressed in the Pretreatment and Conditioning (Sect. 11.2.4).

11.2.3 Solid-Liquid Separation

Continuous centrifugation and filtration are common, scalable methods for solids removal and clarification of plant extracts and homogenates. Centrifugation is somewhat more flexible for use with leaf and cell culture homogenates, which have a broader particle size distribution and densities. Seed extract suspensions have large differential solid-liquid densities and are amenable to clarification by decanter centrifugation as well as dead-end filtration using devices such as, rotary vacuum drum filters, horizontal belt filters, filter presses, and, for smaller suspension volumes, basket centrifuges. The "green juice" from screw-pressed leafy plants can be clarified by disc-stack centrifugation (Bratcher et al. 2005; D'Aoust et al. 2004) or vacuum filtration. Algae, *Lemna minor,* and plant cell cultures are too small for screw pressing and their homogenates are clarified by centrifugation followed by depth filtration or cross-flow microfiltration. For plant cell culture with secreted products, cross-flow filtration is the most suitable method for separation of cell biomass that results in a well clarified feed for packed-bed chromatography (Hellwig et al. 2004). If extract pretreatment is required, polishing filtration can be performed after pretreatment rather than after solid-liquid separation.

11.2.4 Pretreatment and Conditioning

Extraction conditions that maximize the amount of released recombinant protein also results in the release of a variety of indigenous proteins and other plant cell components such as nucleic acids, chlorophyll, alkaloids, phenolics, polysaccharides, and proteases. These compounds can bind to or degrade the recombinant protein, reducing final product quality and yield. Another potential consequence of co-extracting these components is the reduction of process efficiency due to fouling of chromatography resins and cross-flow filtration membranes. To counteract these process inefficiencies, a pretreatment, method is typically included prior to the capture step. The objectives of the pretreatment step are to: (1) reduce recombinant protein modification by plant phenolics or degradation by proteases; (2) minimize interference during purification by removal of impurities; and (3) reduce protein purification burden by removal of native proteins. An overview of selected pretreatment methods is given in Table 11.2. Since seed-based and leafy-based systems have very different compositions, pretreatment methods are typically platform-tailored using conditions based on impurity properties, such as isoelectric point, hydrophobicity, or charge. Impurities of interest for each production system, properties of the impurities, and select pretreatment methods are discussed in detail below.

11.2.4.1 Leaf-Based Systems

The composition of leafy tissue homogenates requires the addition of at least one pretreatment step for nearly all reported purification processes. Leaf extracts prepared by homogenization are typically more complex than the seed ones and contain rubisco, phenolic compounds, proteases, and chlorophyll-derived pigments that could pose difficulties during downstream processing (Barros et al. 2011; Woodard et al. 2009; Yu et al. 2008).

Table 11.2 Preatreatment methods for removal of plant extract impurities

Pretreatment method	Agent	Potential components removed	Example plant system applications	Reference
Precipitation	Ammonium sulfate	Protein, cell debris, phenolics, pigments	Tobacco leaves, tobacco cell suspension culture (intracellular product), rice cell suspension (extracellular product)	Huang et al. (2005), Lai et al. (2010), Peckham et al. (2006)
	Sodium chloride	Protein, cell debris, phenolics, pigments	Tobacco cell culture (intracellular and extracellular products)	Xu et al. (2007, 2010)
	Acid/base	Protein, cell debris, chlorophyll pigments (green tissue)	Tobacco leaves, *Lemna* fronds, rapeseed, rice seed, canola seed, pea	Bendandi et al. (2010), Cox et al. (2006), Garger et al. (2000), Menkhaus et al. (2004b), Pogue et al. (2010), Van Rooijen et al. (2008), Vézina et al. (2009), Wilken and Nikolov (2010), Woodard et al. (2009), Zaman et al. (1999)
Partitioning	Aqueous two-phase (PEG and potassium phosphate)	Phenolics and alkaloids	Tobacco leaves, potato leaves	Miller et al. (2004), Platis et al. (2008), Platis and Labrou (2006, 2009), Ross and Zhang (2010)
Charge neutralization	TRIS	None	Rice seed	Wilken and Nikolov (2010)
Adsorption	Hydrophobic resin	Phenolics, pigments	*Lemna* fronds	Barros et al. (2011), Woodard et al. (2009)
	Anionic resin	Phenolics acids and nucleic acids	*Lemna* fronds	Barros et al. (2011) Woodard et al. (2009)
	Hydrophilic polymers (PVPP)	Phenolics, pigments	*Lemna* fronds, tobacco leaves	Barros et al. (2011), Holler and Zhang (2008)
Filtration	Polymeric membrane (concentration)		Tobacco cell suspension (intracellular product)	Girard et al. (2006)
	Polymeric membrane (diafiltration)	Small molecular weight proteins, salts, sugars	Tobacco leaves, corn seed	Pogue et al. (2010), Zhang et al. (2009b)

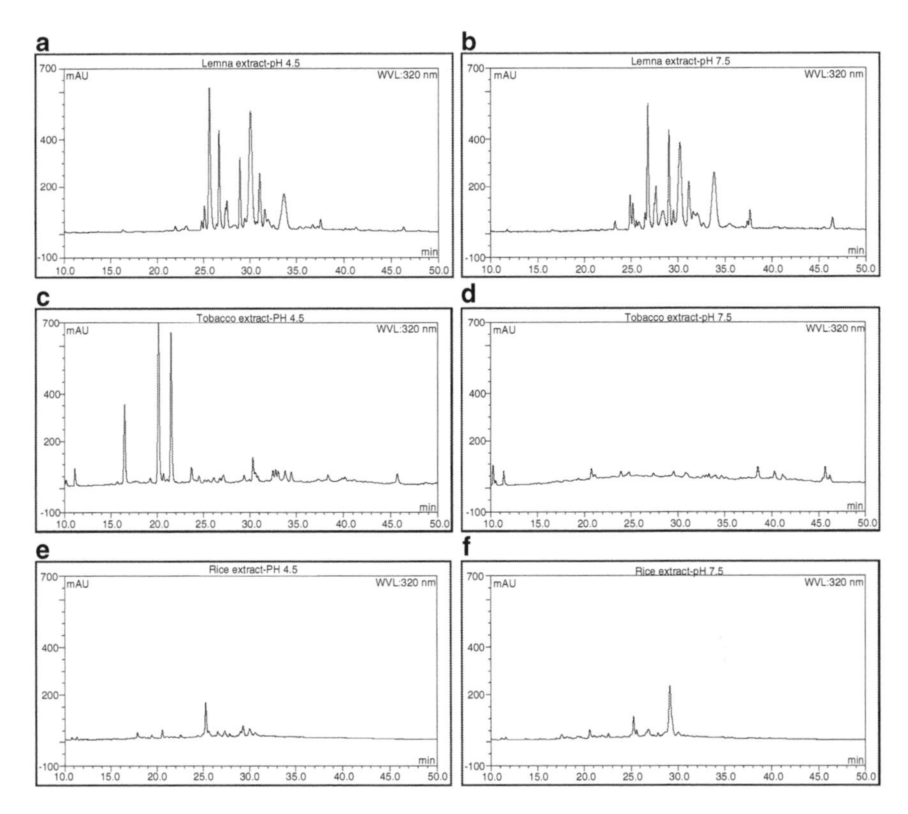

Fig. 11.2 RP-HPLC chromatograms showing 320 nm absorbances of Lemna minor (**a**,**b**), tobacco (**c**,**d**), and rice extracts (**e**,**f**) at pH 4.5 (*left panels*) and pH 7.5 (*right panels*)

Phenolic compounds, including phenylpropanoids and flavonoids, are known to foul chromatography resins and are capable of binding, inactivating, and modifying proteins through various mechanisms (Loomis 1974). The amount of phenolics released during homogenization varies with plant type as shown in the HPLC phenolic profiles in tobacco, *Lemna*, and rice extracts (Fig. 11.2). The chromatograms indicate that extraction pH had little effect on the amount and type of phenolics present in *Lemna* extracts, but did alter the profile for tobacco extracts.

Leafy tissue is composed of active cells containing enzymes needed for metabolic activity and thus, has a significant amount of water-soluble proteins. The chloroplast enzyme, rubisco, can account for as much as 50% of total leaf nitrogen and is the most abundant protein in the world (Spreitzer and Salvucci 2002). The prevalence and well-defined properties of rubisco allows for rather simple removal of this protein by isoelectric precipitation at or below pH 5. Proteolytic enzymes can be particularly challenging in recombinant protein production in plant systems, as degradation can occur posttranslationally or during recovery (Benchabane et al. 2009). Typically, recombinant protein stability in a plant extract is evaluated experimentally using analytical tools such as Western Blot, activity assays, and HPLC. Protease activity in extracts can be evaluated using zymograms. The steps necessary to assess recombinant protein stability and detect a proteolytic problem and selection of a companion protease inhibitor is nicely summarized by Benchabane et al. (2009). The variability and diversity of proteases present in plant systems can make their removal difficult. Therefore, proteases are managed by controlling storage and processing temperatures, selecting appropriate extraction conditions, and adding inhibitors to extraction buffers. An alternative strategy for the protection of recombinant proteins is the co-expression of protease inhibitors.

There are many different strategies used to remove plant impurities that are detrimental to purification methods or recombinant protein quality and yields. Common strategies used for extract pretreament include precipitation, aqueous two-phase partitioning, adsorption, and membrane filtration (Table 11.2). Precipitation is the most common and rather inexpensive approach used for coarse fractionation of proteins that also has the potential to remove other impurities such aggregates, cell debris, phenolics, and pigments. Typically, precipitation is initiated by the addition of ammonium sulphate (Huang et al. 2005; Lai et al. 2010; Peckham et al. 2006), high concentrations of sodium chloride (Xu et al. 2007, 2010), polyelectrolytes (Menkhaus et al. 2002), or by pH adjustment with acid or base (Bendandi et al. 2010; Cox et al. 2006; Garger et al. 2000; Pogue et al. 2010; Vézina et al. 2009; Woodard et al. 2009; Zaman et al. 1999). Rubisco is precipitated by reducing the homogenate pH to 5 or below. Other native leaf proteins that would otherwise not precipitate at $pH<5$ tend to precipitate along with rubisco upon acidification (Pirie 1987). This method, termed isoelectric precipitation, resulted in twofold purification of a mAb from *Lemna* homogenate and also removed chlorophyll pigments (Woodard et al. 2009). Rubisco can also be precipitated from extracts using low concentrations of ammonium sulphate (25–30%), along with cell debris, phenolics, and pigments (Lai et al. 2010; Peckham et al. 2006).

Aqueous two-phase fractionation has been demonstrated to be an effective way to remove phenolics, alkaloids, and native protein from tobacco extracts prior to capture chromatography (Platis et al. 2008; Platis and Labrou 2006, 2009; Ross and Zhang 2010). Aqueous two-phase fractionation of clarified transgenic tobacco (Platis et al. 2008) and corn extracts (Ramessar et al. 2008) allowed the development of non-protein A purification protocols for two mAbs. Careful selection of process conditions (PEG and phosphate concentrations and pH) for the aqueous two-phase system resulted in a twofold purification of two anti-HIV mAbs from unclarified tobacco extracts (Platis and Labrou 2009), which was comparable to the purification level previously achieved with clarified extracts. The flexibility to eliminate both centrifugation and filtration steps provides a distinct advantage over other methods. Additional advantages of using aqueous two-phase partitioning include the relatively low cost and ease of scale-up (Platis and Labrou 2009).

Removal of phenolics prior to capture chromatography is particularly important for expensive affinity resins like Protein A and can be accomplished using adsorption as a pretreatment method. Adsorption methods include the use of hydrophobic resins, anionic resins, and hydrophilic polymers in a batch mode or packed-bed

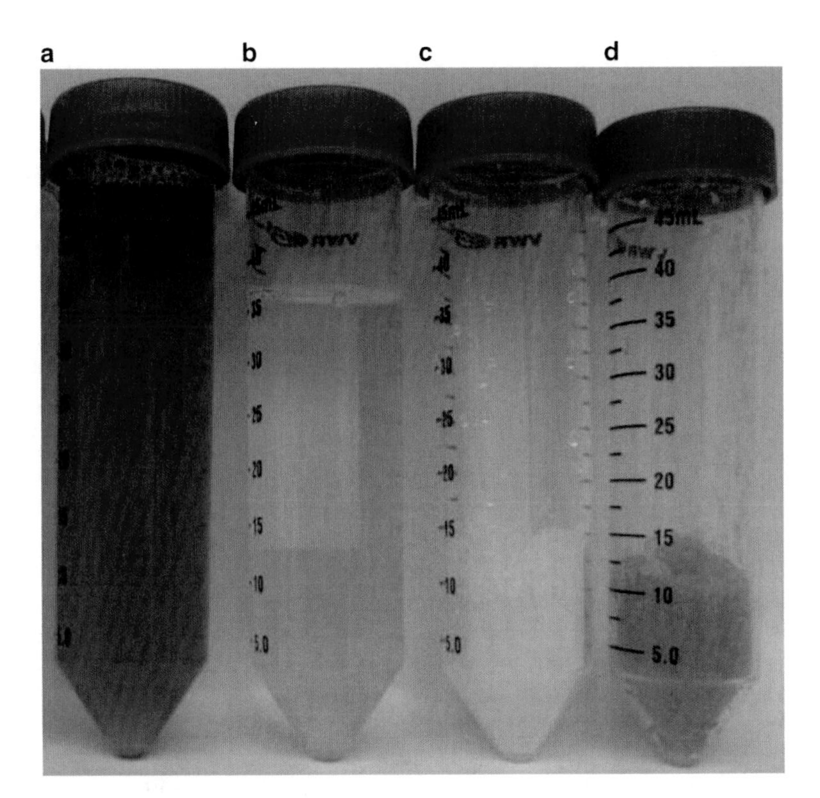

Fig. 11.3 Pretreatment of sugarcane juice with XAD-4: sugarcane juice (**a**), XAD-4 treated juice (**b**), XAD-4 resin before treatment (**c**), XAD-4 resin after treatment (**d**)

configuration to bind phenolics and minimize their interactions with recombinant proteins. The use of polyvinylpyrrolidone (PVPP) to sequester and remove tannins is based on technology used in the fruit juice industry and is not very effective for treating plant extracts and homogenates that contain very little tannic compounds (Barros et al. 2011). The use of hydrophobic and anion-exchange resins have been effectively used for phenolics removal and/or removal of plant pigments. In our experience, non-phenolic color compounds (products of Maillard reactions) could be removed from transgenic sugarcane juice using the hydrophobic resin, Amberlite XAD-4 (Fig. 11.3).

By identifying the amount and type of phenolics compounds present in an extract, a platform-tailored solution can be devised for efficient removal. Barros et al. (2011) identified the types of phenolics present in transgenic *Lemna minor* extracts by using RP-HPLC and evaluated the use of inexpensive adsorption resins for their removal prior to Protein A capture chromatography. The authors found that the added cost of the pretreatment step using Amberlite XAD-4 and IRA-402 could be offset by the extended lifetime (number of cycles) of the Protein A resin.

11.2.4.2 Seed-Based Systems

Aqueous seed extracts inherently lack alkaloids and chlorophyll pigments but do contain small amounts of phytic acid, lipids, and lectins that may interfere with protein purification (Farinas et al. 2005a; Stöger et al. 2002; Wilken and Nikolov 2006). As shown in Fig. 11.2e,f, seed extracts typically contain small amounts of phenolics (primarily ferulic and coumaric acids) compared to leafy extracts. Seed extracts are less complex than leafy-tissue extracts but extracted proteins, phytic acid, and lipids may adversely affect recombinant protein yield and purification efficiency. Pretreatment of seed extracts is not always necessary and is often substituted with a conditioning step. The objective of conditioning is make the clarified seed extract compatible with the subsequent purification steps. Traditional conditioning methods for any protein production platforms include adjustment of extract pH, ionic strength, buffer composition, and volume reduction by cross-flow filtration.

The primary protein classes that would potentially be extracted using typical aqueous buffers would be the water-soluble albumins and salt-soluble globulins. Seed proteins differ by the total protein content and also by protein solubility. For example, the total protein content of brown rice is about 8% and composed primarily of glutelins, which are not readily extracted with common extraction buffers. Corn contains 11% protein with protein pIs and molecular weights similar to rice proteins, implying that purification methods developed for corn-expressed recombinant proteins may also be applicable to rice (Menkhaus et al. 2004a). Soybean contains the highest amount of total protein (40%) compared to canola, corn, and rice, and is composed primarily of globulins and albumins. Therefore, extraction of a soybean-expressed recombinant protein will like have more total soluble protein in the extract than other seed systems. Corn, rice, and soybean proteins are acidic in nature and, thus, low pH extraction minimizes the amount of total native protein in extracts (Farinas et al. 2005a; Robic et al. 2010; Wilken and Nikolov 2006). If high pH extraction is required for optimal recombinant protein extraction, an acidic pH precipitation step can be added to greatly reduce the purification burden. Canola proteins consist of globulins and are more basic than those of corn and rice (Menkhaus et al. 2004a).

Phytic acid, a phosphoric ester of inositol, is a negatively charged molecule in aqueous solutions of pH above 1.0 (Costello et al. 1976). The presence of phytic acid in seed extracts is an important consideration for downstream processing because it can form binary and tertiary complexes with proteins and interfere with protein extraction (Hussain and Bushuk 1992) by shifting the isoelectric point and the solubility profile of proteins (Wolf and Sathe 1998). The type of complex and extent of protein-phytic acid interaction depends on pH, cation concentration and charge, and protein properties (Cheryan 1980). Generally, binary complexes are formed at acidic pH and ternary complexes at neutral pH (Selle et al. 2000). Basic proteins can form binary complexes with phytic acid over a wide pH range and interfere with protein purification (Wilken & Nikolov, 2006, 2010). Another even more common occurrence is the formation of insoluble phytate salts with divalent cations that can precipitate during pH adjustment from acidic to neutral pH (Wilken and Nikolov 2010).

Methods to reduce phytic acid content or interference include extraction of transgenic seed under conditions at which phytic acid is not highly soluble, addition of cations to counteract the negative charges, or using phytase to hydrolyze phytic acid (Wilken and Nikolov 2010).

11.2.5 Purification

Following pre-treatment and conditioning, plant-derived proteins are usually purified using methods developed for existing biopharmaceutical products (Chen 2008; Menkhaus et al. 2004a; Nikolov and Woodard 2004). Because chromatographic resins historically have been developed and optimized for protein purification from microbial and cell-culture systems, some adaptation of loading conditions and resin regeneration protocols may be required. Based on our experience and reported data in Table 11.3, it appears that the current best way to translate previously developed purification and resin regeneration procedures to plant systems is to adequately pre-treat clarified extracts.

Purification of plant-derived recombinant proteins relies primarily on adsorption chromatography because of its superior resolution power and chromatographic resin availability and diversity. Resin selection is determined by recombinant protein properties such as charge, hydrophobicity, and biospecificity. Selecting a resin based on the property most unique to the recombinant protein compared to the plant impurities can improve purification efficiency by increasing binding capacity and/or product purity. Once the resin functionality is selected (ion-exchange, affinity, hydrophobic), particle size, surface area, ligand density and resin backbone can be screened and binding conditions such as pH and ionic strength can be modified for optimal purification.

The function of the first chromatographic step, typically a capture step, is to concentrate the recombinant protein and remove critical feedstock impurities that would be detrimental to protein yield, quality, and/or purification efficiency (Holler and Zhang 2008; Menkhaus and Glatz 2005; Platis et al. 2008; Woodard et al. 2009). Purification methods and selected resin functionalities summarized in Table 11.3 show that the initial capture of recombinant proteins from plant extracts is accomplished primarily by affinity chromatography and ion-exchange adsorption resins. Protein A or G affinity resins are typically used for IgG capture and purification for bench-scale, pilot-scale, and manufacturing-scale processes. Other protein-and tag-specific affinity resins (IMAC, trypsin, ECH-lysine, glutathione, etc.) are available for purification of recombinant proteins intended for characterization and initial clinical trials (Kimple and Sondek 2001). Most affinity resins are very expensive and a pre-treatment of plant extracts to remove impurities that are responsible for resin fouling and loss of resolution is recommended. Ion-exchange resins are heavily relied upon in the biotechnology industry and are very common capture step for plant protein extracts as well. Acidic proteins like β-glucuronidase, chymosin, human serum albumin, and transferrin were captured by anion-exchange adsorption,

Table 11.3 Conditioning, pretreatment, and recombinant protein purification methods (Modified from Wilken and Nikolov 2011)

Plant system	Recombinant protein	pI of recombinant protein	Solid–liquid separation, conditioning and pretreatment	Capture	Intermediate/ Polishing	Purity	Yield	Reference
Seeds								
Corn seed	Aprotinin	10.5	– Dead-end filtration – Adjust to pH 7.8 – Dead-end filtration	Trypsin affinity	IMAC-Cu	79%	49%	Azzoni et al. (2002)
Corn germ	Aprotinin	10.5	– Centrifugation – Dead-end filtration	Weak cation exchange (CM-Sepharose)	HIC (Phenyl-Sepharose)	75%	34%	Zhong et al. (2007)
Corn seed	Collagen	8.3–8.9	– Centrifugation – Dialysis – Dead-end filtration	Strong cation exchange (SP-Sepharose FF)	Gel filtration (Sephacryl 200HR)	>70%	60%	Zhang et al. (2009b)
Corn seed	IgG	>8.0	– Centrifugation – Dead-end filtration (2-step)	PEG/PO_4 separation (10%/15%), 11% NaCl	PEG/PO_4 separation (9%/11%)/ $PEG//PO_4$ (8.2%/14.8%), Precipitation at interface	72%	49%	Lee and Forciniti (2010)
Corn seed	IgG	>8.0	– Centrifugation – Dead-end filtration	Strong cation exchange (SP-Sepharose FF)	IMAC-Zn	90%	50–60%	Ramessar et al. (2008)
Rapeseed	Chymosin	4.6	– Centrifugation – pH 2 acidification – Adjust to pH 5.6 – Centrifugation – Dilution with water to reduce conductivity	Weak anion exchange (DEAE)	HIC (Butyl Sepharose)	NR	NR	Van Rooijen et al. (2008)

Rice seed	Human Lactoferrin	8.2	– Centrifugation or sedimentation – Dead-end filtration	Strong cation exchange (SP-Sepharose FF)	UF/DF	93%	68%	Nandi et al. (2005)
Rice seed	Transferrin	6.3	– Centrifugation	Weak anion exchange (DEAE)	None	>95%	60%	Zhang et al. (2010)
Rice seed	Human Lysozyme	10.2	– Centrifugation and dead-end filtration – pH adjustment from pH 4.5–6 with 1 N NaOH – Dead-end filtration	Strong cation exchange (SP-Sepharose FF)	None	95%	90%	Wilken and Nikolov (2006)
Rice seed	Human Lysozyme	10.2	– Centrifugation and dead-end filtration – pH adjustment from pH 4.5–6 with 1 M TRIS and 1 M NaOH or pH 4.5 precipitation (for pH 10 extracts)	Strong cation exchange (SP-Sepharose FF)	None	98%	97%	Wilken and Nikolov (2010)
Soybean seed	β-Glucuronidase	5.5	– Centrifugation – Dead-end filtration	Weak anion exchange (DEAE)	HIC (Phenyl Sepharose)	NR	110%	Robi et al. (2006)

(continued)

Table 11.3 (continued)

Plant system	Recombinant protein	pI of recombinant protein	Solid–liquid separation, conditioning and pretreatment	Capture	Intermediate/ Polishing	Purity	Yield	Reference
Leafy crops								
Tobacco leaves	Aprotinin	10.5	– Acidification – Centrifugation – Dead-end filtration – Ultrafiltration	Cation exchange	Reverse phase chromatography, UF/DF	>99%	~ 50%	Pogue et al. (2010)
Tobacco leaves	β-Glucuronidase	5.5	– Addition of 2% (w/v) pre-hydrated PVPP – Centrifugation – Dead-end filtration	Polyelectrolyte precipitation (800 mg PEI/g protein)	HIC (Phenyl Sepharose), Ceramic hydroxyapatite, concentration	High purity	40%	Holler and Zhang (2008)
Tobacco leaves	Influenza Virus Like Particles (H5-VLP)	4.8–6.1	– Centrifugation – Clarification (chemical and physical treatments)	Cation exchange	Affinity, cross-flow filtration, formulation	High purity	NR	D'Aoust et al. (2010)
Tobacco leaves	IgG (patient 69)	>8.0	– Precipitation to below pH 5.1 – Filtration	Protein A	Anion exchange (membrane adsorption)	>90%	>50%	Bendandi et al. (2010)
Tobacco leaves	IgG	>8.0	– Dead-end filtration – Centrifugation – 25% ammonium sulfate precipitation – Centrifugation	50% ammonium sulfate precipitation	Protein A (MabSelect), weak anion exchange, flowthrough mode (DEAE)	> 95%	50%	Lai et al. (2010)

Tobacco leaves	IgG	>8.0	– Centrifugation – Adjustment to pH 7.5 – Centrifugation – Dead-end filtration	Protein A	Strong cation exchange (Source 30 S), buffer exchange	99%	7%	Lombardi et al. (2010)
Tobacco leaves	IgG (4E10) IgG (2 G12)	>8.0	– Centrifugation – Dead-end filtration – PEG/Phosphate two-phase partitioning – Dialysis	Strong cation exchange (SP Sepharose FF)	IMAC-Zn	96% 97%	36% 63%	Platis et al. (2008)
Tobacco leaves	IgG	>8.0	– Centrifugation or dead-end filtration – High-speed centrifugation	Protein A (Streamline) EBA	Size exclusion (Sephadex G-25), sterile filtration	95%	44%	Padilla et al. (2009)
Tobacco leaves	IgG (αCCR5)	8.4–9.7	– Horizontal screw press – Acidification – Dead-end filtration	Protein A	Multi-ion exchange, UF/DF	97%	NR	Pogue et al. (2010)
Tobacco leaves	IgG (C5-1)	~8.0	– Precipitation at pH 4.8 – Centrifugation – Adjustment to pH 8 – Centrifugation – Dead-end filtration – Concentration (100 kDa cross-flow filtration)	Protein G	None	>90%	NR	Vézina et al. (2009)

(continued)

Table 11.3 (continued)

Plant system	Recombinant protein	pI of recombinant protein	Solid–liquid separation, conditioning and pretreatment	Capture	Intermediate/ Polishing	Purity	Yield	Reference
Bioreactor-based systems								
Carrot cell suspension	Glucocerebrosidase (intracellular)	7.4	– High-speed centrifugation	Strong cation exchange (Macro-prep High S)	HIC (Phenyl), strong cation exchange	High purity	NR	Shaaltiel et al. (2007)
Lemna	IgG1 (anti-CD 30)	>8.0	– Acidification – Centrifugation – Precipitation at pH 4.5 – Adjustment to pH 7.2 – Dead-end filtration	Protein A (MAb Select SuRe)	Ceramic hydroxyapatite	>90%	NR	Cox et al. (2006)
Rice cell suspension	HSA (extracellular)	4.7–5.0	– 40% ammonium sulfate precipitation – Centrifugation – Dialysis	Weak anion exchange (DEAE-Sepharose)	Dialysis and concentration	>95%	50%	Huang et al. (2005)

Tobacco cell suspension	GFP-fusion protein (intracellular)	5.0	– Centrifugation – Dead-end filtration – 30% ammonium sulfate precipitation – Centrifugation – Dead-end filtration	HIC (Phenyl Sepharose)	Strong anion exchange (Q-Sepharose)	>80%	>70%	Peckham et al. (2006)
Tobacco cell suspension	Human Anti-Rabies Antibody (intracellular)	>8.0	– Fiberglass filtration – 30 kDa membrane concentration	Hi Trap Protein G	None	>95%	25%	Girard et al. (2006)
Tobacco cell suspension	Human Growth Hormone (intracellular)	~5.0	– Filtration – 2 M NaCl precipitation – Centrifugation	HIC (Phenyl Sepharose)	UF concentration	>85%	NR	Xu et al. (2010)
Tobacco cell suspension	Interferon α2b (extracellular)	~5.8	– Filtration – 2 M NaCl precipitation – Centrifugation	HIC (Phenyl Sepharose)	UF concentration	>85%	>60%	Xu et al. (2007)

NR not reported, *UF/DF* ultrafiltration/diafiltration, *HIC* hydrophobic interaction chromatography and basic proteins, aprotinin, collagen, human lysozyme, and human lactoferrin by cation exchange (Table 11.3). Hydrophobic-type resins are often used after ion-exchange steps because recombinant proteins are eluted with high ionic strength buffers that are compatible with subsequent hydrophobic interaction chromatography (HIC). Because of relatively low recovery yields compared to affinity and ion-exchange resins, HIC is not usually used as a capture step unless extract pre-treatment requires the use of high salt concentrations (Peckham et al., 2006; Xu et al., 2011, 2007). Several new purification methods have been proposed including fusion protein, such as elastin-like polypeptides and hydrophobins, as well as induction of protein body formation (Conley et al. 2011). Select cases of more traditional chromatographic and non-chromatographic methods for purification of proteins from leafy and seed tissues are discussed below.

11.2.5.1 Leaf-Based Systems

Chromatographic Methods

Purification of protein products from mammalian cell cultures and yeast rely on a capture step for a quick removal of proteases to protect product integrity. The same is true for leafy extracts which, in addition to proteases, contain chlorophyll pigments and phenolics that should be removed from the extract as early as possible. Irrespective of the protein platform and feedstock origin, process development scientists strive to select inexpensive resins which are resistant to required regeneration chemicals and able to retain capacity and selectivity over multiple cycles. For that reason, ion-exchangers are ideally suited for their robustness, low cost, and availability from several reliable resin manufacturers (Tosoh Bioscience, GE Healthcare, Bio-Rad, Pall, Merck KGaA). Direct application of a crude leafy extract onto an anion-exchange resin is not suitable because rubisco, nucleic acids, and phenolic acids may also bind to the anion ligand and reduce the binding capacity. Extended use of an anion-exchange column without pre-treatment of the crude extract could lead to irreversible fouling and plugging of the column (Holler and Zhang 2008).

Cation exchangers are usually more amenable for direct capture of basic proteins because nucleic acids and phenolics do not preferentially bind to cationic resins. In many situations, simple manipulation of ionic strength and pH provides sufficient protein binding capacity, concentration, and partial purification. (D'Aoust et al. 2010; Pogue et al. 2010; Shaaltiel et al. 2007). Recombinant anti-HIV IgGs were efficiently captured and purified by a strong cation-exchange resin after pre-treatment of the clarified tobacco homogenates by aqueous two-phase partitioning (Platis et al. 2008; Platis and Labrou 2009). Capture and purification of IgG molecules by affinity resins (Protein G or Protein A) still gives superior purification results but implementation of pre-treatment methods discussed previously such as, acidic or salt precipitation (Bendandi et al., 2010; Cox et al., 2006; Pogue et al., 2010;

Vézina et al., 2009), Amberlite IRA-402 adsorption (Barros et al. 2011; Woodard et al. 2009), and aqueous two-phase partitioning (Platis and Labrou 2006) is advisable to increase the lifetime of costly Protein A resins.

Additional purification of eluted proteins from the capture column includes a variety of orthogonal steps such as HIC, IMAC, ion-exchange, and ceramic hydroxyapatite (Table 11.3). The subsequent purification step (intermediate or polishing step) is selected based on the properties of the recombinant protein and residual impurities. If impurities (e.g. DNA and endotoxin) remain after capture chromatography, the subsequent purification step can be performed in a flow-through mode to bind the residual impurities (Bendandi et al., 2010; Lai et al., 2010).

Non-chromatographic Capture Alternatives

Since tobacco impurities made direct capture of the acidic protein, β-glucuronidase, on an anion-exchange column infeasible due to resin fouling, Holler and Zhang (2008) opted for anionic polyelectrolyte precipitation (polyethylenimine) as a capture step. The precipitation β-glucuronidase also removed large amounts of tobacco impurities, concentrated the recombinant protein, and produced an enriched fraction suitable for subsequent purification by HIC (Phenyl-Sepharose). Another interesting non-chromatography option was proposed by Werner et al. (2006) which used *in situ* capture of a plant-expressed mAb by Protein A-engineered viral particles. The Protein A ligand was attached to the C-terminus of the core protein of turnip vein-clearing virus to produce rod-shaped, nanometer-size viral particles in tobacco leaves that displayed protein A on their surface. The high-surface density of Protein A on viral particles allowed the binding of 2 g mAb per g particle. Viral nanoparticles with bound mAb were harvested by centrifugation and mAb was desorbed using pH 2.5 glycine buffer. The released mAb was then precipitated by 15% PEG, resulting in highly purified IgG.

11.2.5.2 Seed-Based Systems

Chromatographic Methods

As discussed above, aqueous seed extracts do not contain alkaloids, chlorophyll pigments, and nucleic acids, but may have small amounts of phenolic acids and phytic acid. Capturing basic and acidic proteins by ion-exchange adsorption is common and usually an effective first step. To achieve required recombinant protein binding and purification one can vary chromatography conditions (pH, ionic strength, and residence time). For example, rice proteins that are extractable in aqueous buffers are acidic in nature (pI < 5) and allowed purification of basic recombinant proteins (pI > 8) to a greater than 90% purity using a single cation-exchange column (Nandi et al. 2005; Wilken and Nikolov 2010). The effect of extraction and adsorption pH on binding capacity and purity of recombinant human lysozyme

Table 11.4 Human lysozyme purification by cation exchange adsorption (Modified from Wilken and Nikolov 2006, 2010)

Extraction pH	pH adjustment	Adsorption pH	Bound lysozyme (mg)	Lysozyme purity (%)
4.5	None	4.5	43	89
6	None	6	24	50
4.5	Phosphate buffer	6	8.6	95
4.5	TRIS buffer	6	25	98
10	Acetic acid	4.5	36	84
10	Acetic acid to pH 4.5 then NaOH to pH 6	6	26	95

from rice is given in Table 11.4. The data clearly exemplify how varying pH can manipulate extract composition and modify interactions of recombinant lysozyme and rice proteins with the strong cation-exchange resin. Extraction and adsorption at pH 4.5 was clearly advantageous in terms of lysozyme binding capacity (43 mg/mL) and resulted in relatively high lysozyme purity. For higher purity, pH 4.5 extraction combined with pH 6 adsorption after pH adjustment using TRIS buffer was the best process option.

Purifications of recombinant proteins with pIs similar to the majority of native seed proteins may require a different approach or intermediate purification step. Acidic proteins can be captured by anion-exchange chromatography but typically require an additional purification step, such as hydrophobic interaction chromatography, to achieve a greater protein purity (Robić et al. 2006; Van Rooijen et al. 2008). Zhang et al. (2010) were able to purify the slightly acidic recombinant human transferrin (pI > 6.3) by a single weak anion-exchange adsorption and achieve >90% protein purity. The selected process was possible because of the high expression level of transferrin (10 g/kg) and weaker adsorption of transferrin at pH 7.5 to the DEAE anion-exchange resin compared to solubilized rice proteins. The weaker transferrin-DEAE interactions made it possible to elute the adsorbed transferrin before native proteins using a low ionic strength buffer (40 mM NaCl). The more acidic and strongly-bound rice proteins eluted during resin regeneration using a high salt buffer.

Non-chromatographic Capture Alternatives

One of the first non-chromatographic methods applied to seeds was the capture and purification of oleosin-mediated fusions by centrifugation (Van Rooijen and Motoney 1995). Since 1999, this technology has been further developed to include several variations. In one case, a ligand for the target protein was fused directly to the oleosin to capture the recombinant protein on extracted and purified oil bodies

(Boothe et al. 2010). An example of this approach was the capture of antibodies on oleosin-Protein A oil bodies which were then released from purified oil bodies by acidic elution. Another variation of oleosin-mediated capture was the fusion of an anti-oleosin ligand to the recombinant protein (e.g. an anti-oleosin scFv). In this case, the anti-oleosin scFv fusion protein was released from the oil bodies by acid or urea, depending on the stability of the specific fusion protein. The fusion partner was then removed from the recombinant protein by either chemical (e.g. acidic cleavage in the case of the scFv-ApoA1 fusion) or enzymatic cleavage (Boothe et al. 2010; Nykiforuk et al. 2011). Standard adsorption chromatographic methods were used for subsequent purification of released recombinant protein.

Several alternative methods for recombinant protein purification developed recently either completely circumvented adsorption chromatography or utilized a non-chromatography capture step followed by traditional intermediate and polishing chromatography. Lee and Forciniti (2010) explored the use of aqueous two-phase (PEG/salt) partitioning as a sole recovery and purification method of a non-glycosylated monoclonal antibody (mAb) from transgenic corn extract. The antibody was purified in a three-stage process by manipulating the two-phase system composition, pH, and ionic strength. The first two stages consisted of a typical two-phase partitioning aimed at increasing the recombinant protein concentration in the bottom (aqueous salt) phase. The third stage consisted of mAb precipitation at the two-phase interface that resulted in tenfold purification. Overall, the three-stage processes delivered 72% pure mAb with 49% overall yield.

Aspelund and Glatz (2010) demonstrated purification of recombinant collagen from low pH corn extracts by cross-flow filtration. Diafiltration of corn endosperm extracts at pH 3.1 by using a 100-kDa MWCO membrane resulted in 89% pure collagen. Further improvement of collagen purity was achieved by protein precipitation of endosperm extracts at pH 2.1 using sodium chloride. Membrane diafiltration of re-suspended precipitate resulted in 99% pure collagen and 87% collagen yield. The favorable sieving coefficient of endosperm proteins and unique properties of collagen (high molecular weight and stability at low pH) permitted the development of this inexpensive purification process.

11.3 Process Economics

The manufacturing cost of plant-produced recombinant proteins consists of upstream production cost, i.e. expenses related to transgenic plant biomass production in open-fields, greenhouses, or contained vessels (bioreactors) and downstream costs. Both, upstream production and downstream processing costs depend on the concentration of extractable recombinant protein, overall process yield, and production scale. Downstream processing costs are affected by extract complexity, required product purity, and intended application of recombinant protein, which in turn determines the extent of process documentation and related cGMP costs.

Table 11.5 Breakdown of direct manufacturing costs[a] of a therapeutic recombinant protein

Biomass production method	Upstream cost ($/g)	Downstream cost ($/g)	Total direct cost ($/g)	Upstream cost (%)	Downstream cost (%)
Open-field	2.5	80	83	3	97
Greenhouse	19	79	98	19	81
Bioreactor	60	78	138	43	57

[a]Direct cost consisted of reagents, consumables, operating labor, supervision, QC/QA and lab charges, utilities and waste treatment

Downstream processing expenses are proportional to the number and type purification stages as the latter directly affect the overall product yield, reagent use, and labor requirements (Nikolov and Hammes 2002). The breakdown of upstream and downstream processing costs depends primarily on product end application with biopharmaceutical and industrial proteins being at the opposite end of the cost and purity spectrum (high to low). Depending on the expression level, purification yield, and annual product output, the upstream production cost for a highly-purified protein (>95%) from a seed crop ranges from 5% to 10% of the total manufacturing cost (Evangelista et al. 1998; Mison and Curling 2000; Nandi et al. 2005; Nikolov and Hammes 2002). For a protein of similar purity and annual throughput produced by leafy tissue grown in a greenhouse, the upstream cost could be as high as 25% of the total manufacturing cost (unpublished estimates). Pogue et al. (2010) compared the production cost ($/g product) of aprotinin transiently expressed in tobacco and reported that greenhouse production cost was five times greater than the open-field cost. There is minimal data available regarding cost of plant cell culture systems, but one could anticipate a significantly higher production costs compared to greenhouse and open-field grown transgenic plants.

To illustrate the manufacturing cost breakdown between upstream and downstream costs for the three biomass production options, we considered a hypothetical case: a recombinant monoclonal antibody (mAb) expressed at 1 g per kg fresh weight in (1) tobacco grown in an open-field; (2) tobacco grown in a greenhouse; and (3) tobacco cell culture. The three production options were analyzed using SuperPro Designer simulation software (Intellgen Inc.) and hypothetical cost numbers were summarized in Table 11.5. The downstream processing trains for the two plants extracts and cell-free homogenate consisted of three chromatographic steps with the same volumetric throughput and product yield and, therefore, have similar cost. The direct upstream cost for the bioreactor-produced mAb ($60/g) is significantly higher than the other two plant production systems because of the bioreactor-related cost, consumables (media, buffer, cleaning agents), and labor associated with bioreactor operation, quality assurance and quality control (QA/QC), and cleaning validation. The upstream cost for an open-field produced mAb is almost an insignificant (3%) fraction of the total manufacturing cost, whereas the greenhouse cost (19% of the total) is still substantially lower than downstream processing cost. The cost breakdown for the bioreactor-produced recombinant protein (43% upstream and 57% downstream) is similar to that for mammalian cell culture systems (Sommerfeld and Strube 2005). We estimate that the capital investment for an open-field biomass production system would be less than 5% of the investment required for building a cell-culture facility, while the investment for a greenhouse would be around 25% of that for cell culture. The capital investment cost could vary because it is dictated by facility size, geographic location, required containment level, and plant growth control (Spök and Karner 2008). This hypothetical study case illustrates the apparent cost advantage of open–field and greenhouse biomass production systems.

Besides the therapeutic protein production example considered here, there is a wide range of recombinant proteins that have been produced in transgenic plants with applications as specialty enzymes, cell-culture media ingredients, in-process reagents, and industrial and food proteins. The breakdown of upstream and downstream processing costs for these products will vary depending on the purification requirements, production scale, and end application. For example, downstream costs for industrial enzymes and oral vaccines could be less than 50% of the total manufacturing costs because the downstream processing trains would consist of inexpensive capture and recovery methods such as membrane filtration and protein precipitation (Arntzen et al. 2006; Nikolov and Hammes 2002).

The decision-making process for choosing the best plant production system is complex and requires case-by-case analyses. From a downstream processing perspective, several evaluation criteria should be applied for selecting the best system that matches product characteristics and allows overall manufacturing cost reduction (Nikolov and Hammes 2002). In addition to protein expression level and stability, the following criteria should be considered: biomass yield and storage stability, access to off-the-shelf purification tools, biomass disposal cost, and byproduct revenues (e.g. biomass and starch conversion to energy). Business drivers for consideration include capital investment, production scale and cost, speed to market, and regulatory requirements (Nikolov and Hammes 2002; Spök and Karner 2008).

11.4 Conclusions

As the transgenic plant technologies continue to mature, it appears that research and development interests in plant biotechnology are slowly shifting toward developing and understanding downstream processing methods. The interest in downstream processing is not surprising, as it follows the natural evolution of mature protein production platforms such as mammalian and insect cell culture systems. In addition, the awareness that downstream processing accounts for a significant portion of the total product manufacturing costs has enticed companies as well as private, state, and federal entities to invest into research and development.

Downstream processing strategies presented in this chapter clearly indicate that recovery and purification of recombinant proteins from seed crops are easier than for metabolically active leafy systems, which contain a number of potentially detrimental impurities. The less complex and more stable seed extracts allow capture chromatography to be carried out without extract pretreatment step(s). Combined with favorable economics for growing seeds in an open-field, transgenic systems, like rice, corn, and safflower, are the best candidates for low-cost protein production. However, the long seed development time and public resistance to use of food and feed crops are barriers to further commercialization of seed-based products. Transient expression offers an advantage for using leafy tissue for production of pandemic and seasonal influenza vaccines, since the short development time of 4–6 weeks cannot currently be fulfilled by other protein production platforms (D'Aoust et al. 2010). In spite of all the progress and positive developments, additional research and technological breakthroughs in downstream processing are needed to capitalize on the lower production cost of transgenic biomass. Future directions to accelerate process development and advance recombinant protein production from plant systems have been outlined by Wilken and Nikolov (2011).

Acknowledgements The authors thank Ms. Georgia Barros of Texas A&M University and Dr. Susan Woodard of Caliber Biotherapeutics for preparation of the chromatograms cited in this chapter as unpublished work.

References

Adkins YB, Lönnerdal B (2004) Proteins and peptides. In: Neeser JR, German BJ (eds) Biotechnology for performance foods, functional foods, and nutraceuticals. Marcel Dekker, Inc., New York

Arntzen C, Plotkin S, Dodet B (2005) Plant-derived vaccines and antibodies: potential and limitations. Vaccine 23(15):1753–1756

Arntzen C, Mahoney R, Elliott A, Holtz B, Krattiger A, Lee CK, Slater S (2006) Plant-derived vaccines: cost of production. The Biodesign Institute at Arizona State University, Tempe

Aspelund MT, Glatz CE (2010) Purification of recombinant plant-made proteins from corn extracts by ultrafiltration. J Membr Sci 353(1–2):103–110

Azzoni AR, Kusnadi AR, Miranda EA, Nikolov ZL (2002) Recombinant aprotinin produced transgenic corn seed: extraction and purification studies. Biotechnol Bioeng 80(3):268–276

Azzoni AR, Farinas CS, Miranda EA (2005) Transgenic corn seed for recombinant protein production: relevant aspects on the aqueous extraction of native components. J Sci Food Agric 85(4):609–614

Bai Y, Nikolov ZL (2001) Effect of processing on the recovery of recombinant β-glucuronidase (rGUS) from transgenic canola. Biotechnol Prog 17(1):168–174

Bai Y, Nikolov ZL, Glatz CE (2002) Aqueous extraction of β-glucuronidase from transgenic canola: kinetics and microstructure. Biotechnol Prog 18(6):1301–1305

Barros GOF, Woodard SL, Nikolov ZL (2011) Phenolics removal from transgenic *Lemna minor* extracts expressing mAb and impact on mAb production cost. Biotechnol Prog 27(2):410–418

Barta A, Sommergruber K, Thompson D, Hartmuth K, Matzke MA, Matzke AJM (1986) The expression of a nopaline synthase: human growth hormone chimaeric gene in transformed tobacco and sunflower callus tissue. Plant Mol Biol 6(5):347–357

Benchabane M, Rivard D, Girard C, Michaud D (2009) Companion protease inhibitors to protect recombinant proteins in transgenic plant extracts. In: Faye L, Gomord V (eds) Recombinant proteins from plants. Humana Press, New York

Bendandi M, Marillonnet S, Kandzia R, Thieme F, Nickstadt A, Herz S, Fröde R, Inogés S, Lòpez-Dìaz de Cerio A, Soria E, Villanueva H, Vancanneyt G, McCormick A, Tusé D, Lenz J, Butler-Ransohoff J-E, Klimyuk V, Gleba Y (2010) Rapid, high-yield production in plants of individualized idiotype vaccines for non-Hodgkin's lymphoma. Ann Oncol. doi:10.1093/annonc/mdq1256

Boothe J, Nykiforuk C, Shen Y, Zaplachinski S, Szarka S, Kuhlman P, Murray E, Morck D, Moloney MM (2010) Seed-based expression systems for plant molecular farming. Plant Biotechnol J 8(5):588–606

Boyhan D, Daniell H (2010) Low-cost production of proinsulin in tobacco and lettuce chloroplasts for injectable or oral delivery of functional insulin and C-peptide. Plant Biotechnol J 9(5):585–598

Bratcher B, Garger SJ, Holtz RB, McCulloch MJ (2005) Flexible processing apparatus for isolating and purifying viruses, soluble proteins, and peptides from plant sources. US Patent 6,906,172

Chen Q (2008) Expression and purification of pharmaceutical proteins in plants. Biol Eng 2(4):291–321

Cheryan M (1980) Phytic acid interactions in food systems. Crit Rev Food Sci Nutr 13(4):297–335

Conley AJ, Zhu H, Le LC, Jevnikar AM, Lee BH, Brandle JE, Menassa R (2010) Recombinant protein production in a variety of Nicotiana hosts: a comparative analysis. Plant Biotechnol J 9(4):434–444

Conley AJ, Joensuu JJ, Richman A, Menassa R (2011) Protein body-inducing fusions for high-level production and purification of recombinant proteins in plants. Plant Biotechnol J 9(4):419–433

Costello AJ, Glonek T, Myers TC (1976) 31P nuclear magnetic resonance-pH titrations of myo-inositol hexaphosphate. Carbohydr Res 46:159–171

Cox KM, Sterling JD, Regan JT, Gasdaska JR, Frantz KK, Peele CG, Black A, Passmore D, Moldovan-Loomis C, Srinivasan M, Cuison S, Cardarelli PM, Dickey LF (2006) Glycan optimization of a human monoclonal antibody in the aquatic plant *Lemna minor*. Nat Biotechnol 24(12):1591–1597

D'Aoust M-A, Lerouge P, Busse U, Bilodeau P, Trepanier S, Faye L, Vezina L-P (2004) Efficient and reliable production of pharmaceuticals in alfalfa. In: Fisher R, Schillberg S (eds) Molecular farming. Wiley-VCH, Weinheim

D'Aoust MA, Couture MMJ, Charland N, Trepanier S, Landry N, Ors F, Vezina LP (2010) The production of hemagglutinin-based virus-like particles in plants: a rapid, efficient and safe response to pandemic influenza. Plant Biotechnol J 8(5):607–619

Daniell H, Singh ND, Mason H, Streatfield SJ (2009a) Plant-made vaccine antigens and biopharmaceuticals. Trends Plant Sci 14(12):669–679

Daniell H, Ruiz G, Denes B, Sandberg L, Langridge W (2009b) Optimization of codon composition and regulatory elements for expression of human insulin like growth factor-1 in transgenic chloroplasts and evaluation of structural identity and function. BMC Biotechnol 9:23–39

Davies HM (2010) Review article: commercialization of whole-plant systems for biomanufacturing of protein products: evolution and prospects. Plant Biotechnol J 8(8):845–861

De Muynck B, Navarre C, Boutry M (2010) Production of antibodies in plants: status after twenty years. Plant Biotechnol J 8(5):529–563

De Zoeten GA, Penswick JR, Horisberger MA, Ahl P, Schultze M, Hohn T (1989) The expression, localization, and effect of a human interferon in plants. Virology 172(1):213–222

Doran PM (2006) Foreign protein degradation and instability in plants and plant tissue cultures. Trends Biotechnol 24(9):426–432

Edelbaum O, Stein D, Holland N, Gafni Y, Livneh O, Novick D, Rubinstein M, Sela I (1992) Expression of active human interferon-β in transgenic plants. J Interferon Res 12(6):449–453

Evangelista RL, Kusnadi AR, Howard JA, Nikolov ZL (1998) Process and economic evaluation of the extraction and purification of recombinant β-glucuronidase from transgenic corn. Biotechnol Prog 14(4):607–614

Farinas CS, Leite A, Miranda EA (2005a) Aqueous extraction of maize endosperm: insights for recombinant protein hosts based on downstream processing. Process Biochem 40(10):3327–3336
Farinas CS, Leite A, Miranda EA (2005b) Aqueous extraction of recombinant human proinsulin from transgenic maize endosperm. Biotechnol Prog 21(5):1466–1471
Fischer R, Liao YC, Drossard J (1999) Affinity-purification of a TMV-specific recombinant full-size antibody from a transgenic tobacco suspension culture. J Immunol Methods 226(1–2):1–10
Fischer R, Schillberg S, Twyman RM (2009) Molecular farming of antibodies in plants. In: Kirakosyan A, Kaufman PB (eds) Recent advances in plant biotechnology. Springer, New York
Fraley RT, Rogers SG, Horsch RB, Sanders PR, Flick JS, Adams SP, Bittner ML, Brand LA, Fink CL, Fry JS, Galluppi GR, Goldberg SB, Hoffmann NL, Woo SC (1983) Expression of bacterial genes in plant cells. Proc Natl Acad Sci USA 80(15):4803–4807
Franconi R, Demurtas OC, Massa S (2010) Plant-derived vaccines and other therapeutics produced in contained systems. Expert Rev Vaccines 9(8):877–892
Garger SJ, Holtz B, McCulloch MJ, Turpen TH (2000) Process for isolating and purifying viruses, soluble proteins and peptides from plant sources. US Patent 6,037,456
Giddings G, Allison G, Brooks D, Carter A (2000) Transgenic plants as factories for biopharmaceuticals. Nat Biotechnol 18(11):1151–1155
Girard LS, Fabis MJ, Bastin M, Courtois D, Pétiard V, Koprowski H (2006) Expression of a human anti-rabies virus monoclonal antibody in tobacco cell culture. Biochem Biophys Res Commun 345(2):602–607
Hassan S, Van Dolleweerd CJ, Ioakeimidis F, Keshavarz-Moore E, Ma JKC (2008) Considerations for extraction of monoclonal antibodies targeted to different subcellular compartments in transgenic tobacco plants. Plant Biotechnol J 6(7):733–748
Hellwig S, Drossard J, Twyman RM, Fischer R (2004) Plant cell cultures for the production of recombinant proteins. Nat Biotechnol 22(11):1415–1422
Hiatt A, Pauly M (2006) Monoclonal antibodies from plants: a new speed record. Proc Natl Acad Sci USA 103(40):14645–14646
Hiatt A, Cafferkey R, Bowdish K (1989) Production of antibodies in transgenic plants. Nature 342(6245):76–78
Holler C, Zhang C (2008) Purification of an acidic recombinant protein from transgenic tobacco. Biotechnol Bioeng 99(4):902–909
Holler C, Vaughan D, Zhang C (2007) Polyethyleneimine precipitation versus anion exchange chromatography in fractionating recombinant β-glucuronidase from transgenic tobacco extract. J Chromatogr A 1142(1):98–105
Hood EE, Howard JA (2009) Over-expression of novel proteins in maize. In: Kriz AL, Larkins BA (eds) Molecular genetic approaches to maize improvement. Springer, Berlin
Hood EE, Witcher DR, Maddock S, Meyer T, Baszczynski C, Bailey M, Flynn P, Register J, Marshall L, Bond D, Kulisek E, Kusnadi A, Evangelista R, Nikolov Z, Wooge C, Mehigh RJ, Hernan R, Kappel WK, Ritland D, Ping Li C, Howard JA (1997) Commercial production of avidin from transgenic maize: characterization of transformant, production, processing, extraction and purification. Mol Breed 3(4):291–306
Hood EE, Love R, Lane J, Bray J, Clough R, Pappu K, Drees C, Hood KR, Yoon S, Ahmad A, Howard JA (2007) Subcellular targeting is a key condition for high-level accumulation of cellulase protein in transgenic maize seed. Plant Biotechnol J 5(6):709–719
Howard JA, Nikolov ZL, Hood EE (2011) Enzyme production systems for biomass conversion. In: Hood EE, Nelson P, Powell R (eds) Plant biomass conversion. John Wiley & Sons Inc., Ames
Huang L-F, Liu Y-K, Lu C-A, Hsieh S-L, Yu S-M (2005) Production of human serum albumin by sugar starvation induced promoter and rice cell culture. Transgenic Res 14(5):569–581
Huang T-K, Plesha MA, Falk BW, Dandekar AM, McDonald KA (2009) Bioreactor strategies for improving production yield and functionality of a recombinant human protein in transgenic tobacco cell cultures. Biotechnol Bioeng 102(2):508–520

Hussain A, Bushuk W (1992) Interference of phytic acid with extraction of proteins from grain legumes and wheat with acetic acid. J Agric Food Chem 40(10):1938–1942

Joensuu JJ, Conley AJ, Lienemann M, Brandle JE, Linder MB, Menassa R (2010) Hydrophobin fusions for high-level transient protein expression and purification in *Nicotiana benthamiana*. Plant Physiol 152(2):622–633

Karg SR, Kallio PT (2009) The production of biopharmaceuticals in plant systems. Biotechnol Adv 27(6):879–894

Kimple ME, Sondek J (2001) Overview of affinity tags for protein purification. Curr Protoc Protein Sci 36(suppl):9.9.1–9.9.19

Kusnadi AR, Nikolov ZL, Howard JA (1997) Production of recombinant proteins in transgenic plants: practical considerations. Biotechnol Bioeng 56(5):473–484

Kusnadi AR, Evangelista RL, Hood EE, Howard JA, Nikolov ZL (1998) Processing of transgenic corn seed and its effect on the recovery of recombinant β-glucuronidase. Biotechnol Bioeng 60(1):44–52

Lai H, Engle M, Fuchs A, Keller T, Johnson S, Gorlatov S, Diamond MS, Chen Q (2010) Monoclonal antibody produced in plants efficiently treats West Nile virus infection in mice. Proc Natl Acad Sci USA 107(6):2419–2424

Lamphear BJ, Streatfield SJ, Jilka JM, Brooks CA, Barker DK, Turner DD, Delaney DE, Garcia M, Wiggins B, Woodard SL, Hood EE, Tizard IR, Lawhorn B, Howard JA (2002) Delivery of subunit vaccines in maize seed. J Control Release 85(1–3):169–180

Lee S-M (1989) The primary stages of protein recovery. J Biotechnol 11(2–3):103–117

Lee JW, Forciniti D (2010) Purification of human antibodies from transgenic corn using aqueous two-phase systems. Biotechnol Prog 26(1):159–167

Ling HY, Pelosi A, Walmsley AM (2010) Current status of plant-made vaccines for veterinary purposes. Expert Rev Vaccines 9(8):971–982

Lombardi R, Villani M, Di Carli M, Brunetti P, Benvenuto E, Donini M (2010) Optimisation of the purification process of a tumour-targeting antibody produced in *N. benthamiana* using vacuum-agroinfiltration. Transgenic Res 19(6):1083–1097

Loomis WD (1974) Overcoming problems of phenolics and quinones in the isolation of plant enyzmes and organelles. Methods Enzymol 31:528–544

Ma JKC, Lehner T, Stabila P, Fux CI, Hiatt A (1994) Assembly of monoclonal antibodies with IgG1 and IgA heavy chain domains in transgenic tobacco plants. Eur J Immunol 24(1):131–138

Ma JKC, Hiatt A, Hein M, Vine ND, Wang F, Stabila P, Vandolleweerd C, Mostov K, Lehner T (1995) Generation and assembly of secretory antibodies in plants. Science 268(5211):716–719

Ma JKC, Drake PMW, Christou P (2003) The production of recombinant pharmaceutical proteins in plants. Nat Rev Genet 4(10):794–805

Mason HS, Lam DMK, Arntzen CJ (1992) Expression of hepatitis B surface antigen in transgenic plants. Proc Natl Acad Sci USA 89(24):11745–11749

Mayfield SP, Franklin SE, Lerner RA (2003) Expression and assembly of a fully active antibody in algae. Proc Natl Acad Sci USA 100(2):438–442

Menkhaus TJ, Glatz CE (2005) Antibody capture from corn endosperm extracts by packed bed and expanded bed adsorption. Biotechnol Prog 21(2):473–485

Menkhaus TJ, Eriksson SU, Whitson PB, Glatz CE (2002) Host selection as a downstream strategy: polyelectrolyte precipitation of β-glucuronidase from plant extracts. Biotechnol Bioeng 77(2):148–154

Menkhaus TJ, Bai Y, Zhang CM, Nikolov ZL, Glatz CE (2004a) Considerations for the recovery of recombinant proteins from plants. Biotechnol Prog 20(4):1001–1014

Menkhaus TJ, Pate C, Krech A, Glatz CE (2004b) Recombinant protein purification from pea. Biotechnol Bioeng 86(1):108–114

Miller KD, Gao J, Hooker BS (2004) Initial clarification by aqueous two-phase partitioning of leaf extracts from *Solanum tuberosum* plants expressing recombinant therapeutic proteins. Bioprocess J 3(2):47–51

Mison D, Curling J (2000) The industrial production costs of recombinant therapeutic proteins expressed in transgenic corn. Biopharm 13(5):48–54

Moeller L, Taylor-Vokes R, Fox S, Gan Q, Johnson L, Wang K (2010) Wet-milling transgenic maize seed for fraction enrichment of recombinant subunit vaccine. Biotechnol Prog 26(2):458–465

Nandi S, Suzuki YA, Huang JM, Yalda D, Pham P, Wu LY, Bartley G, Huang N, Lonnerdal B (2002) Expression of human lactoferrin in transgenic rice grains for the application in infant formula. Plant Sci 163(4):713–722

Nandi S, Yalda D, Lu S, Nikolov Z, Misaki R, Fujiyama K, Huang N (2005) Process development and economic evaluation of recombinant human lactoferrin expressed in rice grain. Transgenic Res 14(3):237–249

Nikolov ZL, Hammes D (2002) Production of recombinant proteins from transgenic crops. In: Hood EE, Howard JA (eds) Plant as factories for protein production. Kluwer Academic Publishers, Dordrecht

Nikolov ZL, Woodard SL (2004) Downstream processing of recombinant proteins from transgenic feedstock. Curr Opin Biotechnol 15(5):479–486

Nikolov ZL, Regan JT, Dickey LF, Woodard SL (2009) Purification of antibodies from transgenic plants. In: Gottschalk U (ed) Process scale purification of antibodies. Wiley, Hoboken

Nykiforuk CL, Boothe JG, Murray EW, Keon RG, Goren HJ, Markley NA, Moloney MM (2006) Transgenic expression and recovery of biologically active recombinant human insulin from *Arabidopsis thaliana* seeds. Plant Biotechnol J 4(1):77–85

Nykiforuk CL, Shen Y, Murray EW, Boothe JG, Busseuil D, Rhéaume E, Tardif J-C, Reid A, Moloney MM (2011) Expression and recovery of biologically active recombinant Apolipoprotein AI_{Milano} from transgenic safflower (*Carthamus tinctorius*) seeds. Plant Biotechnol J 9(2):250–263

Padilla S, Valdés R, Gómez L, Geada D, Ferro W, Mendoza O, García C, Milá L, Pasín L, Issac Y, Gavilán D, González T, Sosa R, Leyva A, Sánchez J, LaO M, Calvo Y, Sánchez R, Fernández E, Brito J (2009) Assessment of a plantibody HB-01 purification strategy at different scales. Chromatographia 70(11):1673–1678

Paraman I, Fox SR, Aspelund MT, Glatz CE, Johnson LA (2010a) Recovering corn germ enriched in recombinant protein by wet-fractionation. Bioresour Technol 101(1):239–244

Paraman I, Moeller L, Scott MP, Wang K, Glatz CE, Johnson LA (2010b) Utilizing protein-lean coproducts from corn containing recombinant pharmaceutical proteins for ethanol production. J Agric Food Chem 58(19):10419–10425

Peckham GD, Bugos RC, Su WW (2006) Purification of GFP fusion proteins from transgenic plant cell cultures. Protein Expr Purif 49(2):183–189

Pirie NW (1987) Leaf protein and its by-products in human and animal nutrition, 2nd edn. Cambridge University Press, Cambridge

Platis D, Labrou NE (2006) Development of an aqueous two-phase partitioning system for fractionating therapeutic proteins from tobacco extract. J Chromatogr A 1128(1–2): 114–124

Platis D, Labrou NE (2009) Application of a PEG/salt aqueous two-phase partition system for the recovery of monoclonal antibodies from unclarified transgenic tobacco extract. Biotechnol J 4(9):1320–1327

Platis D, Drossard J, Fischer R, Ma JKC, Labrou NE (2008) New downstream processing strategy for the purification of monoclonal antibodies from transgenic tobacco plants. J Chromatogr A 1211(1–2):80–89

Pogue GP, Vojdani F, Palmer KE, Hiatt E, Hume S, Phelps J, Long L, Bohorova N, Kim D, Pauly M, Velasco J, Whaley K, Zeitlin L, Garger SJ, White E, Bai Y, Haydon H, Bratcher B (2010) Production of pharmaceutical-grade recombinant aprotinin and a monoclonal antibody product using plant-based transient expression systems. Plant Biotechnol J 8(5):638–654

Ramessar K, Rademacher T, Sack M, Stadlmann J, Platis D, Stiegler G, Labrou N, Altmann F, Ma J, Stoger E, Capell T, Christou P (2008) Cost-effective production of a vaginal protein microbicide to prevent HIV transmission. Proc Natl Acad Sci USA 105(10):3727–3732

Ramírez N, Oramas P, Ayala M, Rodríguez M, Pérez M, Gavilondo J (2001) Expression and long-term stability of a recombinant single-chain Fv antibody fragment in transgenic *Nicotiana tabacum* seeds. Biotechnol Lett 23(1):47–49
Robic G, Farinas CS, Rech EL, Miranda EA (2010) Transgenic soybean seed as protein expression system: aqueous extraction of recombinant beta-glucuronidase. Appl Biochem Biotechnol 160(4):1157–1167
Robić G, Farinas CS, Rech EL, Bueno SMA, Miranda EA (2006) Downstream process engineering evaluation of transgenic soybean seeds as host for recombinant protein production. Biochem Eng J 32(1):7–12
Robić G, Farinas C, Rech E, Miranda E (2010) Transgenic soybean seed as protein expression system: aqueous extraction of recombinant β-glucuronidase. Appl Biochem Biotechnol 160(4):1157–1167
Ross KC, Zhang C (2010) Separation of recombinant β-glucuronidase from transgenic tobacco by aqueous two-phase extraction. Biochem Eng J 49(3):343–350
Sala F, Manuela Rigano M, Barbante A, Basso B, Walmsley AM, Castiglione S (2003) Vaccine antigen production in transgenic plants: strategies, gene constructs and perspectives. Vaccine 21(7–8):803–808
Schillberg S, Twyman RM, Fischer R (2005) Opportunities for recombinant antigen and antibody expression in transgenic plants – technology assessment. Vaccine 23(15):1764–1769
Selle PH, Ravindran V, Caldwell A, Bryden WL (2000) Phytate and phytase: consequences for protein utilisation. Nutr Res Rev 13(02):255–278
Shaaltiel Y, Bartfeld D, Hashmueli S, Baum G, Brill-Almon E, Galili G, Dym O, Boldin-Adamsky SA, Silman I, Sussman JL, Futerman AH, Aviezer D (2007) Production of glucocerebrosidase with terminal mannose glycans for enzyme replacement therapy of Gaucher's disease using a plant cell system. Plant Biotechnol J 5(5):579–590
Sharma AK, Sharma MK (2009) Plants as bioreactors: recent developments and emerging opportunities. Biotechnol Adv 27(6):811–832
Shepherd CT, Vignaux N, Peterson JM, Johnson LA, Scott MP (2008a) Green fluorescent protein as a tissue marker in transgenic maize seed. Cereal Chem 85(2):188–195
Shepherd CT, Vignaux N, Peterson JM, Scott MP, Johnson LA (2008b) Dry-milling and fractionation of transgenic maize seed tissues with green fluorescent protein as a tissue marker. Cereal Chem 85(2):196–201
Sommerfeld S, Strube J (2005) Challenges in biotechnology production–generic processes and process optimization for monoclonal antibodies. Chem Eng Process 44(10):1123–1137
Spök A, Karner S (2008) Plant molecular farming: opportunities and challenges. European Commission: The Institute for Prospective Technological Studies, Seville
Spreitzer RJ, Salvucci ME (2002) RUBISCO: structure, regulatory interactions, and possibilities for a better enzyme. Annu Rev Plant Biol 53(1):449–475
Stöger E, Vaquero C, Torres E, Sack M, Nicholson L, Drossard J, Williams S, Keen D, Perrin Y, Christou P, Fischer R (2000) Cereal crops as viable production and storage systems for pharmaceutical scFv antibodies. Plant Mol Biol 42(4):583–590
Stöger E, Sack M, Perrin Y, Vaquero C, Torres E, Twyman RM, Christou P, Fischer R (2002) Practical considerations for pharmaceutical antibody production in different crop systems. Mol Breed 9(3):149–158
Stöger E, Ma JKC, Fischer R, Christou P (2005) Sowing the seeds of success: pharmaceutical proteins from plants. Curr Opin Biotechnol 16(2):167–173
Tran M, Zhou B, Pettersson PL, Gonzalez MJ, Mayfield SP (2009) Synthesis and assembly of a full-length human monoclonal antibody in algal chloroplasts. Biotechnol Bioeng 104(4):663–673
Tremblay R, Wang D, Jevnikar AM, Ma S (2010) Tobacco, a highy efficient green bioreactor for production of therapeutic proteins. Biotechnol Adv 28(2):214–221
Turpen TT, Garger SJ, McCulloch MJ, Cameron TI, Samonek-Potter ML, Holtz RB (2001) Method for recovering proteins from the interstitial fluid of plant tissues. US Patent 6,284,875

Twyman RM, Stoger E, Schillberg S, Christou P, Fischer R (2003) Molecular farming in plants: host systems and expression technology. Trends Biotechnol 21(12):570–578
Van Rooijen GJH, Motoney MM (1995) Plant seed oil-bodies as carriers for foreign proteins. Nat Biotechnol 13(1):72–77
Van Rooijen G, Glenn KR, Shen Y, Boothe J (2008) Commercial production of chymosin in plants. US Patent 7,390,936
Vaquero C, Sack M, Chandler J, Drossard J, Schuster F, Monecke M, Schillberg S, Fischer R (1999) Transient expression of a tumor-specific single-chain fragment and a chimeric antibody in tobacco leaves. Proc Natl Acad Sci USA 96(20):11128–11133
Vézina L-P, Faye L, Lerouge P, D'Aoust M-A, Marquet-Blouin E, Burel C, Lavoie P-O, Bardor M, Gomord V (2009) Transient co-expression for fast and high-yield production of antibodies with human-like N-glycans in plants. Plant Biotechnol J 7(5):442–455
Weathers PJ, Towler MJ, Xu JF (2010) Bench to batch: advances in plant cell culture for producing useful products. Appl Microbiol Biotechnol 85(5):1339–1351
Werner S, Marillonnet S, Hause G, Klimyuk V, Gleba Y (2006) Immunoabsorbent nanoparticles based on a tobamovirus displaying protein A. Proc Natl Acad Sci USA 103(47):17678–17683
Wilken LR, Nikolov ZL (2006) Factors influencing recombinant human lysozyme extraction and cation exchange adsorption. Biotechnol Prog 22(3):745–752
Wilken LR, Nikolov ZL (2010) Evaluation of alternatives for human lysozyme purification from transgenic rice: impact of phytic acid and buffer. Biotechnol Prog 26(5):1303–1311
Wilken LR, Nikolov ZL (2011) Recovery and purification of plant-made recombinant proteins. Biotech Adv doi:10.1016/j.biotechadv.2011.07.020
Wolf WJ, Sathe SK (1998) Ultracentrifugal and polyacrylamide gel electrophoretic studies of extractability and stability of almond meal proteins. J Sci Food Agric 78(4):511–521
Woodard SL, Mayor JM, Bailey MR, Barker DK, Love RT, Lane JR, Delaney DE, McComas-Wagner JM, Mallubhotla HD, Hood EE, Dangott LJ, Tichy SE, Howard JA (2003) Maize (*Zea mays*)-derived bovine trypsin: characterization of the first large-scale, commercial protein product from transgenic plants. Biotechnol Appl Biochem 38(2):123–130
Woodard SL, Wilken LR, Barros GOF, White SG, Nikolov ZL (2009) Evaluation of monoclonal antibody and phenolic extraction from transgenic *Lemna* for purification process development. Biotechnol Bioeng 104(3):562–571
Xu JF, Tan L, Goodrum KJ, Kieliszewski MJ (2007) High-yields and extended serum half-life of human interferon alpha 2b expressed in tobacco cells as arabinogalactan-protein fusions. Biotechnol Bioeng 97(5):997–1008
Xu J, Okada S, Tan L, Goodrum K, Kopchick J, Kieliszewski M (2010) Human growth hormone expressed in tobacco cells as an arabinogalactan-protein fusion glycoprotein has a prolonged serum life. Transgenic Res 19(5):849–867
Xu J, Ge X, Dolan MC (2011) Towards high-yield production of pharmaceutical proteins with plant cell suspension cultures. Biotechnol Adv 29(3):278–299
Yang LJ, Wakasa Y, Takaiwa F (2008) Biopharming to increase bioactive peptides in rice seed. J AOAC Int 91(4):957–964
Yu D, McLean MD, Hall JC, Ghosh R (2008) Purification of a human immunoglobulin G1 monoclonal antibody from transgenic tobacco using membrane chromatographic processes. J Chromatogr A 1187(1–2):128–137
Zaman F, Kusnadi AR, Glatz CE (1999) Strategies for recombinant protein recovery from canola by precipitation. Biotechnol Prog 15(3):488–492
Zambryski P, Joos H, Genetello C, Leemans J, Vanmontagu M, Schell J (1983) Ti-Plasmid vector for the introduction of DNA into plant-cells without alteration of their normal regeneration capacity. EMBO J 2(12):2143–2150
Zeitlin L, Olmsted SS, Moench TR, Co MS, Martinell BJ, Paradkar VM, Russell DR, Queen C, Cone RA, Whaley KJ (1998) A humanized monoclonal antibody produced in transgenic plants for immunoprotection of the vagina against genital herpes. Nat Biotechnol 16(13):1361–1364

Zhang CM, Glatz CE (1999) Process engineering strategy for recombinant protein recovery from canola by cation exchange chromatography. Biotechnol Prog 15(1):12–18

Zhang C, Lillie R, Cotter J, Vaughan D (2005) Lysozyme purification from tobacco extract by polyelectrolyte precipitation. J Chromatogr A 1069(1):107–112

Zhang C, Glatz CE, Fox SR, Johnson LA (2009a) Fractionation of transgenic corn seed by dry and wet milling to recover recombinant collagen-related proteins. Biotechnol Prog 25(5):1396–1401

Zhang C, Baez J, Pappu KM, Glatz CE (2009b) Purification and characterization of a transgenic corn grain-derived recombinant collagen type I alpha 1. Biotechnol Prog 25(6):1660–1668

Zhang D, Nandi S, Bryan P, Pettit S, Nguyen D, Santos MA, Huang N (2010) Expression, purification, and characterization of recombinant human transferrin from rice (*Oryza sativa* L.). Protein Expr Purif 74(1):69–79

Zhong Q, Xu L, Zhang C, Glatz C (2007) Purification of recombinant aprotinin from transgenic corn germ fraction using ion exchange and hydrophobic interaction chromatography. Appl Microbiol Biotechnol 76(3):607–613

Zhu Z, Hughes KW, Huang L, Sun B, Liu C, Li Y (1994) Expression of human α-interferon cDNA in transgenic rice plants. Plant Cell Tiss Org 36(2):197–204

Chapter 12
Biosafety of Molecular Farming in Genetically Modified Plants

Didier Breyer, Adinda De Schrijver, Martine Goossens, Katia Pauwels, and Philippe Herman

Abstract The use of genetically modified plants for large-scale production of recombinant compounds for pharmaceutical or industrial use, known as plant molecular farming, holds several promises. However, any development in this field must be counterbalanced by a thorough evaluation of risks to human health and the environment. The possible impact of accidental contamination of the food and feed chain or of transgene spread in the environment, in particular when major food/feed crops grown in open fields are involved, highlights the need to carefully address some important issues during the safety assessment of genetically modified farming plants, such as the choice of the production platform, the implementation of containment or confinement measures and the adoption of other relevant management strategies. In this chapter, we report on the applicability of the current risk assessment methodology and principles, outline some important issues linked to the assessment of environmental and health risks, and comment in more detail general management strategies that could be applied to limit potential environmental and human health impacts linked to plant molecular farming.

12.1 Introduction

Plant molecular farming (PMF) offers attractive perspectives to produce compounds for pharmaceutical or industrial purposes on a large scale at low costs. Benefits associated with the use of plants as production platforms also include rapid scaling up, convenient storage of raw material and less concern of contamination with human

---

D. Breyer (✉) • A. De Schrijver • M. Goossens • K. Pauwels • P. Herman
Biosafety and Biotechnology Unit, Scientific Institute of Public Health,
Rue J. Wytsmanstraat 14, B-1050 Brussels, Belgium
e-mail: didier.breyer@wiv-isp.be

A. Wang and S. Ma (eds.), *Molecular Farming in Plants: Recent Advances and Future Prospects*, DOI 10.1007/978-94-007-2217-0_12,

or animal pathogens during downstream processing. Worldwide PMF has enabled the development of thousands of recombinant biotechnology products and some of them are expected to be marketed soon (see *e.g.* Basaran and Rodríguez-Cerezo 2008; Ahmad et al. 2010).

As with current genetically modified (GM) plants, all plants engineered to produce pharmaceutical or industrial compounds must go through a thorough health and environmental risk assessment before they can be used. The risk assessment of GM farming plants is currently supported in most countries by the same procedures and policies that are used for first-generation GM plants (see *e.g.* Spök et al. 2008; Breyer et al. 2009).

In the European Union (EU), the legislation applicable for the risk assessment of GM farming plants cultivated in open fields is Directive 2001/18/EC on the environmental release of GMOs (EC 2001). When the GM product is a pharmaceutical, the safety aspects of the product and its uses are also assessed according to Regulation (EC) No 726/2004 on medicinal products (EC 2004) under the coordination of the European Medicines Agency (EMA, formerly EMEA). The provisions of Regulation (EC) No 1829/2003 on genetically modified food and feed (EC 2003) need to be followed when residual biomass derived from GM farming plants are used as food/feed, and could also be relevant when GM farming plants or their derived products may enter the food/feed chain. In this case, the European Food Safety Authority (EFSA) plays a central role in the risk assessment, in close collaboration with the Member States. Under these three regulatory frameworks, authorizations for commercialization have to be granted at the EU level, involving all Member States and the European Commission (Fig. 12.1).

For activities involving GM microorganisms and conducted under strict containment (e.g. greenhouses or laboratories) Directive 2009/41/EC (EC 2009) applies. Contained use is under the regulatory oversight of each particular EU Member State. It must be noted that in most of the Member States, the scope of activities falling under this regulatory framework has been broadened to GM organisms and therefore also covers the contained use of GM farming plants.

In the United States (US), the institutional structure established for regulating biotechnology products also oversees the risk assessment of GM farming plants. The Animal and Plant Health Inspection Service (APHIS) of the US Department of Agriculture (USDA) regulates GM plant trials. The Food and Drug Administration (FDA) is responsible for regulating pharmacological and safety aspects when the end product is a pharmaceutical. It should be noted that, since 2003, APHIS regulatory requirements have been strengthened with regard to the cultivation of GM plants producing pharmaceutical or industrial compounds. APHIS requires for these GMOs a more constraining "permit" procedure with specific confinement measures and procedures to verify compliance (Federal Register Notice 2003; NARA 2005).

Specific issues associated with the risk assessment of PMF have been discussed in several national and international regulatory bodies resulting in the development of guidance, standards and procedures. In the EU specific aspects related to PMF are being addressed in a guidance document for the risk assessment of GM plants used for non-food or non-feed purposes (EFSA 2009). This guidance document

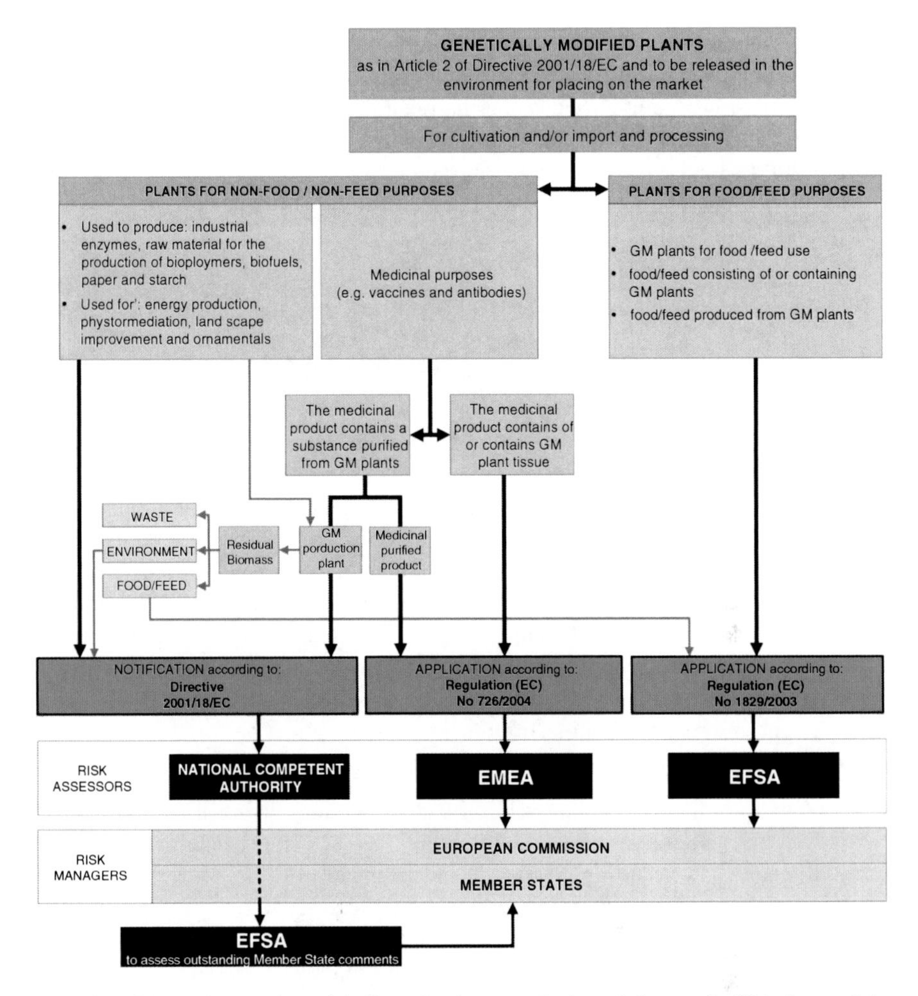

Fig. 12.1 Schematic overview of the interplay between the intended uses of a GM plant and the respective EU legislation applicable. The flowchart also gives an overview of the regulatory bodies that are involved in scientific risk assessment and the ones that are responsible for risk management and authorization decisions (Source: EFSA 2009)

supplements the more general EFSA guidelines for the risk assessment of GM plants and derived food and feed (EFSA 2006, 2010). The US authorities have issued specific guidance to cover the risks associated with PMF (FDA 2002; USDA 2008). These documents inform on elements to consider when addressing containment (of a facility such as a laboratory or greenhouse or during movement), confinement (of the field test site), and environmental issues. The Canadian Food Inspection Agency (CFIA) has also developed several additional rules, terms and conditions to address the environmental and human and livestock health concerns associated with the use of GM farming plants (CFIA 2004a).

Safety issues associated with PMF and possible mechanisms to limit the risks have also been addressed in research programmes such as the project *Pharma-Planta* funded by the EU (see http://www.pharma-planta.org/), and in several reports and reviews in scientific literature including Commandeur et al. (2003), Mascia and Flavell (2004), Liénard et al. (2007), Murphy (2007), Sparrow et al. (2007), Wolt et al. (2007), Spök and Karner (2008), Breyer et al. (2009), Sparrow and Twyman (2009) and Obembe et al. (2010).

12.2 Biosafety and Risk Assessment Methodology

There is no generic or internationally accepted definition of "biosafety" but for the purpose of this chapter, it can be defined as the safety for human health and the environment associated with the use of GM farming plants. Biosafety is determined as a result of a risk assessment process. The objective of risk assessment is to identify and evaluate on a case-by-case basis potential adverse effects of a GM plant on the receiving environment(s) and human health. Risk assessment is based on a multi-step methodology and a comparative approach. Through this approach, the GM plant is compared with its non-GM counterpart which is considered having a history of safe use for the average consumer or animals and familiarity for the environment, in order to identify differences. Assessment is performed principally according to the following steps (SCBD 2000; EC 2001; EFSA 2010):

1. Problem formulation, including hazard identification. This step considers the biological characteristics of the recipient and donor organism, the genetic modification, the resulting GM plant, the intended use, the potential receiving environment and the interactions between those. It takes into account the differences identified between the GM and its non-GM comparator. Relevant differences potentially leading to harm are determined and subsequently assessed in the next steps for their potential adverse effect(s) on human health and the environment (Codex 2003; EFSA 2009). Problem formulation also includes the identification of exposure pathways through which the GM plant may interact with human health or the environment and of the corresponding levels of exposure. Problem formulation further aims at explicitly stating the assumptions (context and scope) underlying the risk assessment, taking into account the protection goals set out by existing policies;
2. Hazard characterization. Through this step, the potential consequences of each hazard identified in the first step (provided that it does occur) are evaluated, by defining the magnitude (qualitative and/or quantitative) of the associated health or environmental harm;
3. Exposure assessment. This step aims at estimating (qualitatively or when possible quantitatively) the exposure (likelihood of the occurrence) for each hazard identified and characterized;

4. Risk characterization. During this step, the risk posed by each hazard is estimated by combining the magnitude of consequences of each hazard and the likelihood of its occurrence. This step will also result in the identification of areas of uncertainties;
5. Identification of risk management strategies. This step aims at reducing risks identified in the previous step (in most cases by minimizing the likelihood of adverse effects occurring) while at the same time considering areas of uncertainties;
6. Overall risk evaluation and conclusions. This step evaluates the overall risk of the GM plant taking into account the proposed risk management strategies. Where such strategies are applied, their reliability and efficacy need to be assessed. In this way, the effect of failure of these strategies can be estimated (EFSA 2009). The risk assessment finally leads to a conclusion as to whether the overall health and environmental impact of the GM plant can be accepted or not.

12.3 Considerations for the Risk Assessment of GM Farming Plants

The risk assessment methodology described above seems appropriate and robust enough to support the evaluation of most applications of PMF (Peterson and Arntzen 2004; EFSA 2009). However, we highlight hereafter some specific issues related to risk assessment that will deserve special attention when performing safety evaluation of GM farming plants (Spök 2007; Rehbinder et al. 2009).

Firstly, the applicability of the comparative approach principle could be challenged due to the difficulty to find appropriate non-GM comparators when plants having less or no history of safe use or little familiarity for the environment are involved. This approach could also be challenged when several extensive genetic modifications are carried out at once in the GM farming plant to obtain the intended property and/or biological confinement. This may lead to substantial – intended or unintended – changes in the original metabolism and composition of the GM plant, thereby making the comparative analysis more difficult (EFSA 2009).

Secondly, the biological activity of the compound(s) produced will need to be considered carefully in the risk assessment of GM farming plants. This is particularly relevant for pharmaceutical plants as they are used specifically to produce substances that have an effect on humans or animals, very often at low concentrations. As a result, small amounts of pharmaceutical compounds may harm people or animals that would accidentally consume GM farming plants or their products (Shama and Peterson 2008a). Given the unintentional character of oral intake, assessment will focus on the impacts of acute and/or short term exposure (EFSA 2009). It should also be noted that most of the plant-made pharmaceutical products currently in the pipeline are not anticipated to have any pharmacological activity when ingested or are expected to be rapidly degraded to innocuous peptides or amino acids upon accidental ingestion (Goldstein and Thomas 2004).

Thirdly, the risk assessment of GM farming plants will also need to consider the potential for GM farming plants, plant parts or products to inadvertently enter the food or feed chain, thereby exposing consumers or livestock to potentially toxic compounds. This potential will largely depend on the characteristics of the production platform (see Sect. 12.4.1). Accidental intake by humans or livestock could arise from unintentional admixture of GM farming plant products with food or feed products or from transfer of transgenes from molecular farming plants to non-transgenic crops due to gene flow. The Prodigene incident in the US, where conventional soybean for human consumption had been accidentally mixed with Prodigene's GM maize seeds genetically engineered to produce trypsin, gives an indication of the potential risks of inadvertent contamination of the food/feed chain, in particular if more pharmaceutical/industrial crops are grown in open field (Fox 2003).

Besides the potential for oral intake, the potential dermal, ocular or inhalatory exposure of people participating in the production and processing of the GM farming plant will need to be considered (Wolt et al. 2007; EFSA 2009). In the case of farming plants, risk for occupational toxicity or allergenicity relates not only to the pharmaceutical or industrial product but also to the plant itself (tobacco, for instance, is well known to produce toxins).

Fourthly, when GM farming plants are grown in open field, risk assessment needs to address the potential environmental impacts. This would include on the one hand the impact of the bio-active recombinant compound synthesized by the GM farming plant on animals (such as mammals, birds, insects) or microorganisms in the environment (Shama and Peterson 2008b) and on the other hand concerns related to changes in the persistence or invasiveness of the GM farming plant. It must be noted that similar impacts could arise from compatible wild relatives in the environment that would capture and express the transgene due to gene flow.

Last but not least, the final use of residual biomass must be carefully assessed and documented, taking into account relevant Good Manufacturing Practices (*e.g.* EC 2006; WHO 2003). If such material is handled as waste, appropriate measures will be taken in order to ensure that the material will not enter the food or feed chain or be released in the environment. If the remaining biomass is used in the environment (e.g. as fertilizer) or as food or feed (e.g. starch produced from GM potato tubers), as proposed by some companies, this approach must be assessed for the additional risk it could pose to the environment and food/feed safety.

12.4 Interplay Between Risk Assessment and Risk Management

When risks have been identified, relevant protective measures need to be implemented in order to minimize the likelihood of adverse effects occurring. In the context of GM farming plants, an important issue to be considered is how far unintended commingling with conventional food or feed is acceptable. In most countries, like in Canada and in the EU, no precise thresholds have been defined. In the US, there is a strict requirement that plants grown for pharmaceutical or industrial compounds (and not approved for food and feed use) must stay clear of the food system under a zero-tolerance standard (USDA 2006). Depending on the contamination level set up as protection goal, several (not mutually exclusive) risk management options can be envisaged.

Some of them clearly fall within the competences of decision-makers, such as the complete ban of farming applications involving the use of food/feed crops (Union of Concerned Scientists 2006; Murphy 2007), the mandatory pre-marketing approval of the GM farming plant and products under the food and feed regulation (Becker and Vogt 2005), or the adoption of threshold limits for adventitious or technically unavoidable presence of molecular farming products in non-GM products at levels low enough that risks are minimal (Moschini 2006; Spök 2007).

Other risk management strategies will aim directly at reducing hazard and/or exposure identified during risk assessment. These include the choice of the production host, the application of physical containment and/or biological confinement strategies or the implementation of wide-ranging monitoring systems, and are described below.

12.4.1 Characteristics of the Host Plant

The choice of the host plant in which the recombinant compounds are produced has critical implications. On the one hand, the production strategy needs to comply with technical factors such as the required level of expression, the purification process or the quality of the end product (see e.g. Vancanneyt et al. 2009). On the other hand, from a biosafety viewpoint, the host plant should be chosen taking into account the potential for and impact of exposure of the environment or the food/feed chain. The choice of the host will therefore be considered on a case-by-case basis, taking into account the potential impact of all aspects of the manufacturing process, including cultivation, harvest, transport, processing, purification, packaging, storage and disposal. Table 12.1 summarizes some of the main potential advantages and disadvantages for different host categories (Breyer et al. 2009).

The use of food/feed crops as production platforms for pharmaceuticals or industrial compounds is a controversial issue. As shown in Table 12.1, there are several arguments in favor of using food/feed crops for PMF (Streatfield et al. 2003; Hennegan et al. 2005; Sparrow et al. 2007; Ramessar et al. 2008). However, as mentioned before, there are also concerns about the risks such GM crops would pose in case they would inadvertently enter the food/feed chain.

It seems obvious that selecting non-food or non-feed crop plants or even non-crop plants for molecular farming would provide containment advantages. The use of such plants would reduce the possibility of admixture in the food/feed chains. Canadian authorities are explicitly recommending the use of non-food/non-feed crop species for the production of pharmaceuticals in plants (CFIA 2004b). Among this category, tobacco is a very efficient production system in which a wide range of

Table 12.1 Overview of host systems for plant molecular farming (Adapted from Breyer et al. 2009)

	Food/feed plants	Non-food/non-feed plants	Non-crop plants	Plant cells in culture
Examples	Maize, rice, potato, soybean, oilseed rape, banana, tomato	Tobacco	Duckweed, mosses, *Arabidopsis*, microalgae	Tomato, tobacco, carrot, rice
Main advantages	Good knowledge of genetics, biology and cultivation practices In most cases, efficient transformation procedures History of safe use Options for tissue-specific expression	Not part of the food/feed chain In some cases, good knowledge of genetics, biology and cultivation practices. In some cases, efficient transformation procedures	Not part of the food/feed chain Some of these plants can easily be grown in containment	Not part of the food/feed chain Propagated under containment In some cases, easier recovery and purification of the product Easier maintenance of quality standards
Main disadvantages	Part of the food/feed chain Risk of gene transfer to cross-compatible crop and wild relatives	In some cases, production of naturally occurring toxins that could have an impact on safety	In some cases, less knowledge of genetics and biology Very often, little experience with propagation	Technical drawbacks (scaling-up…) Higher costs to maintain plant cells in culture

pharmaceutical products is currently being tested for production. Non-crop plants such as the monocot *Lemna* (duckweed), *Arabidopsis*, microalgae or mosses are also tested as production platform. The latter species offer the possibility of being grown easily in contained facilities which gives many advantages in terms of biosafety (see below). However, the use of non-food, non-feed or non-crop plants will in some cases be difficult and might challenge the risk assessment due to the lack of knowledge about the genetics and biology, the lack of history of safe use and/or the limited experience with cultivation (Murphy 2007; Sparrow et al. 2007).

The safety of PMF can even be made greater by producing the recombinant compound in cell cultures of transgenic plants (Plasson et al. 2009). From a biosafety viewpoint, the main advantage is that cell cultures can be grown under contained conditions in bioreactors, avoiding risks associated with gene flow in the environment and reducing potential contamination of the food/feed chain. In some cases, the recombinant compound can even be secreted directly into the culture medium, allowing easier recovery and purification of the product and reducing further the possibility of product contamination with cell debris. Despite the fact that the adoption of cell culture technology has made progress during the last years, this strategy remains limited for the time being to a small number of well-characterized plant cell lines (see Table 12.1) and still needs improvement before it can be used in routine on a commercial scale.

12.4.2 Containment and Confinement Measures

Several physical containment or biological confinement methods are available to limit food/feed chain contamination or environmental impact of PMF. The application of such methods will be decided on a case-by-case basis with the aim to reduce hazard and/or exposure identified during the risk assessment and/or to consider areas of uncertainties. One should also take into account that these methods present different levels of effectiveness, that many of them are not mutually exclusive and that probably none of them will be able to achieve full protection of the environment or a zero level contamination of the food/feed chain.

12.4.2.1 Physical Containment

Physical containment of GM farming plants is a first strategy that can help avoiding contamination of the environment or of the food/feed chain. Potato, tobacco and other leafy crops such as alfalfa, lettuce and spinach are examples of plants that can be grown in contained facilities such as plastic tunnels, greenhouses, growth cabinets, or even large-scale underground facilities such as mines. Production of pharmaceutical or industrial compounds in contained systems also applies to plant cell suspensions, hairy root cultures, microalgae, *lemnaceae* or mosses (Franconi et al. 2010).

In the case of open field cultivation, local physical containment can also be applied, such as physical removal or bagging of flowers to prevent pollen release. In addition, spatial containment can also be implemented to minimize gene flow from GM farming plants to cross-compatible crop species or wild relatives. It includes different approaches such as: the implementation of minimum isolation distances ("buffer zones") between fields of GM farming plants and fields of the same species intended for food, feed or seed production; the planting of a border of non-transgenic "trap" plants around the field to capture the pollen emitted by the GM farming plants; the growing of GM farming plants in geographical areas where neither cross-compatible food/feed crops nor their wild relatives are present; or the restricted planting of GM farming plants for a defined number of growing seasons before the field can be used for production of food/feed crops (FDA 2002; Howard and Hood 2007).

Temporal containment is another strategy consisting in planting and harvesting GM farming crops at different periods than food/feed crops to make sure they flower at different times, which decreases the potential for pollen transfer (Spök 2007). This option could however be difficult to implement in practice because of the difficulty to control environmental factors influencing the timing of flowering in plants.

12.4.2.2 Biological Confinement

Biological confinement strategies are based on many different principles (Daniell 2002; Dunwell and Ford 2005; de Maagd and Boutilier 2009; Ahmad et al. 2010; Hüsken et al. 2010) and new technologies are still being developed and explored (see *e.g.* the Transcontainer project – http://www.transcontainer.org/UK/). It is important to realize that most of the biological confinement mechanisms described below are far from being used for commercial production.

Plastid transformation consists in inserting the transgene into the plant chloroplast genome instead of the plant nuclear genome. Given the maternal inheritance of plastids and their genomes, this strategy has great interest at biosafety level by preventing the risk of transfer of the transgene to other species through pollen (Ruf et al. 2007; Verma and Daniell 2007; Meyers et al. 2010). However, chloroplast genetic engineering remains to be achieved in several major crop species and does not always offer complete confinement.

Male sterility is another option for preventing gene flow through pollen and can be achieved through a great variety of approaches (*e.g.* inhibiting pollen formation or killing off the cells that are involved in the development of the male flower). Very few of these approaches have however been extensively tested to date in the field for their efficacy, apart from the Barstar/Barnase system (Kobayashi et al. 2006).

Other biological confinement strategies being pursued include apomixis (asexual reproduction through seed), cleistogamy (self-fertilization before flower opening), genomic incompatibility (placing the transgene on a genome of a polyploid plant species that is not compatible with related wild species), transgene removal from pollen using a site-specific recombination system (Moon and Stewart 2010), transgenic mitigation which consists in the inclusion of a transgene that confers competitive disadvantage to hybrids or volunteers (Gressel and Valverde 2009), or auxotrophy (inability of the plant to synthesize a particular organic compound required for normal growth). Some of these mechanisms are however complex in nature and many of the genes controlling them have still to be identified or characterized.

Production of sterile seeds (also known as the "terminator" technology) is the only strategy at present aiming at preventing transgene movement through seed (Hills et al. 2007). The original concept, involving inducible expression of a seed-lethal gene, was developed initially as trait protection for seed companies. But it is also envisaged as a mean to engineer sterility into farming plants. However, due to the limited information published on this approach, its reliability still needs to be proven particularly in field experiments.

Almost all of the abovementioned confinement strategies aim at restricting spread of transgenes by targeting flowering, pollen production, seed production, fertility or a combination of those. Auxotrophy is the only strategy targeting vegetative reproduction.

12.4.2.3 Targeted Expression and Temporal Confinement

The unintended exposure to a pharmaceutical or industrial compound can be reduced by restricting its production to a few specific plant parts such as roots, leaves, fruits or edible parts, or subcellular compartments. Expression in seeds has also been presented as a promising strategy (Lau and Sun 2009). Such targeted expression is generally achieved by the use of tissue specific promoters. From a biosafety perspective, production in plant parts that can be efficiently harvested can limit environmental exposure to the molecular farming compound.

Temporal confinement strategies can also be applied such as the post-harvest inducible expression. In this case, the molecular farming compound is not produced at all in the plants in the field, but will only be formed when the plant material is harvested and exposed to a chemical or environmental trigger that permits expression of the transgene (Corrado and Karali 2009).

Another form of temporal confinement, which is applied in contained environments, consists in transient expression systems. These are used more and more extensively by companies and allow fast and high level production of vaccines or other products. They include the agroinfiltration and virus infection methods (Komarova et al. 2010; Pogue et al. 2010). From a biosafety viewpoint, these techniques have the advantage that the transgene is only present temporarily in the plant cells and cannot be inherited by the next generation.

12.4.3 Other Risk Management Measures

Given that confinement- or containment-oriented risk management measures cannot always guarantee full protection against gene flow and possible admixture of GM farming plants with the food/feed chain, application of other risk management measures can be envisaged on a case-by-case basis.

The implementation of a production management system supported by appropriate protocols is a first example (Spök et al. 2008). Such system is likely to be based on or complement the production protocols designed to maintain a high level of batch to batch quality and prevent contamination of the plant-made product during all stages of production. Management measures that could contribute to mitigate the potential contamination of the food/feed chain include the cleaning of equipment and storage facilities, the use of specific equipment and even of specific facilities, the strict control over the inventory and disposition of viable material, or the clear labeling of containers of harvested material (indicating that the material is not to be used for food or feed purposes). If the molecular farming plant is grown in the field, monitoring of the production site, supported by an appropriate inspection plan, will be required as for other GM plants (see *e.g.* EC 2001).

In addition, post-marketing management measures could be implemented to check the efficiency of the confinement and containment strategies and/or to allow detection and identification of the farming plants in case of accidental admixture in food or feed. The use of molecular tools has been proposed in that respect. For instance, GM farming plants could be tagged with a specific DNA sequence identifier, preferably being devoid of an open reading frame, to allow straightforward screening of their appearance in food and feed (EFSA 2009; Alderborn et al. 2010). The introduction of morphological genetic markers, through which the GM plant is made visually distinctive from its food or feed counterpart, is another strategy that could help in the identification and traceability of GM farming plants (FDA 2002; Commandeur et al. 2003). It should be stressed that such measures would imply adding new heterologous genes in the plant genome which could make the risk assessment of such GMOs even more complex.

12.5 Observations and Conclusions

During the past years, plant-based expression systems have emerged as a powerful tool in the production of medicinal and industrial products. In addition to solving technical drawbacks, the success of plant molecular farming rests also on addressing appropriately biosafety issues associated with the technology.

Agricultural-scale cultivation of GM farming plants faces strong reluctance among regulators and the general public. Over the past decade, there have been several cases of unexpected contamination of the food/feed chain by transgenic crops to question the strategy of using major food/feed crops as production vehicles in plant molecular farming. To address these concerns, alternative strategies are intensively developed involving non-food, non-feed crops or cell cultures, the implementation of physical containment and/or biological confinement methods, and the adoption of strict management practices.

We have shown that any development in this field should take into account the results of a prior case-by-case risk assessment, framed according to defined protection goals. In any case, identification and assessment of biological risks will be achieved only if an appropriate regulatory framework supported by relevant guidance is in place.

References

Ahmad A, Pereira EO, Conley AJ, Richman AS, Menassa R (2010) Green biofactories: recombinant protein production in plants. Recent Pat Biotechnol 4(3):242–259

Alderborn A, Sundström J, Soeria-Atmadja D, Sandberg M, Andersson HC, Hammerling U (2010) Genetically modified plants for non-food or non-feed purposes: straightforward screening for their appearance in food and feed. Food Chem Toxicol 48(2):453–464

Basaran P, Rodríguez-Cerezo E (2008) Plant molecular farming: opportunities and challenges. Crit Rev Biotechnol 28(3):153–172

Becker G, Vogt D (2005) Regulation of plant-based pharmaceuticals. CRS report for congress. Order Code RS21418, March 8

Breyer D, Goossens M, Herman P, Sneyers M (2009) Biosafety considerations associated with molecular farming in genetically modified plants. J Med Plant Res 3(11):825–838

CFIA (2004a) Directive 94–08. Assessment criteria for determining environmental safety of plants with novel traits. http://www.inspection.gc.ca/english/plaveg/bio/dir/dir9408e.shtml. Accessed 28 Feb 2011

CFIA (2004b) Directive Dir2000-07: conducting confined research field trials of plant with novel traits in Canada. http://www.inspection.gc.ca/english/plaveg/bio/dir/dir0007e.shtml. Accessed 28 Feb 2011

Codex (2003) Codex Alimentarius Commission 2003 (ALINORM 03/34A). Guideline for the conduct of food safety assessment of foods derived from recombinant DNA plants. Annex on the assessment of possible allergenicity, Rome, Italy. Codex Alimentarius Commission, Yokohama

Commandeur U, Twyman RM, Fischer R (2003) The biosafety of molecular farming in plants. AgBiotechNet 5:1–9

Corrado G, Karali M (2009) Inducible gene expression systems and plant biotechnology. Biotechnol Adv 27:733–743

Daniell H (2002) Molecular strategies for gene containment in transgenic corps. Nat Biotechnol 20:581–586

de Maagd RA, Boutilier K (2009) Efficacy of strategies for biological containment of transgenic crops. Plant Research International B.V., Wageningen

Dunwell JM, Ford CS (2005) Technologies for biological containment of GM and non-GM crops. Defra contract CPEC 47. Final report. http://www.gmo-safety.eu/pdf/biosafenet/Defra_2005.pdf. Accessed 28 Feb 2011

EC (2001) Council Directive 2001/18/EC of 12 March 2001 on the deliberate release into the environment of genetically modified organisms and repealing Council Directive 90/220/EEC. Off J Eur Union L106:1–38

EC (2003) Regulation (EC) No 1829/2003 of the European Parliament and of the Council of 22 September 2003 on genetically modified food and feed. Off J Eur Union L268:1–23

EC (2004) Regulation (EC) No 726/2004 of the European Parliament and of the Council of 31 March 2004 laying down Community procedures for the authorisation and supervision of medicinal products for human and veterinary use and establishing a European Medicines Agency. Off J Eur Union L136:1–33

EC (2006) Commission Regulation (EC) No 2023/2006 of 22 December 2006 on good manufacturing practice for materials and articles intended to come into contact with food. Off J Eur Union L384:75–78

EC (2009) Directive 2009/41/EC of the European Parliament and of the Council of 6 May 2009 on the contained use of genetically modified micro-organisms. Off J Eur Union L125:75–97

EFSA (2006) Guidance document of the scientific panel on genetically modified organisms for the risk assessment of genetically modified plants and derived food and feed. EFSA J 99:1–100

EFSA (2009) EFSA scientific panel on Genetically Modified Organisms (GMO); Scientific opinion on guidance for the risk assessment of genetically modified plants used for non-food or non-feed purposes, on request of EFSA. EFSA Journal 1164. [42 pp.]. Available online http://www.efsa.europa.eu

EFSA (2010) EFSA scientific panel on Genetically Modified Organisms (GMO); guidance on the environmental risk assessment of genetically modified plants. EFSA Journal 8(11):1879. [111 pp.]. doi:10.2903/j.efsa.2010.1879. Available online http://www.efsa.europa.eu/efsajournal.htm

FDA (2002) Draft guidance for industry. Drugs, biologics, and medical devices derived from bioengineered plants for use in humans and animals. http://www.fda.gov/downloads/Drugs/GuidanceComplianceRegulatoryInformation/Guidances/ucm124811.pdf. Accessed 28 Feb 2011

Federal Register Notice (2003) 68 FR 11337–11340. http://www.aphis.usda.gov/brs/pdf/7cfr.pdf. Accessed 28 Feb 2011

Fox JL (2003) Puzzling industry response to prodigene fiasco. Nat Biotechnol 21:3–4

Franconi R, Costantina O, Demurtas OC, Massa S (2010) Plant-derived vaccines and other therapeutics produced in contained systems. Expert Rev Vaccines 9(8):877–892

Goldstein DA, Thomas JA (2004) Biopharmaceuticals derived from genetically modified plants. QJM 97:705–716

Gressel J, Valverde BE (2009) A strategy to provide long-term control of weedy rice while mitigating herbicide resistance transgene flow, and its potential use for other crops with related weeds. Pest Manag Sci 65:723–731

Hennegan K, Yang DC, Nguyen D, Wu LY, Goding J, Huang JM, Guo FL, Huang N, Watkins S (2005) Improvement of human lysozyme expression in transgenic rice grain by combining wheat (*Triticum aestivum*) puroindoline b and rice (*Oryza sativa*) Gt1 promoters and signal peptides. Transgenic Res 14:583–592

Hills MJ, Hall L, Arnison PG, Good AG (2007) Genetic use restriction technologies (GURTs): strategies to impede transgene movement. Trends Plant Sci 12:177–183

Howard JA, Hood EE (2007) Methods for growing nonfood products in transgenic plants. Crop Sci 47:1255–1262

Hüsken A, Prescher S, Schiemann J (2010) Evaluating biological containment strategies for pollen-mediated gene flow. Environ Biosafety Res. doi:10.1051/ebr/2010009

Kobayashi K, Munemura I, Hinata K, Yamamura S (2006) Bisexual sterility conferred by the differential expression of barnase and barstar: a simple and efficient method of transgene containment. Plant Cell Rep 25:1347–1354

Komarova TV, Baschieri S, Donini M, Marusic C, Benvenuto E, Dorokhov YL (2010) Transient expression systems for plant-derived biopharmaceuticals. Expert Rev Vaccines 9(8):859–876

Lau OS, Sun SS (2009) Plant seeds as bioreactors for recombinant protein production. Biotechnol Adv 27(6):1015–1022

Liénard D, Sourrouille C, Gomord V, Faye L (2007) Pharming and transgenic plants. Biotechnol Annu Rev 13:115–147

Mascia PN, Flavell RB (2004) Safe and acceptable strategies for producing foreign molecules in plants. Curr Opin Plant Biol 7:189–195

Meyers B, Zaltsman A, Lacroix B, Kozlovsky SV, Krichevsky A (2010) Nuclear and plastid genetic engineering of plants: comparison of opportunities and challenges. Biotechnol Adv 28:747–756. doi:10.1016/j.biotechadv.2010.05.022

Moon HS, Li Y, Stewart CN Jr (2010) Keeping the genie in the bottle: transgene biocontainment by excision in pollen. Trends Biotechnol 28(1):3–8

Moschini G (2006) Pharmaceutical and industrial traits in genetically modified crops: co-existence with conventional agriculture. Working paper 06-WP 429. Center for Agricultural and Rural Development

Murphy DJ (2007) Improving containment strategies in biopharming. Plant Biotechnol J 5:555–569

NARA (2005) Introductions of plants genetically engineered to produce industrial compounds. Fed Regist USA 70(85):23009–23011

Obembe OO, Popoola JO, Leelavathi S, Reddy SV (2010) Advances in plant molecular farming. Biotechnol Adv. doi:10.1016/j.biotechadv.2010.11.004

Peterson RKD, Arntzen CJ (2004) On risk and plant-based biopharmaceuticals. Trends Biotechnol 22:64–66

Plasson C, Michel R, Lienard D, Saint-Jore-Dupas C, Sourrouille C, de March GG, Gomord V (2009) Production of recombinant proteins in suspension-cultured plant cells. Methods Mol Biol 483:145–161

Pogue GP, Vojdani F, Palmer KE, Hiatt E et al (2010) Production of pharmaceutical-grade recombinant aprotinin and a monoclonal antibody product using plant-based transient expression systems. Plant Biotechnol J 8(5):638–654

Ramessar K, Sabalza M, Capell T, Christou P (2008) Maize plants: an ideal production platform for effective and safe molecular pharming. Plant Sci 174:409–419

Rehbinder E, Engelhard M, Hagen K, Jørgensen RB, Pardo Avellaneda R, Schnieke A, Thiele F (2009) Pharming. Promises and risks of biopharmaceuticals derived from genetically modified plants and animals. Ethics of science and technology assessment, Vol 35. Springer, p 334. ISBN: 978-3-540-85792-1

Ruf S, Karcher D, Bock R (2007) Determining the transgene containment level provided by chloroplast transformation. Proc Natl Acad Sci USA 104(17):6998–7002

SCBD (2000) Secretariat of the convention on biological diversity. Cartagena protocol on biosafety to the convention on biological diversity: text and annexes. Secretariat of the Convention on Biological Diversity, Montreal

Shama LM, Peterson RKD (2008a) Assessing risks of plant-based pharmaceuticals: I. Human dietary exposure. Hum Ecol Risk Assess 14:179–193

Shama LM, Peterson RKD (2008b) Assessing risks of plant-based pharmaceuticals: II. Non-target organism exposure. Hum Ecol Risk Assess 14:194–204

Sparrow PA, Twyman RM (2009) Biosafety, risk assessment and regulation of plant-made pharmaceuticals. Methods Mol Biol 483:341–353

Sparrow PA, Irwin JA, Dale PJ, Twyman RM, Ma JK (2007) Pharma-Planta: road testing the developing regulatory guidelines for plant-made pharmaceuticals. Transgenic Res 16(2):147–161

Spök A (2007) Molecular farming on the rise – GMO regulators still walking a tightrope. Trends Biotechnol 25:74–82

Spök A, Karner S (2008) Plant molecular farming: opportunities and challenges. European commission, joint research centre, institute for prospective technological studies

Spök A, Twyman RM, Fischer R, Ma JKC, Sparrow PAC (2008) Evolution of a regulatory framework for pharmaceuticals derived from genetically modified plants. Trends Biotechnol 646:1–12

Streatfield SJ, Lane JR, Brooks CA, Barker DK, Poage ML, Mayor JM, Lamphear BJ, Drees CF, Jilka JM, Hood EE, Howard JA (2003) Corn as a production system for human and animal vaccines. Vaccine 21:812–815

Union of Concerned Scientists (2006) UCS position paper: pharmaceutical and industrial crops. UCS. http://www.ucsusa.org/food_and_environment/genetic_engineering/ucs-position-paper.html. Accessed 28 Feb 2011

USDA (2006) Permitting genetically engineered plants that produce pharmaceutical compounds. BRS factsheet. http://www.aphis.usda.gov/publications/biotechnology/content/printable_version/BRS_FS_pharmaceutical_02-06.pdf. Accessed 28 Feb 2011

USDA (2008) Guidance for APHIS permits for field testing or movement of organisms intended for pharmaceutical or industrial use. http://www.aphis.usda.gov/brs/pdf/Pharma_Guidance.pdf. Accessed 28 Feb 2011

Vancanneyt G, Dubald M, Schröder W, Peters J, Botterman J (2009) A case study for plant-made pharmaceuticals comparing different plant expression and production systems. Methods Mol Biol 483:209–221

Verma D, Daniell H (2007) Chloroplast vector systems for biotechnology applications. Plant Physiol 145:1129–1143

WHO (2003) WHO guidelines on good agricultural and collection practices (GACP) for medicinal plants. http://whqlibdoc.who.int/publications/2003/9241546271.pdf. Accessed 28 Feb 2011

Wolt JD, Karaman S, Wang K (2007) Risk assessment for plant-made pharmaceuticals. CAB Rev Perspect Agric Vet Sci Nutr Nat Resour 2(12):1–9

Index

A. Wang and S. Ma (eds.), *Molecular Farming in Plants: Recent Advances and Future Prospects*, DOI 10.1007/978-94-007-2217-0,